CURRENT PERSPECTIVES
ON AGING AND
THE LIFE CYCLE

Volume 3 • 1989

PERSONAL HISTORY
THROUGH THE LIFE COURSE

CURRENT PERSPECTIVES ON AGING AND THE LIFE CYCLE

A Research Annual

PERSONAL HISTORY THROUGH THE LIFE COURSE

Editors: **DAVID UNRUH**
Office of Instructional Development
University of California, Los Angeles

GAIL S. LIVINGS
Department of Sociology
University of California, Los Angeles

VOLUME 3 • 1989

 JAI PRESS INC.

Greenwich, Connecticut   *London, England*

CONTENTS

LIST OF CONTRIBUTORS

Katherine R. Allen

Department of Family and Child
    Development
Virginia Polytechnic Institute and
    State University

Denise D. Bielby

Department of Sociology
University of California
Santa Barbara

Diane Bjorklund

Department of Sociology
University of California
Davis

W.R. Bytheway

School of Social Studies
University College of Swansea
United Kingdom

Jaber F. Gubrium

Department of Sociology
University of Florida

Patricia J. Gumport

School of Education
Stanford University

N. Laura Kamptner

Department of Psychology
California State University
San Bernardino

Jean R. Kayano

Department of Psychology
California State University
San Bernardino

Hannah S. Kully

Department of Sociology
University of California
Los Angeles

Gail S. Livings

Department of Sociology
University of California
Los Angeles

Judy Long

Department of Sociology
Syracuse University

Joan L. Peterson

Department of Psychology
California State University
San Bernardino

Beverly J. Robinson

Department of Film and Television
University of California
Los Angeles

Johannes J. F. Shroots

TNO Institute of Preventive Health
    Care
Leyden, The Netherlands

Corine A. ten Kate

TNO Institute of Preventive Health
    Care
Leyden, The Netherlands

David Unruh

Office of Instructional Development
University of California
Los Angeles

Meira Weiss

Department of Sociology and
    Anthropology
Tel Aviv University

INTRODUCTION

The papers included in this volume of *Current Perspectives on Aging and the Life Cycle* are original research reports and theoretical essays which address, in various ways, the uses of "personal history through the life course." Research matters which relate to the issues of personal history through the life course is accumulating within the fields of anthropology, gerontology, psychology, and sociology. The specific emphases and issues addressed by researchers in these fields may vary, and the concepts developed and theories applied may also vary enormously. Anthropologists, for example, may attend to the differences across cultures when analyzing oral histories, while psychologists may be more interested in processes of the life review or with the differences between reminiscence and long term memory. Sociologists may be more interested in fluidic changes of the self over time than would the oral historian whose goal is to facilitate the reconstruction of an individual's past which has been interpreted and reinterpreted many times.

The notion of "personal history through the life course," for the purpose of this volume, has been interpreted quite broadly so that we could demonstrate the breadth and originality of current research. We have included original papers from scholars representing many different disciplines and perspectives which address these issues. It would have been as easy to develop a single volume around research on reminiscence, or the life review, or even changes in the self over time. We have chosen, instead, to bring together papers which draw from a variety of perspectives and which focus on a number of different issues so that some possibilities for further collaboration might be explored.

Nearly all of the papers explicitly or implicitly deal with methodological issues. The researchers' attempts to interpret the data from the stories, narratives, and even surveys they used were necessarily fraught with uncertainty and ambiguity. This should not be surprising since nearly all of the papers are based on empirical data which must be interpreted within a particular context, based on data which could be quite different if the people had been interviewed five years earlier, or even later that same day. These are some of the issues and problems which these papers address.

Overview

The paper by Bielby and Kully begins by noting that life stories, life histories, autobiographies, diaries, and the like are all "self-narratives" which are used by individuals to establish coherency in their lives. These accounts of one's life history are, in their presentation, complete and continuous as they present lives which seem rational, stable, and directed. Bielby and Kully have sought to understand *how* people assign meaning to their lives and they do so by examining empirical evidence of the processes which underlie the social construction of self-narratives.

George Herbert Mead's *The Philosophy of the Present* provides the framework for theoretical analysis. The accomplishment of meaning, in the here and now, according to Mead is achieved by bridging present contingencies through a restructuring of the past. The authors correctly note that the process of interpretation, reinterpretation, and resolution is inherently social, but that there is little empirical research which explores the connections between these processes and the development of self-narratives.

The personal documents of British socialist Beatrice Potter Webb (1852–1943) provide case study materials for some first steps toward a better understanding of the social production of self-narratives. The personal documents of Webb's life seem perfectly suited for in-depth analysis since she was a remarkably introspective, observant, and articulate woman. Her life spanned a period of considerable social change which presented her with considerable discontinuities which were addressed in the construction of the many personal documents which she left behind.

Bielby and Kully have made a number of interesting observations regarding the strengths and weaknesses of Mead's theory as they seek to understand the construction of self-narratives. Perhaps the most relevant and important implication for the theme of this volume is that self-narratives are public accounts of identity, the construction of which requires a measure of acquiescence among interactants for them to be successful. In ongoing relationships, significant others must agree to one's interpretation if the reconstructed identity is to survive. The negotiated nature of these processes is one of the major threads which runs through all of the papers in this volume.

The essay by Unruh on the "Social Psychology of Reminiscence" continues this theme by looking at the possibility of a sociological social psychology of reminiscence, which would entail a shift away from a focus on personal adaptation and psychological well-being toward analysis of reminiscence as a social process. This kind of analysis emphasizes that people explicitly and implicitly seek to manage the content, occurrence, imagery, and effects of reminiscence. That is, episodes of reminiscence do not simply emerge from some benign psychological state, but they are to some extent influenced and managed by the actors themselves.

It is assumed throughout that reinterpretation of the past is a continual human endeavor, but that there are certain times and specific places where the process is magnified in social psychological importance, where it is marked by shifts in personal awareness, and where such reinterpretation is characterized by a focus on matters of personal identity. The analysis begins by making the distinction among three different "orders" of reinterpretation. These orders are differentiated and marked by the degree to which the reinterpretation is self-conscious, controlled, and managed by the individual.

After a brief discussion of some methodological issues surrounding the study of reminiscence, the paper concludes with a discussion of four areas which a sociological social psychology of reminiscence must address. These areas are the: (1) Social process by which reminiscences are stimulated and structured by social situations; (2) Ways reminiscence is a personal accomplishment with people constructing, stimulating, and controlling its content and occurrence; (3) Importance of place in affecting reminiscence episodes through the life course; and (4) Role of emotional states in reminiscence and emotional outcomes as a way of understanding the affects, functions, and social psychological importance of reminiscence.

The paper by Bjorklund continues in the sociological tradition which influenced the previous two papers by looking at yet another way by which individuals seek to make sense of their lives. The focus here is on written autobiographies as documents which "force" individuals to selectively sort through their pasts on their way toward developing a narrative which portrays their lives as they prefer. Authors of autobiographies must, for example, deal with questions of how to structure such a narrative, what events to select out of the countless ones possible, and the extent of explanations or accounts that should be offered to the future audience of readers. Bjorklund also notes how the authors of autobiographies, in managing what is essentially "a story of self," display their own assumptions of self and self development.

The data for Bjorklund's analysis are derived from American autobiographies published between 1800 and 1982. One theme which emerges from this analysis is that autobiographers in the late nineteenth and early twentieth centuries tended to focus much more on the development of the self than did those who wrote earlier autobiographies. As the twentieth century progressed, the focus on self development became much more intense, and it became more complicated. This historical shift, according to Bjorklund, seems to mirror the growth of the many disciplines which study human behavior. The author demonstrates a number of avenues for further study, and she demonstrates the usefulness of autobiographies as rich sources of data on socialization, conceptions of self development, and the life course.

The study by Kamptner, Kayano, and Peterson continues the theme of how people construct meaning in their lives by drawing from their pasts. The focus here, however, is on the meaning and importance of inanimate objects after childhood. Object possession may serve a variety of functions which include serving as a source of social status, providing effectance motivation, defining one's self, providing security or solace, representing ties with others, serving a utilitarian function, and as a source of memories. The authors provide an overview of how these functions change over the life course and how the meanings of various objects may change over time.

The exploratory study described in this paper sought to understand the role of objects in life course development, and to find out how the objects named, and the meanings associated with them, varied with life stage. The research presented here is, in a sense, the study of human psychosocial development in a microcosm. The themes of attachment and separation, of identity and sense of self, and of seeking security and mastery over aspects of one's environment are all evident in the study of objects and their meanings, as well as being primary themes of human life course development.

The essay by Bytheway chronicles the shift in one researcher's emphasis away from a relatively structured approach to interviewing and a shift toward a more flexible approach. The author is quick to note that a change in interview structure necessarily alters the relationship between interviewer and interviewee. The "telling of stories" allows informants to participate as subjects and objects in the construction of knowledge about life histories. The process of "telling one's life story" is a social act in and of itself. The interviewer, therefore, becomes a participant observer who is simultaneously observing and affecting the interview. This process is also situated in a context which is itself always changing. In interviewing members of various households, for example, Bytheway came to the realization that households (as well as the individuals located therein) are in a state of flux most of the time wherein specific events do not occur at a perceivable rate. The process of collecting life histories requires the interviewer to be sensitive to the tendency for subjects to divorce life history from present circumstances.

The focus of the volume begins to shift toward the analysis of life narratives and oral history per se with Robinson's paper on black women's culture. The approach utilized here is one rooted firmly in a folkloric tradition where individuals are encouraged to describe themselves in terms of the past and, as a result, facilitate the preservation of the past—both in terms of one's own personal history and the preservation of certain cultures. The author effectively argues that it is not enough simply to facilitate and listen to people's descriptions of the past, but one must locate those narratives within the appropriate historical, cultural, and sociological context. This process requires knowing the communities (past and present) within which the person has lived and of piecing together these seemingly disparate element into a cohesive and understandable whole.

The research article by Livings expands the use of oral history methods by grounding the analysis of data in a dialectical perspective. Her analysis of the ways by which working-class women dealt with their participation in the 20th century U.S. labor force is one which focuses on the social construction of meaning on the part of a subordinate class. The expansion of union activity prior to World War II provided a common referent for the older women who were interviewed. The primary research question, then, is "How did the working-class woman deal with the conflicts between the traditional woman's role and the necessity to engage in paid labor outside the home?" Livings has examined, through a series of lengthy interviews and an analysis of archival testimony, the interpretations of the past from waitress union leaders focused on the period between the 1930s and the present. The expansion of union activity is an important referent for defining the period in which to investigate working women's consciousness through oral histories. Union activity, as Livings notes, clearly indicates significant identification with work, and such involvement is likely to give these women opportunities to articulate their ideas. Perhaps most important is the fact that organizationally active women are more likely to leave some record of their consciousness than women who are more isolated and detached.

This essay makes a contribution not only to our substantive knowledge of women's meanings of self and work, but it advances and clarifies the importance of oral history methods as well. Livings argues for heightened sensitivity toward the oral history interview as a communicative experience so that the greatest potential use and value of the "strips" against which the oral historian/ethnographer is testing knowledge.

The paper by Gumport explores the ways by which one organizational setting, higher education, not only affects the actions and experiences of individuals but also how the setting frames the possibilities within which individuals construct their future actions. The tone of the essay is one of encouraging those who study higher education organizations to consider seriously the use of life histories as a way of glimpsing beneath the veneer of organizational structure and hierarchical arrangements to see the effects of life within on individuals.

The substantive grounding of Gumport's methodological discussion is the emergence of feminist scholarship from the late 1950s to the present. The three generations of women interviewed were historically situated within different periods of an emerging feminist scholarly discourse. The use of generational analysis emerged from the data and eventually facilitated contextualization of the data derived through life history interviews. The researcher (much like Bytheway) found herself analyzing the interview process itself—in addition to analyzing the tape-recorded conversations which were originally conceived to be the primary source of data.

The essay by Long, to some degree, combines the approaches of Bjorklund and Gumport as she seeks to understand "the content that females seek to communicate in autobiographies and the institutional context that provides her pattern, her audience, and her judges. The context for later analysis is set by discussing the "generic model of autobiography" which has as its defining characteristic the interpretation of life in terms of a pattern. Throughout the essay, Long is sensitive to the concerns of women autobiographers—their daily life, relationships with others, a yearning to be understood, the fear of masculine ridicule. These concerns are not abstract themes, but real reflections of the subject's social world. Three strategies are outlined by which women "tell their lives" within a societal context which has not been all that receptive to women's voices. "Telling it slant," "telling it messy," and "telling it straight" serve most usefully as sensitizing concepts which facilitate an understanding not only of the broad range of women's autobiographies, but to facilitate understanding of certain shifts in tenor and tone found in a single work.

Allen's study focuses on the life histories of four women and the ways by which they organized their lives around their families throughout the life course. The analysis of these women's experience has not been "flattened" by an emphasis on "the women" or some monolithic definition of "family." The emphasis, however, is on the particular rather than the general so that the commonalities of life course experiences may be understood without sacrificing the integrity of individual lives. The hardships and losses endured by the four women profiled in this chapter were dealt with in very different ways. Some individuals managed to remain focused on the family through a continuing and long-lasting relationship, while others refocused their attachments on previously less important family members as a way to cope with a loss. Implicit throughout these four narratives, as Allen later notes, is the tension between individual assertion and family need. This tension was reflected in the two "types" of memory which emerged in the life history data. The first type was a "core" memory which interviewees wove throughout their interviews as themes to which they returned time and time again. The second type of memory was an "overall" picture or assessment of where they fit in the family structure and the need for unit stability. That greater awareness of family need seemed to help these women accept their lives in the here and now, while, at the same time, remembering the discontinuities of their lives.

Gubrium's paper represents a shift in focus away from the analysis of individual lives and toward a theoretical understanding of the social nature of emotions. The focus on "emotion work" may, at first blush, seem somewhat distant from the analysis of personal history through the life course. The focus on the effects of Alzheimer's disease on the family and the construction of emotions by family members, however, allows us to see some of the social psychological aspects one's life history undergoing dramatic change. This disease significantly alters familial relationships in the here and now at a time when the patient needs kindness and attention the most. Implicit within Gubrium's discussion is the notion that the Alzheimer's experience is such a devastating experience that the continuity of one's life history is destroyed. The analysis of feelings involved in the disease experience, therefore, allow us to glimpse the fragility of one's social relationships through the life course, and to home in on the essence of human emotions and the interconnectedness between individuals in families.

The social discontinuity brought to a parent's life by the death of a child is the focus of Weiss' study. We here see the tendency of bereaved parents to use the loss as a new starting point in their life. The circumstances surrounding the loss seem to encourage the survivor to examine ongoing relations in various spheres, and to seek opportunities which previously had been perceived to be blocked or, in some instances, unobtainable. The death of a child, as with all deaths of significant others, stimulates among survivors a review of their lives before the death, but they do not seem to question the basic assumptions upon which their lives had been based. The profiles of bereaved parents presented by Weiss, however, would seem to indicate that the institutions of the family and of work are not broken down by intense analysis and review to the same extent as other aspects of survivor's lives. The challenge of the life review, as Weiss reminds us, is not to actually destroy the existing social order, but to challenge one's position within it.

The essay by Schroots and ten Kate is a methodological piece which introduces the "Life-Line Interview Method." The authors believe that metaphor is implicit in all theory development and that their use should be an informed and conscious act rather than a hidden assumption which implicitly affects one's interpretation. Existing metaphors used by various theorists to encapsulate and describe certain qualities of aging are discussed as a way of laying the groundwork for introduction of the method. The interviewee is asked to convey his or her life story utilizing particular spatial metaphors and to plot the "ups and downs" of one's life history on a graph. This interviewing method would seem to facilitate more sophisticated analyses of life history data and it would also encourage more systematic comparisons among many different life histories.

We hope that readers of this volume will enjoy these research reports, theoretical essays, and methodological statements focused on various aspects of personal history through the life course. This collection of papers is not intended to serve either as a definitive statement of the field, or as a circumscribed view all possibilities for research. This volume should be taken, instead, as a stepping stone toward more and better interdisciplinary research which would extend the thoughts and ideas of the authors represented here. The study of personal histo-

ry—with all of its attendant uncertainties, ambiguities, and interpretational problems—is an exciting and important scholarly endeavor which should enrich the
lives of one's subjects while extending our collective knowledge of the field.

David Unruh
Gail S. Livings

SOCIAL CONSTRUCTION OF THE PAST:
AUTOBIOGRAPHY AND THE THEORY
OF G.H. MEAD

Denise D. Bielby and Hannah S. Kully

INTRODUCTION

In the last ten years, sociologists have returned to the study of the life cycle. Influenced by the life span perspective in developmental psychology which seeks explanation of the "average expectable life cycle" (Butler, 1968; Baltes and Brim, 1980), sociologists have generated a primarily theoretical literature which articulates the impact of cohort and social structure on the course of individual lives (Elder, 1974, 1981; Featherman, 1983; Marini, 1984, 1985; Ryder, 1965; Riley, 1976). Recent sociological work has focused upon articulation of a life

Current Perspectives on Aging and the Life Cycle,
Volume 3, pages 1–24.

course perspective (Featherman, 1981), structural and contextual constraints upon the life cycle (Dannefer, 1984), and the form and content of the life course (Sorensen, Weinert, and Sherrod, 1986). This recent attention by sociologists represents a renewal and extension of early, innovative research which concentrated on the study of social determinants of personality (Mannheim, 1928; Thomas and Znaniecki, 1918–1920).

Increased awareness of personal documents—life stories, life histories, autobiographies, and diaries—as sources of life history data has drawn attention to the similarities and differences among them.[1] All are self-narratives, that is, individual accounts of the relationship among self-relevant events across time in order to establish coherency in one's life (Gergen and Gergen, 1983). What is most notable about recollected personal accounts of a life (i.e., life stories, life histories, and autobiographies) is that they usually conform to a narrative structure. That is, they are completed and continuous, and they yield lives portrayed as full of purpose and direction and minimally chaotic and random (Bertaux and Kohli, 1984; Cohler, 1982). These features prescribe an identity that appears to be relatively rational, stable, and directed in its development.

While similarities and differences are apparent, social scientists are still searching for clarification of what is learned from individuals when they speak or write about their lives (Crapanzano, Ergan, and Modell, 1986; Kohli, 1981; Gottschalk, Kluckhohn, and Angell, 1945). Within sociology, a relatively limited amount of attention has been devoted to analysis of the construction of self-narratives, especially autobiographies. To what extent do autobiographies account for an individual's life? Why do narratives about lives have a distinctive beginning, middle, and end? Why do lives recounted unfold in a progressive, purposeful manner, in contrast to their seemingly stochastic manner as we live them? Why are individuals compelled to assign meaning to their lives?

In the research reported here we seek an understanding of one part of this ambitious agenda: We address how individuals assign meaning to their lives. To accomplish this, we seek empirical evidence of the sociological process underlying the social construction of self-narratives. Our purpose in this research is to qualitatively analyze autobiography as it develops out of particular social conditions and is achieved through social psychological processes.

The Problem

Contemporary sociologists have largely ignored analysis of self-narratives for the study of the life course for two reasons. First, the use of autobiographical material is constrained by the relative absence of a theoretical framework that can be used for interpretation of autobiographically generated life course data. Second, autobiographies are recognized as "metaphors of self" (Olney, 1972) and this has inhibited their use as objective sources of life history data. Thus, a closed theoretical circle exists because of inadequate sociological knowledge about the process of reinterpreting the past. If the process of self-narrative construction was more clearly understood, autobiographies as well as other personal documents should be able to yield valuable insights to the researcher about the life course (Crapanzano, Ergas, and Modell, 1986).

In the work reported here, we explore the sociological contribution of George Herbert Mead's *The Philosophy of the Present* (1932) to the analysis of auto-biography in particular and personal documents in general. In this work, Mead articulated the process whereby the present is used to reconstruct the past, a process through which individuals assign meaning to their lives. According to Mead, the passage of time is experienced because novelty constantly arises in the present. Order, or personal continuity, is maintained through an interaction of present events with selected meaningful past events, taking into account the structural conditions of the past. Thus, the accomplishment of meaning in the present is dependent upon continuity, which is achieved by bridging present contingencies through a restructuring of the past.

Mead's central argument is that although the present implies a past and a future, we always experience past and future through the present. Mead's thesis was a response to prevailing beliefs, primarily psychological in origin, that the remembered past is comprised of mental images stored unitarily but retrieved in clusters comprising an event or experience. Central to psychologists' thinking at that time (and even today) is the assumption that the past overwhelmingly defines the present, that ". . . everyone uses the past to define themselves . . . (it) defines me, together with my present and the future that the past leads me to expect" (Neisser, 1982, p. 13).

While Mead does acknowledge existence of immutable facts that transpired previously, he argues that their subjective interpretation may vary, depending upon the existing present. Mead did not discuss how the social and cultural context of the present influences the negotiation of the meaning of the past and the setting of agendas for the future. By ignoring the emergent present, Mead in effect limited himself to discussion of the function of interpreting the past—the past provides continuity in the present. However, the process by which life material is reviewed by the individual occurs in the context of the present and is interpreted, reinterpreted, and resolved within a (present) social context, so that meanings attached to events, and thus the significance of events themselves are subject to much variation, depending upon the circumstances of the present. The process of interpretation, reinterpretation, and resolution is inherently social, yet little empirical research exists exploring its connection to construction of self-narrative.

To examine the utility of Mead's contribution to understanding the social production of self-narrative, we analyze as a case study the personal documents, including the autobiography, diaries, and letters, of British Socialist Beatrice

Potter Webb (1852–1943). Webb was one of the first quantitative sociologists in England, and in partnership with her husband Sidney, was a central figure in the Fabian Socialist movement in the late 19th and early 20th centuries. One of their many accomplishments was founding the London School of Economics. Webb was an articulate and analytical political activist whose life intersected with the end of the Victorian Era, the rise of industrialization, and the first wave of feminism. Thus, Webb's life spanned a period of considerable social change. Using her life documents as a case study, we examine both the contributions and limitations of Mead to the study of the life course. However, before discussing Mead in detail and applying his concepts to self-narrative, we turn to the work of historians and students of literary forms to explain the historical and cultural distinctiveness of autobiographies.

The Emergence of Autobiography

Historians argue that the turn of the 18th century marked the rise of a rapidly growing number of published autobiographies, reflecting the emergence of a consciousness of humankind's place in historical time (Weintraub, 1975; Pomerleau, 1980, p. 21). The production of consciously written life stories was possible only with the abandonment of a pre-modern view of life which held that human development occurred through either: (1) the unfolding of an inner natural logic; or (2) a potentiality of personal characteristics which reveal themselves in due course if conditions are favorable. Instead, the idea took hold in the 18th century that life was a developmental process arising through interaction with the world: The individual experiences the world in relation to his or her own capacities and historically present contingencies.

Also emerging in the 18th century was a conception of the self as an individual personality in the developmental process of life. Each human life came to be valued as unique, and the purpose of existence then was to make manifest that life for the enrichment of the world. The search for the individuality of each person is guided by a maxim to be true to the self. Deviations from the "true" self become points of clarification in autobiographies, events which must be seen as wrong turns or false steps, but perhaps necessary lessons to learn before the return to the ideal.

Thus, the cultural ideal of self as an individual emerged. This developed against a larger historical backdrop of secularization of life and a celebration of human diversity, made possible by the fading of absolutist standards and providential explanations for human existence. The structural development of political states and the spread of trade led to discoveries of cultures outside the Western European sphere and eventually brought a reinterpretation of the Bible as only one of many orientations to human history. While Christianity had previously provided to most humankind a purpose to pain and suffering, ". . . in the modern form it is the individual himself or herself who must, in the very writing of autobiography, both detect and create life's shape, progress and meaning" (Weintraub, 1975).

The Construction of Autobiography

The construction of autobiography took on a unique form as a result of these societal-level developments. Belief in individuality and personal determination made possible the phenomenon of autobiographies, and it followed that individuals would seek a "looking back" upon one's personally influenced course of life. And, once it was believed that individuals contributed to life's determination, individuals became capable of seeking purpose in its process, in part in an attempt to anticipate and control the future. With meaning "found," the author is then capable of acting back upon the world from this viewpoint.

According to Weintraub (1975), an essential feature of autobiographies is that they are written from a point of view located in time beyond a break in continuity of action—a crisis or a series of events in life, that is, a break from one's "true" self, which is assigned meaning in retrospect. If this feature is central, then autobiographical content should include evidence of a shift in life perspective, and this opens the possibility of Mead's phenomenon of the past being reinterpreted in the context of the present. With the past reinterpretable, meanings attached to past life events, and thus the significance of events themselves, are open to change.

Finally, reinterpretation does not occur in a social vacuum, especially in autobiographical construction of self-narrative. Mills (1959, p. 175) noted that an individual's life can be understood only when one makes "reference to the historical structures in which the milieux of their everyday life are organized." It is also the case that self-narratives in general and autobiographies in particular are the outcome of specific structural and cultural conditions and reflecting themes of a given historical period. "At this point, personal life can be understood only by viewing it under its historical dimension" (Weintraub, 1975, p. 833; cf. also Bertaux, 1981; Nord, 1985).

Although steps have been taken to understand the construction of autobiographies, there is limited empirical evidence explaining autobiographies as a social product. At the micro-sociological level, social psychological explanations of the development of the self only partly illuminate the process whereby a life becomes an event with a history and a purpose. They do not examine how an individual goes about reweaving life's random events into a unified narrative rooted in prevailing cultural themes. Below, we discuss the utility of Mead's theory of temporality for explaining the gathering in and recording of a set of life events.

THEORETICAL BACKGROUND: MEAD'S CONCEPTUAL FRAMEWORK

In 1932, George Herbert Mead's *Philosophy of the Present* was published in an effort to anchor the concept of temporality and its boundaries to time and space. Mead's argument was that although the present implies a past and a future, "reality is always that of a present" (1929, p. 235). In Mead's words,

> The specious present is, then, that within which are present not only the immediate abstracted sense data but also the imagery of past and future experiences taken out of their place in the act which they imply. . . . These experiences belong to the reconstruction to which a later response will take place. They belong to the beginning of a later act. As such they are in the present (1938, p. 221).

Thus, Mead asserts that the past is comprised of previous presents, in contrast to philosophical beliefs during that period of an irrevokable and unchanging past. As Mead said, "we speak of the past as final and irrevocable. There is nothing that is less so. . . . " (1932, p. 95). Rather, ". . . the long and the short of it is that the past (or some meaningful structure of the past) is as hypothetical as the future" (1932, p. 12).

While Mead does acknowledge existence of the objective past, i.e., the immutable facts that transpired previously, the subjective interpretation may vary, depending upon the existing present (1938, p. 614). The rationale for this is continuity of behavioral action. "We are engaged in spreading backward what is going on so that the steps we are taking will be a continuity in advance to the goals of our conduct" (Mead, 1929, p. 237). The past ". . . connects what is unconnected in the merging of one present into another" (Mead, 1929, p. 240). And, while the past does not exist apart from the present, it is the present that interprets the meaning of the past (Mead, 1938, p. 615). Subjective interpretation is the ongoing, continued, never-ceasing effort to give meaning and sense to our experiences.

According to Maines, Sugrue, and Katovich (1983), Mead's theory of the past is based upon four implied dimensions: (1) the symbolically reconstructed past, (2) the social structural past, (3) the implied objective past, and (4) the mythical past. The first, the symbolically reconstructed past, is most salient to Mead's assertions about the past. This dimension clarifies the process whereby individuals in the present selectively draw from past events so the present may be understood and the future anticipated. The second, the social structural past, is that which objectively influences the past and thus "structures and conditions the experiences found in the present" (Maines et al., 1983, p. 163). While influential, this past is not fully deterministic, but merely predisposing. The third dimension, the implied objective past, according to the authors, is the occurance of previous events which made the existing present possible. By virtue of certain past events transpiring, the present is "in place" and possible. The fourth dimension is the mythical past, a non-factual recall of previous events. The mythical past is based, according to the authors, upon creation, not re-creation of the past, and this kind of past is overt manipulation of the present.

Mead's assertions about the past are only indirectly addressed in existing empirical research on self-narrative. It is well known, for example, that the review and certification of one's life is an important personal task prior to death (Butler, 1968; Lieberman and Falk, 1971), and in this process, as in others involving self-narrative, one resolves anomalies and disjunctures as well as progressive change in one's life. It is also generally acknowledged that individuals reconstruct their lives in the final telling of it, omitting inconsistencies, resolving conflicts and heightening successes. Others have hinted at some of Mead's concepts in the study of autobiographical memory (Freud, 1982; Ferrarotti, 1981; Linton, 1982; Kohli, 1981).

Although Mead has captured the process whereby the past achieves meaning, he does not articulate, except for the concept of discontinuity, its social construction, that is, the social or behavioral elements through which it occurs. Life is experienced as a "continuity of presents" and is uninterrupted process. Discontinuity of process is that which creates novelty, departure, or the unexpected, and is that which must be integrated with the past. Mead says: "The social nature of the present arises out of its emergence. I am referring to the process of readjustment that emergence involves. . . . I am here using the term 'social' with reference not to the new system, but to the process of readjustment" (1932, p. 47). He offers us little about relevant content, although we need look no further than his concept of identity. In interpreting the past, individuals maintain presentation of a consistent, public, socially constructed self-concept. As Mead and others have shown, identity is derived through social interaction ". . . characterized by the relation of an organic individual to his environment or world. The world, things, and the individual are what they are because of this relation" (Mead, 1938, p. 215). We recognize that all forms of self-narrative involve the process of constructing identity, and that the inherently social process of establishing a plausible and meaningful continuity between present and past should be empirically observable. In the analysis presented below, we focus upon several aspects of social construction of self-narrative, including:

1. societal shifts in social and moral codes or belief systems;
2. social origins;
3. personal tragedies and triumphs, including death, political defeat, successes, and accomplishments;
4. unfulfilled ideals or goals; and,
5. unresolved conflict.

These factors include several sources of possible interruption in the continuity of identity which typically precipitate explanation in self-narrative. While not inclusive of all sources of disruption, they encompass both societal and individual-level elements. We examine the life of Beatrice Webb in detail by focusing upon each of Maines, et al.'s four dimensions, and considering the impact of the above factors upon her self-narrative.

A CASE STUDY OF BEATRICE WEBB

Beatrice Potter Webb (1858–1943) was born into an upper middle-class English family as the second to the youngest of nine girls. She led an early life detached from maternal affection, full of rigorous self-directed scholarship, analytical insights about her life, and social comforts and societal benefits of living within an economically prosperous family. Between 1883 and 1890, she chose to train as a quantitative sociologist under Charles Booth. In 1891 Beatrice became engaged to Sidney Webb, a Fabian socialist and by 1892 they were married. By 1893 Beatrice formally joined the Fabian Society and began with Sidney what was for the remainder of her life a partnership and marriage dedicated to social analysis and reform. During her life, Webb published numerous treatises on trade unionism, the Poor Law, and labor reform, co-founded the London School of Economics with Sidney Webb and was the first female elected to the British Academy in Economics. In the last decades of her life she endorsed Communism as a model political state but ultimately qualified her embracing of it. She died in 1943, having retired to a life of interview-granting in her home at Passfield Corner.

These events comprise the objective structure of her life, and her telling of it at age 68 in her two-part autobiography *My Apprenticeship* and *Our Partnership* reveals a discernable rational narrative structure to her life, with purpose, intention, and above all progressive change. The introduction to *My Apprenticeship* concentrates on the theme of struggle between the "ego that affirms and the ego that denies." She faces her double self straight on, and in fact makes this schism the narrative line of her story. Webb's self is caught between feeling and independent rational thinking, in the form of prayer versus science, as she strives to find both satisfaction and fulfillment in a vocation of work. She is also caught between vanity (both class and personal) and work in the form of egoism and women's place in Victorian England. Her story is one of direct engagement with the crises resulting from her struggle, and she seeks resolution through her adoption of Fabianism and through her marriage and partnership to Sidney Webb. She states, "At last I became a socialist. . ." (1926) as if joining an organized religion. Publically, she presents a coherent whole to life, one which achieves satisfaction though faith in Fabian socialism. Her solution was to combine an ideology with a vocation, and simultaneously a career with marriage.

But, the personal story of Webb did not achieve the total coherence and resolution of conflicting themes she attempted to portray. "The text she produced redeemed and celebrated the notion of progress, at least in the individual life. Without violating the documentation of her past that she had in the form of her diary, she managed to create out of her personal history the story of a quest that reached a felicitous end" (Nord, 1985, p. 236). In her diaries, kept since the age of 15, she reveals a life that was always searching and unresolved. We can see evidence of some satisfactions but also manifestations of depression and limited fulfillment of her life's aims. Instead of a life successfully integrating her dual nature through Fabianism, Webb was painfully torn among the conflicts of faith, science, work, and the limited public freedom allowed to women for articulation of private concerns, and she expressed these pains freely to herself in her diaries particularly toward the end of her life. At age 68 she notes her "inability to make clear even to myself, let alone others, why I believe in religious mysticism—why I hanker after a Church—with its communion of the faithful, with its religious rites, and its religious discipline and above all with its definite code of conduct (Diary,[2] April 14, 1928). And in 1931 she notes: "Sometimes I cursed myself for becoming entangled in the self-conscious scribblings of a woman" (as Sidney once called the diary). The diary is nonetheless, her "pet pasttime . . . gossip with an old friend and confidant . . ." (Diary, December 29, 1933).

In the remainder of this chapter, we analyze the content of Beatrice Webb's life, using the four dimensions of Mead's theory of the past as developed by Maines et. al, for uncovering reinterpretation and meaning Webb assigned to her life. We then assess the extent of Mead's contribution to interpreting autobiographical self-narratives by extending our analysis of Webb to additional points of sociological insight.

DATA ANALYSIS

1. Symbolic Reconstruction of the Past

When Beatrice Webb, a trained social investigator, wrote her autobiography she was well aware of the process of reinterpretation of past events for the purpose of self-narrative. She says, "Memory is a risky guide in tracing the ups and downs of belief and unbelief; gaps in the argument are apt to be filled in, and the undulating line of feeling becomes artificially straightened. As being free from the fallacy of 'being wise after the event', I prefer the contemporary entries in the MS. diary" (1926, p. 80). Indeed, she notes how the training she received as a social investigator instilled the importance of achieving objectivity and validity in observations (Diary, May 24, 1886). Despite her level of awareness, trunctions in autobiographical accounting occurred, making synthesis seem smoother than was actually the case.

Religion was a recurring feature in Beatrice Webb's life, and many of her early writings were about its place in her life. Indeed, Webb ties the struggle between her two egos to the validity of religious mysticism (1926, p. 90). Beatrice Webb was drawn to belief in the mystical but rejected it for its inherent restriction upon intellectual freedom, and she struggled throughout her life to find a replacement and to reconcile her need for faith with her intellectual beliefs.

As a sixteen-year-old adolescent, Beatrice Potter struggled against ''. . . self-satisfaction which I consider is one of my worst faults. . . . the only way to cure myself of it is to go heart and soul into religion. It is a pity I ever went off the path of orthodox religion; it was a misfortune that I was not brought up to believe that to doubt was a crime'' (Diary, September 1874). But she remains uneasy with the tenet of Atonement, thinking it, in retrospect at the writing of her autobiography more than irrational, but immoral (1926, p. 67). Having vowed to ''become a true Christian . . . making Him my sole aim in life'' she receives the Great Sacrament as affirmation of her resolve to remain religious.

She is disappointed, however, with the fact that the ''high resolutions'' that come with religion do not rid her of the ''world of vanity and vexation of spirit'' (Diary, July 1875). At the publication of her autobiography, she interprets this as ''a struggle . . . with my own lack of morality'' (1926, p. 67). By 1876, Beatrice Potter had lost her hold on orthodox Christianity. She notes its disappearance as one simultaneously occurring with the emergence of eastern religions and the ''. . . 'religion of science': that is, an implicit faith that by the methods of physical science, and by these methods alone, could be solved all the problems arising out of the relation of man to man and of man towards the universe'' (1926, p. 72).

In retrospect, Webb herself identifies for us the social opportunity for her acceptance of the religion of science. ''But it so happened that during these very months intellectual curiosity swept me into currents of thought at that time stirring the minds of those who frequented the outer, more unconventional and, be it added, the more cultivated circles of London society; movements which, though unconnected with and in some ways contradictory to each other, had the common characteristic of undermining belief in traditional Christianity'' (1926, pp. 71–72). As an aspiring participant in these social circles, Webb was predisposed to their influence.

But in fact, her movement toward science as a replacement for religion was somewhat less linear, and in fact never fully achieved as asserted in her autobiography. By August 16, 1875, she noted in her diary how her religious beliefs have been transformed. Two years later after September, 1877, she recollects that she rejected all traditional religions, including Buddhism, with which she had been experimenting, as well as Christianity. She noted in *My Apprenticeship* that by 1876 she had found the religion of science: ". . . it was during the autumn of 1876 that I thought I had reached a resting place for the soul of man, from which he could direct his life according to the dictates of pure reason, without denying the impulse to reverence the Power that controlled the Universe. This resting place was then termed . . . the Religion of Science (1926, p. 77).

However, a year later in the March 19, 1877 entry to her diary, Webb despairs about her continuing nonreligious state, for she notes ". . . I am afraid I have no religion whatever, for I have not yet grasped the religion of science." She remained ambivalent about foregoing religion altogether. "Of one thing I am quite certain, that no character is perfect without religion . . .", and in that same year restates her beliefs and interpretations of Christianity. She had yet to fully let it go.

It is most interesting to discover in later writings that her rejection of religion was ultimately only an intellectual one and that religion's emotional appeal was never fully excised. Her replacement of religion by Fabian socialism, once accomplished, was never complete. Nearly a half century later, in 1925 and 1926, she reinterprets her belief in the importance of religion, and by adopting it in revised form, reasserts the importance of religion to herself and to society. Publicly, in her autobiography, she only minimally asserts her belief, but in private writings her reflections are extensive and troubled.

Several societal transformations, including increased absence of definite rules of conduct, particularly sexual morality, the use of science in the service of world war, and the predominance of capitalism and the failure of Fabianism to revolutionize social structure (Diary, October 29, 1925) troubled Webb deeply and precipitated her return to faith later in life. She names religion as the alternative that "makes for righteousness . . . to raise human values and enoble behavior." But she despairs over her conclusion that one cannot achieve the moral results of a religious faith without dogma—a dogma that offends intellectual integrity and moral sincerity. She concludes, "I have the consciousness of being a spiritual outcast . . . (with) no home for my religious faculty. I wander around disconsolate—that is the root of my indifference to life. In spite of my unusually happy circumstances and keen intellectual interests, I am not at peace with myself. I have failed to solve the problem of life—of man's relation to the Universe and therefore to his fellow man" (Diary, April 14, 1926). The religion of science had not reformed society as she had idealistically hoped 50 years earlier.

We see clearly here the symbolic reconstruction of the past as developed by Mead. Webb does not go so far as to deny past faith in science, but she does announce and accept a shift in her belief in what it can accomplish for humankind based upon her present perspective on society. The precipitate of this shift is that within her lifetime, Webb's intellectual mission appeared to her to be unsuccessful. Once that discontinuity emerged, her quest for solutions came to rest on the only other ideology she knew—religion. According to Mead, when novelty yields perceptable discontinuity in the chain of events, the past must be rein-terpreted in light of the present, so that progress to the present can be understood and intention for the future discerned.

While we are unfortunately unable to uncover any exact situational discon-tinuities or interpersonal interaction that precipitated her reassessment of re-ligion, according to Gergen and Gergen (1983) continuities are dependent upon tacit agreement with one's social interactants. Situational meanings must be negotiated and accepted by significant others. Additionally, significant others must willingly participate in the social interaction; if withdrawn, there is nothing to negotiate. Nord (1985) points out Webb's realization within a year of her marriage that Sidney lacked a need similar to hers for a "faith" in the ideology of socialism. His nonparticipation led her to a withdrawal into her diaries for affirmation of her beliefs. With Webb interacting with herself, the impact of social history becomes an important source of novelty in her life. And it is here we see some limitations to Mead's contribution about causes of reinterpretation.

The influence of social history upon the course of Webb's life—the rejection of a theologically-based belief system and its ultimate rediscovery is very clear-cut. Her search for meaning in life is integrally linked to prevailing system-level events. Linked with this particular backdrop is her almost puritanical need for a vocation that would save her from vanity and fantasy and silliness—Victorian notions of femininity (Nord, 1985). The ultimate failure of her grand social mission undermined her faith in herself and in her identity.

2.   The Social Structural Past

The social structural past, according to Maines, et al. (1983), establishes micro-level probabilities for experience in the present, which in turn affects perceptions about the past and expectations about the future. Webb begins her autobiographical account with this clearly in mind by commencing her auto-biography with vivid observations of structural elements of her childhood: her mother, her father, and family dynamics. As the second youngest of nine girls and the sibling of a much desired and deceased younger brother, she was the recipient of maternal attention almost by default. As a sixteen-year-old she notes, "What is this feeling between Mother and me? It is a kind of feeling of dislike and distrust which I believe is mutual" (Diary, March 24, 1874).

The ascetic parenting of her mother is given particular attention. Mother-daughter estrangement was nearly total. ". . . I am, as Mother says, too young, too uneducated and, worst of all too frivolous to be a companion to her" (Diary, September 27, 1874). When contrasted with that of her indulgent but pater-nalistic father ("Notwithstanding absence, my father was the central figure of the family life—the light and warmth of the home" (1926, p. 8), we see a childhood fairly devoid of maternal affection in the traditional sense. Left on her own to educate herself, Webb begins a self-directed quest for intellectual challenges within a relatively isolated childhood.

Webb's observations illustrate that many familial elements of her social structural past resonate throughout her perceptions of a fundamental issue in her life, which she put as the choice between vocation and "the Victorian code of feminine domesticity" (1926, p. 100). And, we can document similarities between mother and daughter's intellectual interests and life-style which predispose Webb to the life of public work that she chose and, to some extent, the debate within herself over religion.

Webb's mother, Laurencina Potter, had a stern personage and she ruled their household aloofly. She spent much of her life in scholarship, studying Greek Testament, searching for "mystical consolations and moral discipline of religious orthodoxy," and rigorously practising religious rites (1926, p. 12). "She had visualized a home life of close intellectual comradeship with my father, possibly of intellectual achievement, surrounded by distinguished friends, of whom she had many as a girl and young woman" (1926, p. 11). At times this circle extended to her daughters: there were heated discussions about Spiritualism with her daughters (Diary, November 15, 1877). Laurencina Potter was an ardent student of Adam Smith, Malthus, and Herbert Spencer whom she welcomed into her She also pursued foreign languages: "More and more absorbed in her lonely studies and despairing of solving the problems which troubled her, her restless intellect fastened on the acquisition of languages . . . . " (1926, p. 14). If her scholarship kept her aloof from her family, so did personal reasons. "She had been reared by and with men, and she disliked women. She was destined to have nine daughters and to lose her only son. Moreover, her daughters were not the sort of women she admired or approved. . . . (they) refused to be educated and defied caste conventions" (1926, p. 11).

If Laurencina Potter's demeanor was a factor in Beatrice's subsequent personal dilemmas, the ultimate determinant of the influential mother-daughter estrangement was brought on by additional circumstances of family structure. "The birth of an only brother when I was four, and his death when I was seven years of age, the crowning joy and devastating sorrow of my mother's life, had separated me from her care and attention; and the coming of my youngest sister, a few months after my brother's death, a partial outlet for my mother's wounded feelings, completed our separation" (1926, p. 10). Webb's mother pronounced in her diary that Beatrice was "the only one of my children who is below the average in intelligence" and this, Webb thought, explained her mother's indifference. Beatrice Webb states: "I was neither ill-treated nor oppressed: I was merely ignored. For good or for evil I was left free to live my own little life within the large and loose framework of family circumstance" (1926, p. 51).

Within this context, attempts at formal education left Beatrice indifferent to classroom learning and in a continuous state of ill health that proved disruptive to tutoring. Eventually, all formal learning was abandoned. But left to her own devices, Webb became a scholar in her own right. From an early age she read, and when recuperation and all pretense at formal education was abandoned, Webb ". . . invented a device of my own for self-culture—reading the book of my free choice, and in my private manuscript book extracting, abstracting and criticising what I had read" (1926, p. 53). She read from the myriad of available household literature, pamphlets, periodicals, newspapers, and novels, and thrived in the attention and sponsorship of Herbert Spencer. And from that knowledge and exposure grew her intellectual agenda and her observant mind.

From within the more general context of maternal rejection and self-education emerged Webb's adult dilemma over public vocation versus the Victorian code of feminine domesticity, and in this too Laurencina Potter had a hand. Webb was at a crucial point in her own intellectual development, when in 1882, Laurencina Potter died. "To win recognistion (sic) as an intellectual worker was, even before my mother's death, my secret ambition. I longed to write a book that would be read; but I had no notion about what I wanted to write" (1926, p. 94). Ironically, Webb had developed a quest for knowledge surprisingly similar to the mother who ignored her. The similarity occurred to Beatrice within a year of her mother's death. "When I work, with many odds against me, for a distinct and perhaps unattainable end, I think of her and her intellectual strivings, which we were too ready to call useless, and yet will be the originating impulse of all my ambition, urging me onward towards something better in action or thought" (Diary, August 27, 1882). She reiterates this in retrospect, when at the writing of her autobiography at age 68 she says, ". . . as we eventually discovered, we had the same tastes, we were puzzling over the same problems; and she had harboured, deep down in her heart, right up to middle life, the very ambition that I was secretly developing, the ambition to become a publicist" (1926, p. 10).

The death of her mother brought the intervention of a new familial circumstance to condition Webb's present. Webb, as the oldest unmarried daughter, became responsible for managing her father's household. "My duty now lies clearly before me—to Father and Rosy first, secondly to the home as a centre for the whole family" (Diary, April 23, 1882). She was also invited by her father, should she not marry, to become his companion in business and travel. Webb had not yet clearly identified her desire to pursue the craft of sociologist, and the more immediate demands of society hostess seriously challenged her ambiguous career goals. It is here that the backdrop of Victorian Era shaped not only Webb's choices but her perception of those choices. "Now my life is divided sharply into the thoughtful part and the active part, completely unconnected one with the other. They are in fact an attempt to realize the different and almost conflicting ideals necessitating a compromise as to energy and time which has to be perpetually readjusted. My only hope is that the one ideal is hidden from the world, the truth being that in my heart of hearts I'm ashamed of it and yet it is actually the dominant internal power" (Diary, April 24, 1883).

Her separation into two selves became her "dead point," constrained by a socially dictated dichotomy: "the normal woman seeking personal happiness in love . . ." and "the right to the free activity of 'a clear and analytic mind' " (1926, p. 239). Resolution came gradually and by default. Her disappointment in love with Joseph Chamberlain who demanded "intelligent servility" from women (Diary, January 12, 1884), sponsorship by Charles Booth, and volunteer activity in the Charity Organization Society reinforced her predisposition to "intellectual individuality" (Diary, March 16, 1884). By 1887 she had granted herself permission for the latter, embarking on further training for her vocation. In 1888, she had already begun a symbolic reconstruction of her relationship with her mother in support of her intended career. "So Mother seems to stand by my side, to be watching me, anxious to reach out to me a helping hand, at any rate to bless me. I have been wounded, horribly wounded, and the scar can never leave me, but I can fight through the rest of the bullets of life with courage. And perhaps, when it is over, I shall know that she has been by my side" (Diary, June 29, 1888).

What is clear analytically is that Webb sought the perceived intellectual comradeship and passion shared by her parents but in her life, due to both an evolution of family circumstances in interaction with Victorian constraints upon women, Webb was unable to achieve an easy integration of the two. Her less than complete resolution was partially due to what are now clearly recognizable Victorian system-level factors which not only constrained her options by forcing a dichotomy between femininity and vocation but also shaped her perception of those options. Her resolution was agreeing to an initially passionless marriage and partnership with Sidney Webb, a male who did not question the expression of intelligence in women. She wrote to him, "In the future, my life will be one life only—you and my work bound together" (Letter, January, 1892). As she put it publicly: "Here ends 'My Apprenticeship' and opens 'Our Partnership': a working comradeship founded in a common faith and made perfect by marriage; perhaps the most exquisite, certainly the most enduring, of all the varieties of happiness" (1926, p. 354).

However, Webb's degree of resolution can be called into question, and in one final way, this illustrates further the importance and validity of Mead's social structural past. While Webb's public autobiographies reveal little of her disappointment with Sidney's disinterest with spirituality, there are key passages in her diary which belie the near perfect marriage and partnership she publically portrays, and instead underscore the influence of her past. Webb married Sidney to escape the duality of her life created by family structure and Victorian society, but within two years of her marriage she questions her decision to remain childless (Diary, July 28, 1894), and by 1900 she was haunted by a continuing interest in Chamberlain. Nord (1985) explains the link between the two as ". . . imply(ing) the acceptance of a mind/body split that Webb could not seem to escape. If Sidney Webb was the man with whom to write books, then perhaps Joseph

Chamberlain, for whom her feelings were explicitly sexual, would have been the right man with whom to conceive a child" (1985, p. 224). Thus, her past was never fully resolved in the present and as a result the present was continuously resonating throughout the past, seeking resolution for the future.

3.   The Implied Objective Past

The implied objective past refers to the "obdurate realities in the past, e.g., since I am here, I must have arisen this morning and driven here" (Maines, 1983, p. 164). Citing Mead, Maines et al. continue definition of this dimension: "The past is what must have been before it is present in experience as a past" (Mead, 1929, p. 238). According to Maines et al. (1983), unlike other dimensions of Mead's theory, reference to this element of the past does not include the meaning that the past has for the present. Instead, this dimension "is referring to the existence of previous events, and is proposing a situational ontology pertaining to consensus about the facts of the past."

Further definition of the implied objective past clarifies its relevance and application for our use here. First, these pasts are implied and exist only through memory. In this regard, they come closest to the pasts that psychologists acknowledge, that is, events stored and unitarily retrieved. Second, "memory selects out of the past those events which only could have occurred, given the present structure or arrangement of events" (Maines et al., 1983, p. 164). Thus, the second aspect does include an element of the present influencing the past. The present is selecting from memory only those events which could have occurred, given the existing present. A brief example will be drawn from Webb for illustrative purposes only.

Webb writes: "The first scene I remember was finding myself naked and astonished outside the nursery door, with my clothes flung after me, by the highly trained and prim woman who had been engaged as my brother's nurse. What exactly happened to me on that particular morning I do not recollect" (1926, p. 51). This passage clearly recounts in Webb's present as she wrote her autobiography the occurrence of a past event but not the events leading up to it. Clearly, events in the past preceeded Webb's finding herself naked, even though she cannot recall them for us. She implies an objective past albeit one she does not remember. Webb does not clarify the long term symbolic importance of this incident for us, except as an indication of personal displacement in her family configuration. It stands in contrast to the central importance she had for Sidney Webb and the Fabian Society later in life.

4.   The Mythical Past

The mythical past according to Maines, et al. (1983) is the non-factual recall of previous events for overt manipulation of the past in service of the present.

Based upon creation of the past, not recollection of it, the mythical past is based upon legends and other kinds of apocryphal stories. In general, evidence of the mythical past is difficult to find in Beatrice Webb's writings. It appears that Webb's training as a social investigator made her reluctant to actively engage in ad hoc reinterpretation of her past for the needs of the present, at least as she presents her life in her autobiography. Her preference for historical veracity and distaste of being "wise after the event" is stated clearly in her autobiography (1926, p. 80). While the possibility exists that her autobiography, taken as a whole, is an illustration of the mythical past, we accept for our purposes here her self-description at face value.

However, we are able to identify aspects of the mythical past dimension of Mead's theory by turning to biographical writings of others, writings which include both descriptions and attributions about the impact of the Webbs on British politics. Here we see in rather stark contrast to the Webb's public attributions about themselves the creation of legends about the means of the Webbs' political influence. First, we summarize how they portray themselves. Second, we discuss how others portray them. And third, we discuss the mythical elements in the writing of others, and to the extent that they can be found, in the writings of the Webbs.

Late in life the Webbs, in commentary upon their joint career, often referred to themselves as "servants" to society. "The most honourable of all titles, they used to say, is that of servant; and they lived as they thought" (Tawney, 1953, p. 5). In fact Beatrice Webb liked to refer to the political activity of Sidney and herself as mere "publicists" (Muggeridge and Adam, 1968, p. 245). Sidney and Beatrice have been described as "prolific pamphleteers" (Tawney, 1953, p. 4). Beatrice Webb's self-description as a mere "social investigator" belies the power that any of her collections of evidence had upon a political argument. There are also anecdotal records of her interpersonal style which ranged from warm to exceedingly stern.

However, there is no doubt that the combined influence of Sidney and Beatrice Webb, as central figures among the Fabians, belies the Webbs' public modesty. "The Webbs were generally regarded, both in this country and abroad, as the intellectual leaders of British Socialism" (Proceedings of the British Academy, Volume 29, p. 14). They maneuvered vast and controversial political and policy agendas to completion between 1880 and 1930, including reform of trade unionism, cooperation, and the rise of the Labour Party. They also reformed public attitude on industrial policy, financial policy, public education and public health, unemployment, the Poor Law, municipal enterprise, and local government in general. The social impact of their political reforms had to be reckoned with, and after Beatrice's death in 1943 and Sidney's in 1947, recollections among Fabian society descendants of the Webbs' influence found their way into print and public lecture, explaining for the reader an image of power within a by then disarrayed Fabian society.

Neutral biographical descriptions of Beatrice Webb portray her as a persuasive and influential activist. Earlier in her career, "Beatrice was in her prime, and full of energy and ideas. She was a striking, if austere, hostess, who used her salon so effectively to promote Webbian policies that she deservedly earned a reputation for political intrigue; she was a forceful and uncompromising member of the Royal Commission on the outmoded Poor Law; and it was she, rather than Sidney, who controlled the national campaign which the Webbs subsequently launched to promote their own proposals for the break-up of the Poor Law and the creation of a welfare state" (MacKenzie and MacKenzie, 1982, p. xiii). Beatrice was instrumental in Sidney's winning election to Parliament from the mining community in Seaham in county Durham. Later in life, according to these accounts, she did not hesitate to use her credible reputation to sway opinion through public lectures and radio broadcasts over the BBC. "Mrs. Webb, since her evidence before the Lord's Committee on Sweating, had been known as a woman who could be relied on to handle thorny questions with candor and without sentiment" (Proceedings of the British Academy, Volume 29, p. 17).

Because the Webbs' career accomplishments were extensive, some attribution had to be made about their techniques for success. We find initial evidence of differences between the Webbs' perception of themselves and references to the public's perception in biographical writings of their students. The Webbs create their own mythology by describing their methodology as the "Webb speciality," an approach which they declared brought knowledge simultaneously drawn from historical and analytic data to bear on a social problem. They asserted their strategy was to persuade through appeals to knowledge, not ignorance (Proceedings, pp. 6–7).

In contrast, their political opponents offer a different mythical portrayal of the Webbs' past. The publication *Punch* assigned the derogatory term "Sidneywebbicalism" to the Webbs' vigorous political activism. A Webb Memorial lecturer makes reference to public perceptions of the "furtive paw" and "hidden hand" of the Webbs (Tawney, 1953. p. 5). Beatrice Webb's niece Muggeridge and co-author Adam (1968, p. 172) state: "They set about fostering a socialist evolution by deliberately cultivating the acquaintance of those in positions of power and of authority in order to persuade or bamboozle them into putting socialist legislation onto the statute book." The campaign to change the Poor Law stands out as a period of extreme political activism on Beatrice's part and she understood its function, even if it seemed tiresome at the time. As Muggeridge and Adam state: ". . . for the first time she had (almost) tasted the heady experience of pushing through a major reform of her own, just as she had conceived it, and after that, the normal slow process of permeation and compromise and manipulation, of fighting every step and giving way when necessary, looked unbearably drab."

Others were more severe. Nord states: "the historian Robert Scally, describing the Webbs' courting of imperialist politicians in the 'Coefficients Club,' writes of the machinations of 'Beatrice Webb, the accomplished hostess,' overseer of 'interesting little dinners,' and of 'Beatrice, the Fabian' whose obsession was the 'permeation of potential leaders.' She commits a double offense in two different but complementary personae. As the Gradgrindism of the Fabians is exaggerated in representations of Beatrice Webb, so is this aspect of manipulation and, appropriately, seduction. Virginia Woolf described Beatrice not as an eagle but as a spider: she recorded going to her first Fabian meeting and seeing 'Mrs. Webb, seated like an industrious spider at the table; spinning her webs (a pun!) incessantly' " (1985, p. 5).

Thus, we exemplify the mythical past dimension of Mead by contrasting the image the Webbs asserted about themselves with others' discussion of the Webbs. The Webbs have two mythical pasts. Depending upon one's purpose either of these pasts can be involved as the "correct, valid, and true" interpretation of the past which is only of use to us in light of the present. In her autobiography, Webb was intent upon portraying her political accomplishments in a manner consistent with the rest of her life: moral, informed, and without compromising integrity. Her past successes allowed her to assert a legend that was consistent with her original mission. At least from her viewpoint in the present, she had not been corrupted or destroyed by political compromise. Her opponents and critics, of course, felt otherwise. How could have one achieved such far-reaching political reform on sheer integrity and intelligence, without compromising one's ideals. Which mythical past we accept as observers depends, according to Gergen and Gergen (1983), to some extent on our present, that is, the extent to which in our social context we ought to accede credibility to her narrative. Mead's dimension of the mythical past creates the possibility of considering both narrative accounts for use in our past.

DISCUSSION

With increased awareness of personal documents for the study of the life cycle, we have sought evidence for the sociological process underlying the social construction of self-narratives, with particular attention to its clarification of the role of meaning in autobiographical accounts of individual lives. Our primary interest in exploring Mead's theory of the past is to extent sociological thinking about "current frames of reference," that is, the effect of the present on the remembered past (Strauss, 1979) to discussion of its constituent social processes and elements. Maines et al. explicated four dimensions in Mead's theory, and we in turn explored their utility through the life documents of Beatrice Webb. By comparing her public, autobiographical account of her life with her private diaries, we have uncovered evidence in support of Mead's four dimensions. But we have also uncovered sociological limitations in Mead's theory. Before turning our attention to those limitations, we discuss the sociological contribution of Mead to interpreting autobiographical documents in particular.

Above all, Mead has illustrated the importance of how identity is socially negotiated, and how identity extends to encompass events from our past. Individuals we interact with have memories, and since behavior in our culture is expected to be consistent across time, as well as purposive and true to our "true" self, we seek explanations of past behavior which in some cases involve reconstruction, and sometimes reinterpretation of the past. The more public our identity, the more consistency is expected, and in response to that, the more continuity the author seeks to offer. A public figure like Webb had an enormous audience to whom her identity was accountable. But even individuals leading ordinary lives have audiences to whom they are accountable, even if it is only one's spouse, therapist, child, or creator, if one believes in God. Regardless of the size or importance of one's audience, the self-narrative still has to be negotiated, and the challenge to sociologists is uncovering the event being reconstructed and the occasion that precipitated the renegotiation of the past.

Mead's theory of the past also clarifies the origin and purpose of meaning to an individual's life account. All self-narratives, and especially autobiographies as metaphors of self, seek links across previous events to establish continuity in lines of conduct. Meaning lies between what actually happened the past and what continuity the author is compelled to assert, given that it is the author's "present" that needs explanation. For example, late in life, Webb described her movement toward socialism as something that occurred in stages, as an "evolution" into a socialist (Muggeridge and Adam, 1968, p. 51). Webb's overall success as a socialist *required* that she portray her transition through life as intentional, even when it was not so obvious to herself. Because the phenomenon of reinterpretation underlies the search of meaning in construction of self-narratives, sociologists using personal documents such as autobiographies can only cautiously rely upon them for valid descriptions of a life. Establishing the veracity of public accounts of an individual's life should be a fundamental objective in sociological use of materials such as these.

On the other hand, through our case study of Beatrice Webb's life documents for her use of the past in negotiation of self-narrative, it is apparent that Mead's theory of the past is sociologically limited. Mead has offered an insightful elucidation of the *process* whereby the present brings reinterpretation of the past, but he offers us little to clarify how or under what conditions this happens. As Gergen and Gergen state, "Narrative construction can never be entirely a private matter. In the reliance on a symbol system for relating or connecting events one is engaging in an implicit social act. . . . its position within a meaning system is shared by at least one other person" (1983, p. 268).

What Mead does suggest generally is that novelty, or a break in continuity in the succession of events to the present, precipitates reinterpretation or reconstruction of the past. Without novelty itself, continuity could not be discernable, particularly in establishing one's identity. Webb's documents illustrate this nicely, even if we cannot observe direct interactional precipitates in her life. On several occasions, Webb realized that her expectations were going unfulfilled: her marriage with Sidney, her trust in science, her satisfaction with organized religion. In each situation, we see a growing awareness, often in the form of dissatisfaction, with the turns her life was taking. The break in continuity brought with it the realization that the events she had experienced were not progressively moving her toward the goal to which she aspired.

What are the behavioral and situational elements which contribute to the construction of self-narrative? At an intrapersonal level self-awareness is essential for the conscious construction of any form of self-narrative. It is often assumed that personal identity achieved in self-narrative is simply an expected maturational accomplishment. Psychologists attribute self-awareness to "the mature individual, (who) on this account is one who has 'found,' 'crystallized,' or 'realized' a firm sense of self or personal identity" (Gergen and Gergen, 1983, p. 256). However, Gergen and Gergen argue it is more likely that the individual does not arrive at a stabilized state of mind. Rather, he or she develops the capacity for understanding him or herself in this manner and for communicating this understanding creditably to others. One does not acquire a state of "true self" but a potential for being "true to one's self, at least one's public self," that is, communicating that self-awareness is possessed. In fact, even self-awareness would seem to be based upon perception of discontinuity across situational events rather than any increase in a sense of one's "true" self or departures from it.

But more important sociologically, self-narratives, including those reconstructing the past in light of the present in support of identity, require acquiescence among interactants for them to be successful. Self-narratives are public accounts of identity and their construction requires social negotiation. For example, Webb's task was to have the account of her life structure accepted by the general public. When it was, she notes that the reviews were "unexpectedly good; and my self-esteem ought to be satisfied" (Webb quoted in Muggeridge and Adam, 1968, p. 223). If interactants are unwilling participants, as in a case of unresolved conflict where interactants are at an impasse, then one's identity is not accepted as established across time, at least with that individual or group. That is, in an ongoing relationship, significant others must agree to one's interpretation. Had Webb's reviews been negative, her identity would have been questioned. Indeed there is some evidence of this occurring during her lifetime and after her death as seen in our discussion of the implied mythical past. Gergen and Gergen argue that there is a delicate interdependence between constructed narratives, and that reciprocity between interactants is essential for the negotiation of meaning. Without agreement, narratives about identity remain unaccepted and identity remains unestablished, or at least in question.

Finally, Mead does not make direct reference to the impact of societal-level social structural past upon the negotiation of identity in the present. Social and cultural constraints upon Victorian womanhood had a profound impact upon Webb's life, and in retrospect predisposed her to a limited range of life options and limited her perceptions of these options. Nord (1985) carries this issue further: She argues that the basis of Webb's private identity as manifested in her diaries centered on Victorian themes prevalent in the memoirs of Webb's contemporaries. These included the sense of inner division, the elusiveness of integration of self, the struggle against fantasy, the admonishment for vanity and egoism, the difficulties of study and work, and the question of marriage and its tyrannies. "In *My Apprenticeship,* however, these separate problems are raised to the level of structure-giving themes . . . creating a pattern and teleological form" (1985, p. 82). As Webb incorporated these elements into her self-narrative, she was in fact, incorporating direct evidence of societal-level influences in her identity. Nord also argues that social limitations upon notions of work in the lives of Victorian women limited the possibility of integration work with marriage, let alone children. Choosing a vocation raised the issue of one's sexual identity.

In conclusion, Mead has provided an elegant elaboration of a process through which identity is achieved through the use of the past in the present. Through our exploration of his theory of the past, using Beatrice Webb as a case study, we found limitations in its use, centering around the social construction of self-narrative, notably autobiography. We suggest the following agenda to remedy this.

First, there is need for systematic examination of the kinds of events around which the process of modification of the past occurs and whether there are patterns in its public justification. We suggest that one should be able to observe integration and reinterpretation through scrutiny of life documents written by the same individual at different life intervals about the same event. We feel a comparative approach across personal documents will continue to account for intrapersonal distinctiveness.

Second, there is need for systematic examination of how the process of integrating the past with the present varies depending upon the audience for which the personal document is intended. Do public figures who are negotiating identity with the general public simply face a larger task or a different task from a mother negotiating a "good parent" identity with an adolescent daughter? We suggest one ought to be able to observe degrees of veracity across documents regarding a specific event, but which may also vary as a function of time passage. In conclusion, by recognizing how meaning is achieved in self-narratives, we have begun a systematic search for the sociological processes underlying the construction of self-narratives.

ACKNOWLEDGMENT

We wish to thank Tim Silva for bibliographic assistance at the early stages of this project.

NOTES

1. Life stories, life histories, and autobiographies are recollected structured personal accounts of one's life. A life story is an oral account of a person's life, a life history is an oral account supplemented by biographical information from other individuals. An autobiography is a self account written in established literary form. Biographies are the structured telling of another's life. In contrast to structured forms are diaries which are ongoing unstructured personal records kept by an individual for private use. Diaries are usually not intended for public audiences.
2. Citations to "Diary" are from MacKenzie and MacKenzie, 1982–85.

REFERENCES

Baltes, Paul B. and O.G. Brim. 1980. *Life-Span Development and Behavior.* Vol.3. New York: Academic Press.

Bertaux, Daniel. 1981. "Introduction." Pp. 5–15 in *Biography and Society,* edited by D. Bertaux. Beverly Hills: Sage.

Bertaux, Daniel and Martin Kohli. 1984. "The Life Story Approach: A Continental View." Pp. 215–237 in *Annual Review of Sociology.* Palo Alto: Annual Inc.

Butler, Robert N. 1968. "Toward a Psychiatry of the Life Cycle: Implications of Socio-Psychological Studies of the Aging Process for Policy and Practice of Psychotherapy." Pp. 233–248 in *Aging and Modern Society: Psychosocial and Medical Aspects,* edited by A. Simon and L.J. Epstein. Washington, DC: American Psychiatric Association.

Cohler, Bertram. 1982. "Personal Narrative and the Life Course." Pp. 205–241 in *Life-Span Development and Behavior.* Vol.4, edited by P.B. Baltes and O.G. Brim. New York: Academic Press.

Cole, Margaret. 1946. *Beatrice Webb.* New York: Harcourt Brace.

Crapanzano, Vincent, Yasmine Ergas, and Judith Modell. June 1986. "Personal Testimony: Narratives of the Self in the Social Sciences and the Humanities." *Items.* New York: Social Science Research Council 40 (2):25–30.

Dannefer, Dale. 1984. "Adult Development and Social Theory: A Paradigmatic Reappraisal." *American Sociological Review* 49:100–116.

Elder, Glen. 1974. *Children of the Great Depression.* Chicago: University of Chicago Press.

Elder, Glen. 1981. "History and the Life Course." Pp. 77–115 in *Biography and Society,* edited by D. Bertaux. Beverly Hills CA: Sage.

Featherman, David L. 1981. "The Life-Span Perspective." Pp. 621–648 in The National Science Foundation, *The 5-Year Outlook on Science and Technology* 2. Washington, DC: USGPO.

Featherman, David. 1983. "Life-Span Perspectives in Social Science Research." Pp. 1–57 in *Life-Span Development and Behavior* 5, edited by P.B. Baltes and O.G. Brim. New York: Academic Press.

Ferrarotti, Franco. 1981. "On the Autonomy of the Biographical Method." Pp. 19–27 in *Biography and Society,* edited by D. Bertaux. Beverly Hills: Sage.

Freud, Sigmund. 1982. "An Early Memory from Goethe's Autobiography." Pp. 64–72 in *Memory Observed,* edited by Ulric Neisser. San Francisco: Freeman.

Gergen, Kenneth J. and Mary Gergen. 1983. "Narratives of the Self." Pp. 254–389 in *Studies in Social Identity,* edited by T. Sarbin and K. Scheibe. New York: Praeger.

Gottschalk, Louis, Clyde Kluckhohn, and Robert Angell. 1945. *The Use of Personal Documents in History, Anthropology, and Sociology.* Bulletin 53. New York: Social Science Research Council.

Kohli, Martin. 1981. "Biography: Account, Text, Method." Pp. 61–75 in *Biography and Society,* edited by D. Bertaux. Beverly Hills: Sage.

Lieberman, Morton and Jacqueline Falk. 1971. "The Remembered Past as a Source of Data for Research on the Life Cycle." *Human Development* 14:132–141.

Linton, Marigold. 1982. "Transformations of Memory in Everyday Life." Pp. 77–91 in *Memory Observed,* edited by Ulric Neisser. San Francisco: Freeman.

MacKenzie, Norman. 1978. *The Letters of Sidney and Beatrice Webb.* Volumes 1–3. Cambridge: Cambridge University Press.

MacKenzie, Norman and Jeanne MacKenzie. 1982–1985. *The Diary of Beatrice Webb.* Volumes 1– 4. Cambridge, MA: Harvard University Press.

Maines, David R., Noreen Sugrue, and Michael Katovich. 1983. "The Sociological Import of G.H. Mead's Theory of the Past." *American Sociological Review* 48:161–173.

Mannheim, Karl. 1928. *The Problem of Generations: Essays on the Sociology of Knowledge.* New York: Oxford University Press.

Marini, Margaret M. 1984. "Age and Sequencing Norms in the Transition to Adulthood." *Social Forces* 63:229–244.

Marini, Margaret M. 1985. "Determinants of the Timing of Adult Role Entry." *Social Science Research* 14:309–350.

Mead, George H. 1929. "The nature of the past." Pp. 235–242 in *Essays in Honor of John Dewey,* edited by John Coss. New York: Henry Holt.

Mead, George Herbert. 1932. *The Philosophy of the Present.* Chicago: Open Court Publishing Company.

Mead, George H. 1938. *The Philosophy of the Act.* Chicago, University of Chicago Press.

Mills, C. Wright. 1959. *The Sociological Imagination.* New York: Oxford University Press.

Muggeridge, Kitty and Ruth Adam. 1968. *Beatrice Webb: A Life 1858–1943.* New York: Knopf.

Neisser, Ulric. 1982. "Memory: What Are the Important Questions?" Pp. 3–19 in *Memory Observed,* edited by U. Neisser. San Francisco: Freeman.

Nord, Deborah Epstein. 1985. *The Apprenticeship of Beatrice Webb.* Amherst, MA: University of Massachusetts Press.

Olney, James. 1972. *Metaphors of Self.* Princeton, NJ: Princeton University Press.

Pomerleau, Cynthia. 1980. "The Emergence of Women's Autobiography in England." Pp. 21–46 in *Women's Autobiography,* edited by E. Jelinek. Bloomington, IN: Indiana University Press.

Proceedings of the British Academy. n.d. "Beatrice Webb: 1858–1943." Vol XXIX. London: Humphrey Milford.

Riley, Matilda. 1976. "Age Strata in Social Systems." In *Handbook of Aging and the Social Sciences,* edited by R. Binstock and E. Shanas. New York: Van Nostrand Reinhold.

Ryder, Norman B. 1965. "The Cohort as a Concept in the Study of Social Change." *American Sociological Review* 30:843–861.

Sorensen, Aage, Fr. Weinert, and Lonnie Sherrod. 1986. *Human Development: Interdisciplinary Perspectives.* Hillsdale, NJ: Lawrence Erlbaum.

Strauss, Anselm. 1979. "Mead's Multiple Conceptions of Time: Context and Consequence." Paper prepared for Colloquium on Mead, University of Constance, June.

Tawney, R.H. 1945. "The Webbs and Their Work." *Webb Memorial Lecture,* May 11, 1945. London: Fabian Publications Ltd.

Tawney, R.H. 1953. "The Webbs in Perspective." *Webb Memorial Lecture,* December 9, 1952. London: Athlone Press.

Thomas, W.I. and Florian Znaniecki. 1918–1920. *The Polish Peasant in Europe and America.* Chicago: University of Chicago Press.

Webb, Beatrice. 1926. *My Apprenticeship.* London: Longmans, Green and Co., Ltd.

Weintraub, Karl. 1975. "Autobiography and Historical Consciousness." *Critical Inquiry* 1:821– 848.

TOWARD A SOCIAL PSYCHOLOGY
OF REMINISCENCE

David Unruh

INTRODUCTION

In recent years, the phenomenon of reminiscence and its utility for adaptation, coping, and maintaining a sense of self into old age has emerged as a major focus of interdisciplinary research on aging. However, despite widespread theorizing (cf. Butler, 1963; Erickson, 1950; Marshall, 1980) and much empirical research, a number of theoretical, conceptual, and methodological problems remain. Paramount among these problems is that the phenomenon of reminiscence remains imprecisely defined and its nature unclear. Specifically, some researchers have defined it as "the act or process of recalling the past" (Butler, 1963), "linguistic acts referring to the remote past" (Coleman, 1974), "retrospection both purposive and spontaneous" (Havighurst and Glasser, 1972), or simply "the remembered past" (Lieberman and Falk, 1971). Many others have proceeded in the

Current Perspectives on Aging and the Life Cycle,
Volume 3, pages 25–46.
ISBN 0-89232-739-1

direction of measurement without defining it at all (cf. Boylin et al., 1976; Costa and Kastenbaum, 1967; Fallot, 1980; Lewis, 1971; McMahon and Rhudick, 1967).

Beyond definitional imprecision, the relationship among reminiscence, the life review, long-term memory, and other concepts remains unclear, as does the amount of time one must go back to for an episode of memory and recall to be termed "reminiscence" (see Merriam, 1980). With research on reminiscence dominated by psychiatric, psychological, or social work orientations, the social and situational nature of the process has been downplayed in deference to matters of psychological adaptation, guilt assuagement, identity continuity, and the universality of the phenomenon. The orientation of many researchers toward personal adjustment and adaptation among the aging has discouraged inquiry into the multifaceted character of the phenomenon (cf. Ebersole, 1978; Lewis and Butler, 1974; Liton and Olstein, 1969; Pincus, 1970; Zeiger, 1976).

The focus of this paper is on the possibility of a sociological social psychology of reminiscence with a shift away from personal adaptation and psychological well-being toward analysis of reminiscence as a *social process*. From a sociological standpoint, we must note that people explicitly and implicitly seek to manage the content, occurrence, imagery, and effects of reminiscences. A social psychology of reminiscence must also attend to the social settings and circumstances wherein reminiscences emerge. Recent conceptualizations of reminiscence will first be discussed and, at the same time, the phenomenon will be developed into a subject worthy of sociological investigation. Finally, the essay will focus on four research areas by which a social psychologist reconceptualization of the phenomenon might occur.

PROBLEMS IN THE CONCEPTUALIZATION OF REMINISCENCE

This section focuses on two conceptual problems which characterize current research on reminiscence: (1) Ambiguities regarding the relationship between reminiscence and related phenomena, conflicting images of what constitutes reminiscence, and factors which influence it; and (2) Measurement techniques which are attuned to limited aspects of the phenomenon which often contradict the social and processual nature of reminiscence in everyday life.

Ambiguities with Related Phenomena

Some observers have not differentiated reminiscence from similar phenomena. The notion of a "life review" has emerged with many of the same images, purposes, and functions associated with reminiscence. Butler (1963, p. 67) has defined the life review as:

. . . a looking-back process that has been set in motion by looking forward to death and potentially proceeds toward personality reorganization. Thus, the life review is not synonymous with but includes reminiscence; it is not alone either the unbidden return of memories, or the purposive seeking of them, although both may occur. . . . Presumably this process is prompted by the realization of approaching dissolution and death, and the inability to maintain one's sense of personal invulnerability.

Other researchers have developed several categories of reminiscence with one being "evaluative" and considered synonymous with the life review (cf. Lieberman and Falk, 1971; Lo Gerfo, 1980). From a social psychological perspective, it is necessary to understand how, when, and why reminiscence is transformed into the life review. We know that unresolved personal conflicts or discontinuities are said to give rise to increased self reflection with the intent of resolving, reorganizing, and reintegrating that which is troubling (cf. Havighurst et al., 1968; Marshall, 1980; Mischel, 1969; Myerhoff, 1978; Rosen and Neugarten, 1964). It is also apparent that unresolved conflicts and discontinuities in people's lives probably increase with age. However, most analysts also agree that reminiscence occurs as early as childhood when people have a stock of experiences and memories which may form episodes of reminiscence.

Orientations toward the past, present, and future change over time as people compare their lives with others and adjust to a life cycle perceived as "predictable" (Neugarten, 1979), engage in age appropriate behavior (Neugarten, et al., 1965), and estimate the course and length of their lives (cf. Cain, 1978; Marshall, 1974; Reynolds and Kalish, 1974; Teahan and Kastenbaum, 1970; Tolor and Murphy, 1967). Havighurst and Glasser note that the ratio of attention given to the past compared to the future through the life course "probably changes systematically, with an increase of the pats/future ratio as a child grows up, with a past/future ratio of approximately one in old age and a ratio exceeding one in the very old person" (1972, p. 245). This statement reflects developmental trends, but we must attend to the highly situational character of orientations toward the past, present, and future.

Mead's theory of time and the past is useful for highlighting the importance of the present in affecting interpretations of the past and the future. Neither the past nor the future exist in and of themselves but are, instead, components of the present. Mead acknowledged that a person's past has a certain irrevocability and that it can have no meaning apart from its relevance for the present (see Maines et al., 1983, p. 162). That is, one's past might be documented on the basis of seemingly unquestioned facts such as place of birth, schools attended, where a lover was found, and so on. The meaning and relevance of these events emerges in the present as the person adjusts to new situations and events (Mead, 1929; 1932). The imagery of Mead's thought is that people continually adjust, reinterpret, and align their pasts so the present becomes comprehensible. Symbolic reconstruction of the past involves redefining the meaning of previous events so as to render them meaningful and useful for the present (Maines et al., 1983, p.

163). The activity's importance is that discontinuous events are assimilated into a comprehensible flow making the life course fluidic, connected, and manageable.

Reminiscence and the life review are extensions of processes undertaken by all people in everyday life. That is, reinterpretation of the past is a continual human endeavor, but there are times and places where it is magnified in social psychological importance, marked by shifts in personal awareness, and characterized by a focus on matters of personal identity. The differences in intensity, degree, and function among the routine reinterpretation of the past in everyday life, reminiscence, and the life review represent different orders of the same process. Of course, these orders are not concrete entities, but "sensitizing concepts" which emphasize changes in a fluidic and amorphous process (see Blumer, 1967).

First order reinterpretation routinely occurs in an often unsystematic and spontaneous way and is related to the notion of selective memory where aspects of the past are relegated to the "far recesses" of the mind, while others are maintained at the forefront of consciousness. It merges the past with the present as previous experiences, feelings, identities, or facts are reinterpreted in light of present circumstances and the projected future. It is likely that the recent past will be the object of reinterpretation, but distant memories may also be included. For example, people reflect upon recent events in order to plot a course of action in the present. As a part of everyday life, then, first order reinterpretation incorporates memories and conceptions of the past into present consciousness. With little conscious effort, a reinterpreted past lives on in the minds of people as they align their actions in the present and plot a course for the future.

An important shift in reinterpretation occurs when images and memories are no longer stable and continuous elements of everyday consciousness. That is, another level of reinterpretation focuses on long-term memories which are not at the forefront of consciousness or integrated into the person's present thoughts and behaviors, but have been brought to mind through particular occurrences. The literature on long-term memory notes the degree to which it is a byproduct of the transaction between the individual and context—with reminiscence heavily relying upon long-term memory (Kvale, 1977; Meacham, 1977). Reminiscence is a *second order reinterpretation* of the past whereby memories of experiences, feelings, identities, or facts not only lack stability and continuity in the stream of everyday consciousness, but are firmly rooted in *situations* which have been invoked in the present. Here reminiscence is distinguished from long-term memory which may involve the recall of simple facts, faces, actions, and the like divorced from the social contexts within which they occurred. Reminiscing, on the other hand, involves the recall of detailed memories which are partial reconstructions of events or interactions. Let us consider the memories an adult may have of a grandparent's garden. The recall of instructions by the grandparent on how to garden, planting techniques, or kind of vegetables grown may be incorporated into current activities and influence current actions. Similarly, mental snapshots of an apricot tree, colors of the flower blossoms, or hot summer days in the garden may be evoked by eating an apricot or seeing similar flowers in the present. These images have been stored in long-term memory and invoked in the present, but they do not constitute episodes of reminiscence as here conceived. Instead, they are first order reinterpretations of the past divorced from their original social situations. These images and memories ease into second order reinterpretation when the actions of individuals, snippets of conversations, and additional visual, auditory, and interactional detail are invoked. Memories of the past become episodes of reminiscence when they are pieced together and form *situations* in which people are active components. The memories might not be accurate representations of what happened in an objective sense, but they are subjective renderings of previous situations reinterpreted through present contingencies, needs, or desires. The shift from first to second order reinterpretation may be gradual, but a qualitatively different orientation toward long-term memories is apparent.

Some analysts have noted that reminiscence may be distinguished from long-term memory when people engage in "linguistic acts" by which the contents are transmitted to others (cf. Coleman, 1974; Costa and Kastenbaum, 1967; McMahon and Rhudick, 1967). However, verbal attempts to "infuse others" with the contents of reminiscences is a way others become aware of people's reminiscences, but not their presence or absence in the minds of individuals. Much reminiscence is internal and unspoken by people often unaware of its occurrence. The definition of reminiscence to be used throughout this essay is as follows: *The recall to mind of experiences, feelings, identities, or facts rooted in situations of the relatively remote past which are not stable and continuous elements of everyday consciousness.*

Another qualitative difference in intensity and content marks transition into *third order reinterpretation.* This level corresponds with most analysts' conception of the life review wherein people reevaluate their lives and seek to make sense of their pasts as death approaches (cf. Butler, 1963; Lieberman and Falk, 1971; Marshall, 1980; Myerhoff, 1978). The shift from first to second order reinterpretation occurred when details of situations were recalled. Emergence of the life review, however, is signified by *identity work.* This term refers to the ways people review their lives, attempt to make sense of the self, and seek to preserve some personal identities while discarding others. People develop many identities and play a number of roles. With increased awareness of impending death, people sort through these identities and allow some to fade away, while others are remembered and preserved. While at least some identity work is involved in first and second order reinterpretation, the life review is marked by an intensification of this work to the degree that it becomes a significant aspect of the self and a focus of much social psychological time and effort.

Identity work may be consciously pursued as in the case of written autobiographies wherein the past is selectively interpreted and communicated to others. According to Dilthy:

> The person who seeks the connective threads in the history of his life has already, from different
> points of view, created a coherence in that life. . . . He has, in his memory, singled out and
> accentuated the moments which he experienced as significant; others he has allowed to sink into
> forgetfulness (1962, p. 86).

Myerhoff has also noted that the search for a sense of meaning is fundamental to autobiographical endeavors and writers may be both subject and object. That is, it is not necessary for people to have a definite audience in mind, but the act of sifting through the past has social psychological benefits even if no one else reads it.

Identity work also includes activities which are less conscious. For example, Butler (1963, p. 68) notes how "mirror gazing" indicates heightened self reflection as people ponder previous experiences. Even the preparation of wills and testaments, or the distribution of personal artifacts to friends and relatives reflects the outcome of the life review process. Through these activities people may reveal unknown qualities of their character, hidden themes in their lives, previously undisclosed opinions, or the changed nature of long-term relationships. Rosenfeld (1979) notes that friendships found by older people in nursing or retirement homes may first be revealed to their children through wills and testaments read after death. Some elderly residents have left substantial sums to nursing home friends or institutions at the expense of other survivors.

Immersion into the life review process is commonly associated with aging, but a number of other factors come into play. Old age may afford people the time and freedom from responsibilities and facilitate self-reflection, but heightened awareness of impending death is also influenced by comparison of one's own age to that at which parents died (Marshall, 1975), life expectancy of an age cohort (Cain, 1978), exposure to the deaths of others (Tolor and Murphy, 1967), occupational risks (Teahan and Kastenbaum, 1970), and self-perceived health (Marshall, 1980).

Methods and Measurement

Many researchers have failed to define reminiscence and distinguish it from related phenomena, yet some have proceeded in the direction of measurement and quantification. Two studies which vary in approach, image, and analytical technique exemplify some of the limitations which haunt this area of inquiry. Havighurst and Glasser (1967) administered a questionnaire to middle class adults in their 70's designed to measure the frequency and affective quality of reminiscences. It was also designed to conceptualize attitudinal, life style, personality, and perception characteristics of respondents. Lieberman and Falk (1971), on the other hand, sought to conceptualize and quantify reminiscence data as part of a larger investigation of the adaptation and survival of the aged under stress. Their sample consisted of 180 aged people (averaging 78 years) and 25 middle aged people (averaging 49 years).

Data on the frequency and content of reminiscences derived from the self report instruments were crucial for both studies in determining the amount of energy directed toward the past, its functions, and changes from middle to old age. Havighurst and Glasser (1967, p. 246) believed that much silent reminiscence might never be shared verbally with researchers. Reminiscences of sexual episodes, childhood memories, embarrassing situations, or distasteful actions may be downplayed or concealed by respondents revealing yet another aspect of social life managed for the presentation of self (see Goffman, 1959). However, they believed a questionnaire would not receive the same resistance or concealment found in interviews.

Sociologists have long debated the discrepancies between verbal and written responses, observational and questionnaire studies, or attitudinal and behavioral emphases (cf. Becker and Geer, 1957; Campbell, 1964; Deutscher, 1973). It is not important to claim the superiority of one approach over the other, but merely to note that interpreting reminiscence data is extremely problematic. While researchers have been able to immerse themselves in the culture of groups to learn of life within them (cf. Gans, 1962; Liebow, 1967), it is impossible to step inside the mind to observe reminiscences which are spontaneous, free flowing, and fraught with quick changes in mood or imagery. Even Myerhoff (1978) could not get inside the minds of the elderly Jews of Venice, California with whom she interacted at great length, but rather learned their life histories through written autobiographies, workshops designed to elicit self reflection, and informal conversations.

In all but the rarest instances, reseachers must ask people to tell them the contents of their reminiscences. Questions must be asked and directions given which become active features of the situation (cf. Gubrium and Lynott, 1983; Schutz, 1967; 1970). The language of any question, to some degree, produces the kind and limits the range of responses given. For example, Havighurst and Glasser (1967, p. 247) asked their subjects the following questions:

> Events or experiences you recall during reminiscence may be pleasant, unpleasant, or neutral in affect. Try to think how often these occur and fill in the table below.
>
> Were your reminiscences pleasant, unpleasant, or neutral at the following ages: 20–30; 40–50; 60–70?
>
> What do you think reminiscence does for you now?
>
> Is there any difference between your retrospective thinking now and in the past?

The assumption is that reminiscences exist beyond the brief moments in which they emerge in the mind and may objectively be evaluated. Focusing on unpleasant, pleasant, or neutral affects may not be the most salient way of categorizing that which is personally relevant since it is doubtful that some reminiscence is recognized as such by people as they engage in it—much less remembered throughout the life course. It is a mental activity in which people become immersed for brief moments and then snap back into the conscious present. The process ebbs and flows shifting from the past to the present. To require an evaluation of its affective qualities after the fact encourages respondents to "ad hoc" evaluations to fit the research exercise (see Garfinkel, 1967). That is, life may be "circumstantially constructed for consideration, understood within the context of the relevancies at hand, or made over the purpose of examination each time one encounters a question about it" (Gubrium and Lynott, 1983, p. 31). This activity is magnified when people are asked to remember the affective qualities of reminiscences 20, 30, or 50 years ago. Asking what reminiscence "does for them" implies that it has functions which should be recognized and prompts people to speculate in a predetermined way. The prompting is also present in Havighurst and Glasser's questionnaire when they provide the categories of "gives me a good feeling," "helps me understand young people better," and "helps me over a lonesome period." Specific functions are created where none had existed.

Lieberman and Falk relied upon structured questionnaires and the coding of verbal life histories. Their request for subjects to "recount the stories of their lives" was believed to stimulate an episode of reminiscence resembling that which occurs in everyday life. However, reminiscences may be cued or evoked by the sights, sounds, and smells of the natural world. Therefore, it is likely that the content and emotional tone of reminiscences in everyday life differ in important ways from those created in artificial settings. Once again, asking for a "life history" implies a tone, evaluation, and set of experiences which respondents believe should be included. For example, when older people were asked to describe present social involvements, most returned to the far past in order to "set the stage" and began the narrative with the date and place of their birth and proceeded through their life histories noting marked changes or great satisfactions (Unruh, 1983a; 1983b). The request suggested that certain events and junctures associated with the life course ought to be included. To quantify responses to these questions is not the equivalent of studying reminiscences emerging in everyday life.

Verbal reports of reminiscence present another problem regarding language and experience. While language creates a certain reality and way of thinking, it cannot represent the real world completely nor convey the full content of human experience (cf. Chomsky, 1968; Luckmann, 1975; Sapir, 1949; Vygotsky, 1962; Whorf, 1956). Whatever verbal reports received cannot be complete portraits of the processes, images, emotions, and judgments involved because it is a process characterized by stream of consciousness thinking and, perhaps, elements of fantasy which cannot easily be communicated to others. Visual and auditory images, personal dramas, reviews of past scenes, subverbal comments, and internal dialogues jostle one another within the mind (cf. Strauss, Lindesmith, and Denzin, 1977). Verbalizing these occurrences for friends, family, or researchers requires an ordering which condenses the images into words and facilitates verbal communication. However, that which is communicated may not be of the same order as the reminiscences themselves. Here a certain logic may unwittingly be imposed on abstruse memories and images lacking a communicable logic of their own.

This essay argues for a broader conception of methodology than that conventionally employed in reminiscence research. The personal experiences and mental processes of those studying reminiscence should not be dismissed as invalid and unscientific. Sociologists have long studied experiences, conditions, and occupations of which they themselves were a part (cf. Becker, 1963; Faulkner, 1971; Irwin, 1977; Johnson, 1977; Riemer, 1977). The benefit has been insight into the phenomenology of processes unavailable through other approaches. It is profitable for researchers to look inside themselves and gain personal insight into the processes, situations, settings, circumstances, and contents of reminiscence. By noting when, where, and how they themselves reminisce, researchers will construct interview techniques, questionnaires, and other means tailored to the natural occurrence of the process rather than artificial reconstruction.

Awareness of activities in everyday life may tell us about reminiscence in natural settings. Researchers inevitably find people who are telling the stories of their lives, recounting past experiences, and the like in restaurants, movie lines, shopping trips, at the beach, or any number of places. Those studying the aged will most likely find themselves in these situations and should pay heed—even though the focus of their current research may be something quite different (cf. Gubrium, 1975; Hochschild, 1978; Marshall, 1975; Matthews, 1979; Myerhoff, 1978). An opportunistic stance might be adopted whereby data on reminiscence will be seized upon in the course of everyday affairs. Guided observations in the natural world for the expressed purpose of gathering reminiscence data is the next step. Retirement settings, hospital waiting rooms, laundromats, beauty parlors, coffee shops, pool halls, and park benches are places where people might be found free from time constraints and occupational pressures which inhibit self reflection. Here are the ways reminiscences are presented and communicated to other might be explored.

The use of personal diaries has also proved useful and nicely adapts to the study of reminiscence (cf. Alport, 1942; Campbell, 1955; Dean et al., 1969; Dollard, 1935; Zimmerman and Wieder, 1977). The diaries in mind are not the "intimate journals" people keep for their own private purposes, but observational logs which serve as the basis for intensive interviewing. Respondents are given diaries and explicit instructions regarding the kind of information and experiences which should be recorded. For a social psychology of reminiscence the times, places, moods, companions, and other circumstances surrounding the occurrence of reminiscences are basic.

The rationale of this approach involves more fully exploiting people as observers and informants (Zimmerman and Wieder, 1977, p. 484). The first step requires subjects to be observers of their own experiences, to record their impressions, and to be introspective regarding the emergence and role of reminiscence in their lives. The second step uses diary entries as the basis for lengthy, detailed interviews. That is, the personal observations of people are used as points of departure to explore the matter with greater depth. Subjects are cast in the role of informant and encouraged to reflect upon their own observations which are in their own words, and to describe their images and feelings. The interviews would not be based on preconceived notions and images imposed by the researcher. A more complete portrait might be developed which would modify existing questionnaire and interview schedules to be attuned to reminiscence in everyday life.

SUGGESTIONS FOR REMINISCENCE RESEARCH

This section focuses on four areas of inquiry which a social psychology of reminiscence must address. The topics are not minor modifications of psychological, psychiatric, or social work orientations, but suggestions for sociological understanding. These areas are the: (1) Social process by which reminiscences are stimulated and structured by social situations; (2) Ways reminiscence is a personal accomplishment with people constructing, stimulating, and controlling its content and occurrence; (3) Importance of place in affecting reminiscence episodes through the life course; and (4) Role of emotional states in reminiscence and emotional outcomes as a way of understanding the affects, functions, and social psychological importance of reminiscence.

The Social Process

In contrast to the conventional image of reminiscence as spontaneous and impulsive, this section emphasizes the social character of the phenomenon. It is logical to begin with social influences on the initiation of reminiscence in everyday life.

As people conduct their affairs they encounter situations which either replicate portions of past experiences, or have acquired symbolic meanings which provide continuity from the past to the present. One elderly widower interviewed described a situation which sparked reminiscences of his deceased wife.

> I have been living in this apartment for five years now. I get lonely now and then. . . . When my wife was here, she wanted to know where I was and what I was doing all the time. She did this with the kids and me for many years. I guess it was sort of a bone of contention for some time, and we argued about it. Now that I'm by myself, I kind of miss it. You know, all she was trying to do was be a good wife. . . . It's funny, but from time to time for several years after she died, I would remember her nearly everytime I left the house to get someplace. . . . I would remember what she said to me when I didn't tell her where I was going, and how I felt angry most of the time.

It is possible to see that repeating an act which once evoked negative feelings toward a now deceased wife now sparks episodes of reminiscence more favorably colored. According to the informant, reminiscences of this kind routinely emerged shortly after his wife's death, but continued sporadically for over five years. The onslaught of reminiscences about the recently deceased is widely noted in the survivorship literature, as is the gradual decrease in those images over the years (cf. Charmaz, 1980; Glick, Weiss, and Parkes, 1974; Lopata, 1979; Unruh, 1983b).

More interesting are circumstances where everyday activities unexpectedly shift attention to the past. These episodes tend to occur at times when people have periods of unstructured time and are able to reflect on the past without the interruption of present contingencies. An excerpt from the diary of the writer Joyce Horner who recorded many thoughts, experiences, and reminiscences while spending her last years in a nursing home is instructive.

> Thinking of Mrs. Danielson who can always pull another name out of the hat. . . . I also think how impossible it would be for me to hang on to all my names—all the lists I had to learn every semester and how there was always a period in which it seemed impossible I should learn them, and then suddenly they were all individuals, with their names fixed for a while. While one goes through the process of losing more and more, *one is still startled when a name from the remote past—Canada or Hood College—comes into one's mind, with the image attached, and sometimes a whole scene, with words.* The name Mary Lyons Agnew arrives from nowhere and I think of taking her home from church over the snow, and she leaning all her weight—considerable—on my arm. She could have grandchildren now. Not worth mentioning if one did not see or hear so many people who have lost their boundaries in time, place (1982, p. 22).

It was thinking about Mrs. Danielson's ability to recall names that began Horner's reminiscence. In the midst of lamenting the loss of previously memorized names and faces, she returned to a situation of the past. It began with a name, a place, and images infused with snippets of conversation and activities. The episode concluded by returning to the present, bringing with her memories of Mary Lyons Agnew and comparing the past with speculations about her lot in the present.

Physical objects are also imbued with personal meaning and stimulate reminiscences. Memories of past accomplishments, talents, journeys, and sentiments are stored in things people have in their environs. Many people interviewed have recalled past events and important junctures in their lives by using physical objects to prompt and stimulate memories and their accompanying stories (Unruh, 1983b). People often walk around their house touching and fondling vases, photographs, and plants as if those objects in fact held memories which had to be coaxed from within. Sherman and Newman (1977) also found that 81 percent of their sample of 94 older people in community senior centers and nursing homes could identify a "most cherished possession." For the elderly, these objects often represent the last symbolic remnants of who and what they once were.

Most of the subjects associated symbolic jewelry with a spouse and used photographs to evoke memories of their children. It is not that the objects were necessary for certain reminiscences to occur, but they provided cues for their emergence. Some people (especially the aged) have decorated their homes with objects which cue and stimulate pleasurable reminiscences. Trophies of past accomplishments, photographs of time when families were intact, and mementos of earlier travels are placed in prominent places so the return to situations of the past is facilitated.

Reminiscences are also initiated by events which symbolize similar occurrences in the past. There are times which encourage, if not require, people to reflect upon life changes. Anniversaries, birthdays, holidays, annual events, and even seasons of the year are symbolic events which encourage people to reflect with depth and detail. The 75th anniversary of the 1906 San Francisco earthquake was a time for reminiscing among survivors like the one in the following excerpt.

> . . . I was only eight at the time, and to me it was a great spectacle. They were dynamiting one building after another. We had a fine time. . . . I was watching people go by. A woman came by in her beautiful evening gown with her Japanese houseboy. They didn't have a thing. A man went by carrying a bird cage with a cat in it. I said, 'Why have you got the cat in the cage?' He said, 'The cat ate the canary' (Caen, 1981).

Similarly, there is a relationship between present events and the proclivity to reminisce.

> The World Series is our bond with the good old days, which possibly might not have been so good at that—unless you were Grover Cleveland Alexander coming in as a reliever to face Tony Lazzeri in the bottom of the ninth of the seventh game of the 1926 Series. . . . In the mind's eye you still can picture Reggie. And Gionfriddo. And Mays. And Rudi and Agee (Spander, 1981).

Reminiscences initiated by symbolic events require a comparison between the past and present and thereby bring about feelings of nostalgia (see Davis, 1979).

The degree to which comparisons between the past and present exist in reminiscence episodes is affected by many social factors—including the span of available time, possible interruptions, social discontinuities, and willingness to resurrect the past. Attention to the course of reminiscences helps highlight the role of these influences. Some reminiscences are little more than momentary flashes back to the past which are barely perceptible and last for a few seconds. People may return briefly to a situation in their adolescence and then be thrust back into the present. These brief flashes do not allow for much reflection and little identity work would be accomplished.

However, reminiscences may also progress so that people continually shift back and forth from past to present comparing, questioning, and striving to understand their lives. This is the point at which reinterpretation of the past shifts from the second to third order. These reminiscences require more free time, as well as mental agility. An entry from Horner's diary illustrates this shift.

> Sudden fit of nostalgia for Maine, not brought on by reading Anne Lindbergh to Miss K. and her mention of a place at North Haven, but by a rambling plan, as I ate my supper. . . . No doubt North Haven began it, with the image of the hills of Islesboro seen from the Head of the Cape or perhaps Orr's Cove. But this was a violent desire for the meadow—I hope Phyllis Riley will never sell any of it off for building—and then the scramble down to the rocks, and the driftwood trees where one sat and looked out to islands. I remember crossing in a boat to pick blueberries on Van Black's "mountain" and saying to K. Wynd (who is dead), "I'd rather be here than anywhere else in the world" (1982, p. 31).

Here we see how Horner's images of Maine rooted in the past continually brought her to the present as she grew concerned about its preservation, and then back to the past again for images of situations with greater detail. She engages in identity work when pieces of conversation were remembered facilitating comparisons between who she was and what she once wanted with her present state of mind. Horner, however, did not allow herself to reminisce at length without adopting a reflexive posture so that memories and current experiences were evaluated in terms of one another. The image portrayed in most studies of reminiscence is that the life review (third order reinterpretation) is something into which the aged spontaneously lapse for long periods of time. In this section, we have begun to see how the social situation, context, and setting influences the course and content of reminiscences.

Personal Accomplishment

As people conduct their daily affairs they come into contact with faces, places, and things which stimulate reminiscences. The importance of this relationship emerges when people are aware of the relationship between their actions and the emergence of reminiscences—thereby seeking to control times, places, frequencies, and contents of reminiscences. Horner also described how reading certain magazines took her back to the past. As she thumbed through *Country Life,* the advertisements and articles reminded her that as a child she looked at auctioneers' catalogues and thought about houses she would like to own. Her recognition of this relationship allowed her consciously to manage the onset of a certain type of reminiscence.

It is not unusual for people to imbue possessions with meanings which represent past accomplishments. Trophies, plaques, scrapbooks, photographs, and the like are obvious examples of objects which may be used to stimulate reminiscences. The literature on death and survivorship is replete with examples of survivors who continue visiting a summer cabin a partner loved, maintain joint group memberships in certain associations, and walk the same routes the deceased once traveled (Lopata, 1973; 1979).

The personal accomplishment of reminiscence by survivors may be an important way of maintaining emotional attachments with the deceased, but it also extends into the mundane minutiae of everyday life. People journey to the places of their birth, visit childhood friends, and watch old television shows hoping to evoke certain kinds of reminiscences. In a nostalgic exercise, one person described his attempts to "set the stage" for a 50th high school reunion as he eschewed modern thoroughfares in favor of primitive roads through backward towns hoping to evoke images, feelings, and memories of the past.

> These and equally sprightly thoughts danced through my Swiss cheese of a brain as I headed
> for the 50th reunion of Sacramento High's Class of '32. To get in the mood, I eschewed
> Interstate 80—the very term lacks romance—and headed along the Sacramento River route
> back and forth across those quaint little drawbridges, feeling the valley heat in towns like
> Locke, Hood, Ryde, Courtland, Walnut Grove. In some of these places, it is still 1932; you
> expect to see the glorious Delta King or Delta Queen sternwheeling around the bend of a great
> river that very much resembles the Mississippi. . . . I stopped at the Ryde Hotel, that
> monument to the 1920's, its downstairs night club still called "The Speakeasy," the peephole
> still in the door. . . . I finally got to a street I recognized—21st—and drove past my old
> fraternity house, Hail, Iota Kappa! All I could remember was being paddled black and blue by
> a guy named Al Minasian. That, and drinking and throwing up a lot (Caen, 1982).

Whether or not the accomplished reminiscences equal the effort expended to encourage them, the preceding excerpt illustrates the activity of facilitating the onset of certain memories.

The Importance of Place

In analyzing the importance of place, it is necessary to begin with the notion of "autobiographical insideness." Rowles (1983, p. 115) has used the term in reference to the "temporal legacy of having lived one's life in an environment. . . . Each location lives as an 'incident place'—where a child was born, a husband was met, or a first job obtained." The importance of place is more than the physical or social influence of the present, but it is also a part of personal history—a series of remembered places which have become a landscape of memories. Places live on in the minds of people even though they may not have visited them for some time or when the places no longer exist. When asked to produce a cognitive map of her neighborhood, one elderly woman included a trolley line which had been missing for 20 years and thereby indicated the relevance of that nonexistent place in her life (Regnier, 1983, p. 69).

Remembered places may affect reminiscences in at least two ways. They serve as background for reminiscences of social situations which occurred in and around those places. One older man interviewed by Rowles (1978, p. 95) reminisced as he toured a former neighborhood.

Ooh, we used to light bonfires. We used to take old tires and put them over hydrants, and set it afire. And the fire team would come, and they couldn't get the water! I'll tell you the funniest one we had. There was an old guy by the name of Baker who was a baker. He had a baker's shop on Durham Street between Cutler and Sentry, and he had some steps. Now we were building a bonfire—not me, the kids—down the corner of Tintering and Cutler Street. They had old boxes. And he came out of his house with a hose. The kids jumped the fence, and they let him get near the fire, start putting the fire out, and they cut the hose.

This reminiscence was sparked by a visitation to the place where the event occurred. In this example, the place itself was incidental to the social interaction which occurred. That is, the bonfires and pranks might have occurred anywhere but the incident was more noteworthy. The spaces and places gave the remembered experience its peculiar character, and provided a route toward preserving memories of social interaction.

Conversely, place may be dominant over any specific episode of social interaction which may have occurred. That is, spaces and places may themselves be objects of reminiscence. The reminiscence, then, involves interaction of self with space. For example, as adults return to places of their childhood and traverse those spaces, it is not unusual for reminiscences to be focused on markets, streets, parks, fences, and hiding places which once were central to their worlds. When walking around a former neighborhood, one older man began to reminisce about how he interacted with certain bushes and fences. He noted the gap in a chain link fence and began to reflect on how he used to crawl through that and similar spaces on his way to school (see Rowles, 1978, p. 140). This reminiscence would not have occurred if the person had not personally confronted the hole in the fence—or least one similar to the original. While reminiscences focused on social interaction include images of the places in which they occurred, so too with places as the object. While spaces and places may begin as the object of reminiscences, memories of specific social interactions may take over. For example, the older man reminiscing about crawling through the hole in the fence later recalled doing it on his way to the railroad tracks where he and his friends flirted with danger by jumping onto the moving trains.

The importance of place extends into the effects of location on the emergence of reminiscences. Some places are more conducive than others. For example, physical relocation is experienced by many older people moving from inappropriate neighborhoods or outsized homes into retirement communities, nursing homes, or the homes of children. With the move, some people lose cues for reminiscences because of the changed surroundings, displaced physical objects, and the like (cf. Lawton et al., 1973; Tobin and Lieberman, 1976). When the transition occurs rapidly, people may lose their position in space, time, and society because they have been unable to remold or transfer pieces of the past into a new habitat. Rowles (1979, p. 90) notes that limitations in the design of many facilities for the aged leave insufficient wall space and shelving for the display of photographs and memorabilia—in addition to many other social and environmental factors which discourage accommodation.

Emotional States and Outcomes

Finally, a social psychology of reminiscence must attend to the emotional states out of which second and third order reinterpretations of the past arise, as well as the emotional outcomes. This area of inquiry is, perhaps, the most crucial for understanding the ways by which reminiscence affects the person. The notion that episodes of reminiscence may result in people feeling content, satisfied, bitter, lonely, alienated, and the like has not been explicitly addressed by many researchers.

Instead, there has been much attention to one particular emotion—nostalgia— and the tendency for many reminiscences to be tinged or colored with this emotion. Davis (1979) notes how the term "nostalgia" first referred to feelings of homesickness among Swiss army soldiers. Nostalgia has since acquired meaning related to special feelings toward the past. It is characterized by comparison of the present with the past. As Davis notes,

> Nostalgic feeling is almost never infused with those sentiments we commonly think of as negative—for example, unhappiness, frustration, despair, hate, shame, abuse. . . . Some will, to be sure, allow that their nostalgia is tinged with a certain sadness or even melancholy but are then inclined to describe it as a "nice sadness" (1979, p. 14).

The relationship between nostalgia and reminiscence is not direct and highly situational. Not all episodes of reminiscence are nostalgic, nor can every nostalgic experience qualify as reminiscence. Instead, there are times, situations, and places where memories of situation rooted in the past sharply contrast with present circumstances. These differences are most apparent during sharp transitions or periods of turmoil. In the realm of everyday life in the present, the anxieties, uncertainties, and feelings of strangeness about the present and future constitute the "object" for the person, while the background is composed of familiar people, places, and identities rooted in the past. According to Davis (1979, p. 58), the nostalgic reaction is an inversion of the "typical" relationship of the two. The richly textured past preserved in long-term memory which was once only the backdrop for action and consciousness in the present emerges as the "object" of attention. The present, then, fades from object to background. With their increased ratio of attention toward the past, the aged are more likely to be immersed in feelings of nostalgia than the young, thereby further separated in mind from succeeding age groups.

Arising out of social losses, grief is an emotion entering the lives of many people—especially the aged. Through the deaths of significant others, divorce, relocation, or even retirement, grief may be experienced as people compare their current and projected lives with the past. The intensity, course, and manifestation of grief is affected by that which was lost since people are linked to others through the roles they play, the help they receive, their network of others, the selves they create, the comforting myths formed, the reality they validate, and the futures they make possible (Lofland, 1982). For periods of time succeeding various losses, reminiscences may be focused on situations of the past where that which was recently lost might momentarily be recaptured and relived. Other emotions may also arise briefly to supplant the grief with satisfaction, joy, levity, contentment, or affection. For example, reminiscences of the ways by which a deceased spouse would inject humor into conversations may, in fact, spark feelings of happiness and even draw a slight smile from survivors (cf. Glick, Weiss, and Parkes, 1974; Lopata, 1979).

Nostalgia and grief are only two emotions which serve as emotional states or outcomes of reminiscence. Many more remain unexplored with greater attention to emotions as the foundation and function of reminiscence will provide a better understanding of the ways by which the process is structured, controlled, managed, and accomplished by people in social situations.

CONCLUSION

This essay has focused on the phenomenon of reminiscence and the possibility of developing a social psychological perspective on its occurrence and management in everyday life. In that direction, it has been noted that previous research generally has failed to define the concept adequately and differentiate it from similar phenomena. Further, the methods and measurement techniques which have been employed have been viewed as restrictive and, perhaps, attuned to only limited aspects of the phenomenon which contradict the spontaneous and processual nature of reminiscence in people's daily lives. Out of this discussion, the relationship of reminiscence to the life review has been addressed, as has the reinterpretation of the past which occurs regularly and routinely in the course of people's lives. In addition to the linkage created among these phenomena by analyzing them as different orders of essentially the same process, a more precise working definition of reminiscence has been offered. In this context, reminiscence refers to the "recall to mind of experiences, feelings, identities, or facts rooted in situations of the relatively remote past which are not stable and continuous elements of everyday consciousness."

The remainder of this paper has focused on suggestions for reminiscence research which might broaden our understanding of the phenomenon and its role in everyday life—especially for the aged. The emphasis has been on reminiscences as a social process which is influenced by social setting and situations. Here it was possible to see how the course of events affects the emergence, course, and content of reminiscence. In addition, when people recognize those relationships, they engage in actions by which the process is managed—in effect, personally accomplished. The importance of place in affecting the course and contents of reminiscences has also been addressed by looking at some examples of how spaces and places serve as both background and object of memories rooted in situations of the past. Finally, emotional states and outcomes must be viewed as a crucial element in any analysis of reminiscence. The moods and sentiments which lead people into a state where reminiscing is encouraged are as important as the emotional outcome of that process. The emotions of nostalgia and grief briefly were discussed to highlight some of the ways by which these factors directly influenced people's lives. In these ways, I have sought to highlight some neglected qualities of reminiscence, illustrate its management by individuals, and offer some insight into the ways by which it might be analytically approached both in studies specifically focused on reminiscence and as a byproduct of other research.

ACKNOWLEDGMENT

This paper was completed with the help of a National Institute of Mental Health postdoctoral traineeship in mental health evaluation research in the Department of Sociology at the University of California, Los Angeles.

REFERENCES

Allport, Gordon. 1942. *The Use of Personal Documents in Psychological Science.* New York: Social Science Research Council.
Becker, Howard S. 1963. *Outsiders.* New York: Free Press.
Becker, Howard S. and Blanche Geer. 1957. "Participant Observation and Interviewing: A Comparison." *Human Organization* 16:28–32.
Blumer, Herbert. 1967. *Symbolic Interactionism.* Englewood Cliffs, NJ: Prentice-Hall.
Boylin, William, Susan Gordon, and Milton Nehrke. 1976. "Reminiscence and Ego Integrity in Institutionalized Elderly Males." *Gerontologist* 16:118–124.
Butler, Robert N. 1963. "The Life Review: An Interpretation of Reminiscence in the Aged." *Psychiatry* 26:65–76.
————. 1970. "Looking forward to What? The Life Review, Legacy, and Excessive Identity versus Change." *American Behavioral Scientist* 14:121–128.
Caen, Herb. 1981. "Days of our Years." *San Francisco Chronicle,* April 19, 1981.
————. 1982. "Reunion Time." *San Francisco Chronicle,* June 12, 1982.
Cain, Leonard. 1978. "Counting Backward from Projected Death." Paper presented to the Policy Center on Aging, Maxwell School, Syracuse University, March 22.
Campbell, Donald T. 1955. "The Informant in Quantitative Research." *American Journal of Sociology* 60:339–342.

_____. 1964. "Social Attitudes and Other Acquired Dispositions." Pp. 195–162 in *Psychology: A Study of Science* 6, edited by Sigmund Koch. New York: McGraw-Hill.

Charmaz, Kathy. 1980. *The Social Reality of Death.* Reading, MA: Addison-Wesley.

Chomsky, Noam. 1968. *Language and The Mind.* New York: Harcourt, Brace and Jovanovich.

Coleman, Peter G. 1974. "Measuring Reminiscence Characteristics from Conversation as Adaptive Features of Old Age." *International Journal of Aging and Human Development* 5:281–294.

Costa, Paul and Robert Kastenbaum. 1967. "Some Aspects of Memories and Ambition in Centenarians." *The Journal of Genetic Psychology* 110:3–16.

Davis, Fred. 1979. *Yearning for Yesterday.* New York: Free Press.

Dean, J.P., R. Eickhorn, and L.R. Dean. 1969. "Fruitful Informants for Intensive Interviewing." Pp. 142–145 in *Issues in Participant Observation,* edited by George McCall and J.L. Simmons. Reading, MA: Addison-Wesley.

Deutscher, Irwin. 1973. *What We Say/What We Do.* Glencoe, IL: Scott Foresman.

Dilthey, W. 1962. *Pattern and Meaning in History,* edited by H.P. Rickman. New York: Harper Torchbooks.

Dollard, John. 1935. *Criteria for The Life History.* New Haven, CT: Yale University Press.

Ebersole, P.P. 1978. "Establishing Reminiscing Groups," in *Working With The Elderly: Group Processes and Techniques,* edited by I.M. Burnside. New York: Duxbury Press.

Erickson, Erik. 1950. *Childhood and Society.* New York: Norton.

Fallot, Roger. 1980. "The Impact on Mood of Verbal Reminiscing in Later Adulthood." *International Journal of Aging and Human Development* 10:385–400.

Faulkner, Robert. 1971. *Hollywood Studio Musicians.* Chicago: Aldine.

Gans, Herbert. 1962. *The Urban Villagers.* New York: Free Press.

Glick, Ira, Robert Weiss, and C. Murray Parkes. 1974. *The First Year of Bereavement.* New York: Columbia University Press.

Goffman, Erving. 1959. *The Presentation of Self in Everyday Life.* Garden City, New York: Anchor/Doubleday.

Gubrium, Jaber. 1975. *Living and Dying at Murray Manor.* New York: St. Martin's.

Gubrium, Jaber and Robert Lynott. 1983. "Rethinking Life Satisfaction." *Human Organization* 42:30–38.

Havighurst, Robert and Richard Glasser. 1972. "An Exploratory Study of Reminiscence." *Journal of Gerontology* 27:245–253.

Havighurst, Robert, Bernice Neugarten and Sheldon Tobin. 1968. "Disengagement and Patterns of Aging." Pp. 161–172 in *Middle Age and Aging,* edited by Bernice Neugarten. Chicago: University of Chicago Press.

Hochschild, Arlie Russell. 1978. *The Unexpected Community.* Berkeley and Los Angeles: University of California Press.

Horner, Joyce. 1982. *That Time of Year.* Amherst, MA: University of Massachusetts Press.

Irwin, John. 1977. *Scenes.* Beverly Hills, CA: Sage.

Johnson, John M. 1977. "Behind the Rational Appearances." Pp. 201–228 in *Existential Sociology,* edited by Jack D. Douglas and John M. Johnson. Cambridge: Cambridge University Press.

Kvale, S. 1977. "Dialectics and Research on Remembering," in *Life Span Developmental Psychology: Dialectical Perspectives on Experimental Research.* New York: Academic Press.

Lawton, M. Powell, M.H. Kleban and D.A. Carlon. 1973. "The Inner City Resident: To Move or not to Move." *Gerontologist* 13:443–448.

Lewis, Charles N. 1971. "Reminiscing and Self-concept in Old Age." *Journal of Gerontology* 26:240–243.

Lewis, Myrna and Robert N. Butler. 1974. "Life Review Therapy: Putting Memories to Work in Individual and Group Psychotherapy." *Geriatrics* 29:420–243.

Lieberman, Morton and Jacqueline Falk. 1971. "The Remembered Past as a Source of Data for Research on the Life Cycle." *Human Development* 14:132–141.

Liebow, Elliot. 1967. *Tally's Corner*. Boston: Little, Brown.
Liton, J. and S.C. Olstein. 1969. "Therapeutic Aspects of Reminiscence." *Social Casework* 50:263–268.
Lofland, Lyn. 1982. "Relational Loss and Social Bonds: An Exploration into Human Connection." Pp. 219–242 in *Personality, Roles, and Social Behavior*, edited by William Ickes and Eric S. Knowles. New York: Springer-Verlag.
Lo Gerfo, Marianne. 1980. "Three Ways of Reminiscence in Theory and Practice." *International Journal of Aging and Human Development* 12:39–48.
Lopata, Helena Z. 1973. *Widowhood in An American City*. Cambridge, MA: Schenkman.
————. 1979. *Women as Widows*. New York: Elsevier.
Luckmann, Thomas. 1975. *The Sociology of Language*. Indianapolis: Bobbs-Merrill.
McMahon, A.W. and P.J. Rhudick. 1967. "Reminiscing in the Aged: An Adaptational Response," in *Psychologynamic Studies on Aging: Creativity, Reminiscing, and Dying*, edited by S. Levin and R. Kahana. New York: International Universities Press.
Maines, David, Noreen Sugrue, and Michael Katovich. 1983. "The Sociological Import of G.H. Mead's Theory of the Past." *American Sociological Review* 48:161–178.
Marshall, Victor W. 1974. "The Last Strand: Remnants of Engagement in the Later Years." *Omega* 5:25–35.
————. 1975. "Socialization for Impending Death in a Retirement Village." *American Journal of Sociology* 80:1124–1144.
————. 1980. *Last Chapters*. Belmont, CA: Brooks/Cole.
Matthews, Sarah. 1979. *The Social World of Old Women*. Beverly Hills, CA: Sage.
Meacham, James. 1977. "A Transactional Model of Remembering," in *Life-Span Developmental Psychology: Dialectical Perspectives on Experimental Research*, edited by N. Datan and H. Reese. New York: Academic Press.
Mead, George Herbert. 1929. "The nature of the past." Pp. 235–242 in *Essays in Honor of John Dewey*, edited by John Coss. New York: Henry Holt.
————. 1932. *Mind, Self, and Society*. Chicago: University of Chicago Press.
Merriam, Sharan. 1980. "The Concept and Function of Reminiscence." *Gerontologist* 20:604–609.
Mischel, W. 1969. "Continuity and Change in Personality." *American Psychologist* 24:1012–1018.
Myerhoff, Barbara. 1978. *Number Our Days*. New York: Simon and Shuster.
Neugarten, Bernice. 1979. "Time, Age, and the Life Cycle." *American Journal of Psychiatry* 136:887–894.
Neugarten, Bernice, Joan Moore, and John Lowe. 1965. "Age, Norms, Age Constraints, and Adult Socialization." *American Journal of Sociology* 70:710–717.
Pincus, A. 1970. "Reminiscence in Aging and its Implications for Social Work Practice." *Social Work* 15:47–53.
Regnier, Victor. 1983. "Urban Neighborhood Cognition: Relationships between Functional and Symbolic Community Elements." Pp. 63–82 in *Aging and Milieu: Environmental Perspectives on Growing Old*, edited by Graham Rowles and Russell Ohta. New York: Academic.
Reynolds, David and Richard Kalish. 1974. "Anticipation of Futurity as a Function of Ethnicity and Age." *Journal of Gerontology* 29:224–231.
Riemer, Jeffrey. 1977. "Varieties of Opportunistic Research." *Urban Life* 5:467–477.
Rosen, J. and Bernice Neugarten. 1964. "Ego Functions in the Middle and Later years." Pp. 90–101 in *Personality in Middle and Late Life: Empirical Studies*, edited by Bernice Neugarten. New York: Atherton.
Rosenfeld, Jeffrey. 1979. *The Legacy of Aging*. Norwood, NJ: Ablex.
Rowles, Graham D. 1978. *Prisoners of Space?* Boulder, CO: Westview Press.
————. 1979. "The Last New Home." Pp. 81–94 in *Location and Environment of Elderly Populations*, edited by Stephan M. Golant. New York: John Wiley and Sons.
————. 1983. "Geographical Dimensions of Social Support in Rural Appalachia." Pp. 111–130 in

Aging and Milieu: Environmental Perspectives on Growing Old, edited by Graham Rowles and Russell Ohta. New York: Academic.

Ruml, Beardsley. 1946. "Some Notes on Nostalgia." *Saturday Review of Literature,* June 22.

Sapir, Edward. 1949. "The Status of Linguistics as a Science," in *Selected Writings in Language, Culture, and Personality,* edited by David Mandelbaum. Berkeley and Los Angeles: University of California Press.

Schutz, Alfred. 1967. *The Phenomenology of The Social World.* Evanston, IL: Northwestern University Press.

———. 1970. *On Phenomenology and Social Relations,* edited by Helmut R. Wagner. Chicago: University of Chicago Press.

Sherman, Edmund and Evelyn Newman. 1977. "The Meaning of Cherished Personal Possessions for the Elderly." *International Journal of Aging and Human Development* 8:181–192.

Spander, Art. 1981. "World Series of Memories." San Francisco Chronicle, October 18.

Strauss, Anselm, Alfred Lindesmith, and Norman Denzin. 1977. Social Psychology. New York: Holt, Rinehart, and Winston.

Teahan, James and Robert Kastenbaum. 1970. "Subjective Life Expectancy and Future Time Perspective as Predictors of Job Success in the Hard-core Unemployed." *Omega* 1:189–200.

Tobin, Sheldon and Morton Lieberman. 1976. *Last Home for The Aged.* San Francisco: Jossey-Bass.

Tolor, Alexander and Vincent Murphy. 1967. "Some Psychological Correlates of Subjective Life Expectancy." *Journal of Clinical Psychology* 23:21–24.

Unruh, David. 1983a. *Invisible Lives: Social Worlds of The Aged.* Beverly Hills, CA: Sage.

———. 1983b. "Death and Personal History: Strategies of Identity Preservation." *Social Problems* 30:340–351.

Vygotsky, L. 1962. *Thought and Language.* Cambridge, MA: MIT Press.

Whorf, Benjamin. 1956. *Language, Thought, and Reality.* Cambridge, MA: MIT Press.

Zeiger, B. 1976. "Life Review in Art Therapy with the Aged." *American Journal of Art Therapy* 15:47–50.

Zimmerman, Don and D. Lawrence Weider. 1977. "The Diary." *Urban Life* 5:479–498.

CHANGING CONCEPTS OF SELF AND SELF DEVELOPMENT IN AMERICAN AUTOBIOGRAPHIES

Diane Bjorklund

INTRODUCTION

Thomas and Znaniecki ([1918–1920]1958, pp. 1832–3) boldly made this assertion about their use of personal letters and a life-history in their massive work *The Polish Peasant in Europe and America:*

> We are safe in saying that personal life records, as complete as possible, constitute the *perfect* type of sociological material, and that if social science has to use other materials at all it is only because of the practical difficulty of obtaining at the moment a sufficient number of such records to cover the totality of sociological problems. . . .

Current Perspectives on Aging and the Life Cycle,
Volume 3, pages 47–68.
Copyright © 1989 by JAI Press Inc.
All rights of reproduction in any form reserved.
ISBN 0-89232-739-1

Their use of the autobiography of a Polish emigré has been deemed the first major sociological use of life-histories (Plummer, 1983). Other studies followed which made use of personal life-stories (Shaw, 1930, 1931, 1938; Sutherland, 1937), and a methodological debate was touched off about the use of such subjective material (Dollard, 1935, 1938; Blumer, 1939; Allport, 1942; Angell, 1945; Becker, 1966; Denzin, 1978; Bertaux, 1981). Critics often point to problems of subjectivity and reliability in these introspective accounts.

In the use of life-history materials, however, the interest predominantly has been in the supplementary information that such documents contain about particular topics such as family dynamics or the career of a delinquent child. Little attention in sociology has been paid to life accounts such as autobiographies for precisely what they are: documents produced by individuals in which they attempt to make sense of self and life experiences. In an autobiography, an author must deal with questions of how to structure such a narrative, what events to select out of the countless ones possible, and the extent of explanations or accounts that should be offered to the future audience of readers. In organizing and managing a story of self, an author will display, at least implicitly, his or her assumptions about the self and self development. By focusing on these assumptions about self and by treating autobiographies as a social presentation of self, a researcher can leave aside the issue of the subjectivity and reliability of these personal accounts.

In this paper, my intent is to demonstrate that autobiographies can be used as a rich source of data to explore questions of historical changes in assumptions and notions about the self. In recent decades, there has been a substantial literature containing assertions about changes in the modern self (e.g., Trilling, 1972; Sennett, 1974; Turner, 1976; Zurcher, 1977; Lasch, 1978). But what evidence can be used to provide support for such a hypothesis? Zurcher (1977), for example, used evidence gathered from the administration of the Twenty Statement Test (TST) developed by Manford Kuhn. But the use of such a structured measuring instrument does not permit forays very far into the past. Autobiographies, however, are abundant. In the United States, there have been 11,385 autobiographies published during the period 1800–1982 (Kaplan, 1961; Briscoe, 1982). If one includes works such as confessions and memoirs, it is possible to go back at least 1500 years. And finally, autobiographers produce these accounts without the direction or intervention of a researcher (Webb et al., 1966).

Many of those who have suggested that there are changes in the modern self point to an increasing amount of focus by individuals on the inner or private self and a lessened concern with public or institutional roles. This changed attitude is described as narcissism, self-absorption, temporary reflexivity, or a search for the true self. In the balance of this paper, I will discuss how autobiographers use concepts and theories of the self in such a manner that it is possible for a researcher to examine closely and systematically the extent to which there have been changes in how modern individuals understand the self.

METHOD

In choosing to focus on autobiographies as a genre, one encounters a problem of sampling autobiographies from among the wealth of such documents available. It would take an inordinate amount of time to read the more than 11,000 auto-biographies that have been published by Americans. For those who use auto-biographies as a source of supplementary information on a particular topic, it is possible to select only works based on their relation to that topic. And literary historians interested in autobiographies usually select just a few classic auto-biographies based on the assumption that such works represent the essence of the genre or perhaps the "spirit of the age" (e.g., Pascal, 1960; Olney, 1972, 1980; Lyons, 1978; Weintraub, 1978). But for this study, a wider range of "informants" was necessary.

To select a sample, I used what Leo Marx (1969, p. 78) has called "qualitative standards of selection" in order to include a variety of autobiographies.[1] For my sample of fifty autobiographies, my primary criterion of selection was the year of publication. I analyzed at least two autobiographies published by Americans from each decade during the period 1800–1982. The continuous production of autobiographies makes it possible to choose such a large time span in which it may be possible to detect changes that have occurred slowly. The date 1800 was chosen based on its proximity to the date (1797) in which the word "autobiography" first appeared in print in the English language (according to the Oxford English Dictionary), and due to the fact that this is the date at which the bibliographical listing of American autobiographies begins. Additionally, the sample includes famous and obscure autobiographers, both sexes, members of various occupations, autobiographers from various regions in the United States, and both young and old autobiographers.

I use Delany's (1969, p. 1) definition of an autobiography as a literary work "*primarily* written to give a coherent account of the author's life. . ., and composed after a period of reflection and forming a unified narrative." Such a definition eliminates diaries, journals, and collections of personal letters which do not include an attempt to provide an overarching framework, to seek a pattern of unity, or to sum up one's life.

For each autobiography, I recorded statements made by the writers concerning topics such as the self, self development, inherited traits, parents, church, school, mentors, relationships, and goals. Any remarks about psychology or sociology as well as the use of concepts and theories from these disciplines were noted. And I paid attention to the construction of the autobiographies including chapter divisions; use of photographs, indices, and forewords; number of pages spent discussing childhood; and the adherence to a chronology.

It must be granted, of course, that only selected life-stories are ever published. There is a lack of representativeness of the "informants." Those in the circle of autobiographers most likely contain an overrepresented number of successful persons who probably had more access to books, education, and leisure time. Thus I cannot straightforwardly generalize these findings to the general population of the United States. But nevertheless, these authors intend to communicate with an audience of readers and assume a considerable amount of intersubjectivity. An autobiography is a story written to be a public presentation of self. Such a performance, as Horowitz (1977, p. 173) calls autobiographies, will take into consideration the expectations and understandings of the expected readers. On the blurbs on bookjackets and in some reviews of particular autobiographies, one can find the comment that the story will "strike a responsive chord in the reader's heart." Furthermore, I believe it possible to presume that these writers share many cultural assumptions with the reading public, e.g., notions of success, commendable goals in life, stages of life, and the extent to which the self is seen as malleable. As Berger (1966, p. 108) puts it, the individual does not have to reflect each time "about the meaning of each step in his unfolding experience. He can simply refer to 'common sense' for such interpretation, at least for the great bulk of his biographical experience."

This is not to argue that all autobiographies are the same stories with just the names and dates changed. Certainly there are many differences among the autobiographies. If this were not so, it is doubtful that readers would continue to find them interesting. For example, there are differences among the works based on gender, occupation, and purpose in writing—differences which might be fruitfully investigated. But for the purposes of this paper, I examined conventions and uniformities in discussions about the self and self development in relation to various time periods in these American autobiographies.

TELLING THE STORY OF A SELF OVER TIME

If you were asked to sit down and recount your life history, how would you go about it? Out of innumerable events, which would be worth mentioning? Would you relate the story of a self that developed over time or, perhaps, a self that never changed? And how reliable would you consider your memory as a source of information? These are the types of questions which autobiographers must confront and which are inextricably bound to their assumptions about self and self development. As Hart (1970, p. 492–3) has pointed out, there are many ways that the story of a self could be told in an autobiography. It could be the story of a self which is developed, found, gained, overcome, lost, tainted, unchanged, or searched for in vain. Each of these implies a different viewpoint of the self. In choosing a theme to structure a story of self, in selecting and arranging the material, and in gathering the data, the autobiographer reveals his or her assumptions about self and self development. Therefore, the construction of the autobiographical account and the authors' explicit discussions of self and self development are used as data for the following discussion of historical changes in notions about self and self development in American autobiographies.

Mention the word "autobiography" and what probably comes to mind for many is a chronological account which begins with the words, "I was born. . . ." Indeed, this is the case with many autobiographies published in the nineteenth century and into the early decades of the twentieth century. However, in many of these nineteenth-century autobiographies, the focus is not really on the "I" so prominently mentioned in the first sentences. Rather, the author focuses on some external historical event which he or she witnessed, or the author recounts a personal view of some aspect of society. For example, Graydon's (1811) *Memoirs of a Life* is an account of the Revolutionary War from the viewpoint of a minor officer. In such memoirs and reminiscences, the authors show little concern about justifying why the reading public should be interested in their life-stories. The historical value of the account is seen as enough justification, or the author also may add that friends and family urged that the autobiography be made available to the public. As Cartwright (1856, pp. 4–5) modestly claims, ". . . I have reluctantly yielded to the many solicitations of my friends. . . . " These authors' accounts are treated as interesting because they were "at the right place and at the right time."

A noteworthy exception to this type of historical memoir is Rousseau's (1781, p. 17) *Confessions* in which he begins with this claim:

> I have resolved on an enterprise which has no precedent, and which, once complete, will have no imitator. My purpose is to display to my kind a portrait in every way true to nature, and the man I shall portray will be myself.

Thus Rousseau asserts that he is setting a precedent by focusing inward on his own subjective self. Shumaker (1954, p. 28), in a history of English autobiography, argues that this type of subjective autobiography began appearing sporadically in the seventeenth century but did not flourish until the nineteenth and twentieth centuries.[2] I found in my sample of American autobiographies that such a trend is not in evidence until the latter half of the nineteenth century.

In the American autobiographies published during the latter half of the nineteenth century, the authors continue to locate themselves historically with discussions of events and dates. Around 1900, some of the autobiographers even begin to include indices which list both personal and historical events. But at the same time, these authors begin to assume that the audience of potential readers is interested in the inward lives of the authors as well as their views of external events. In particular, the autobiographers begin to recount more extensively the history of their own self development.[3] I found this to be the case beginning in the works published in the 1880s. In the earlier autobiographies, such as Graydon's (1811) account, there were some discussions along this theme, e.g., the story of the author's education. But in most cases, such comments could be compressed into a few paragraphs. In the life-stories published around the turn of the century (1900), however, the development of the self becomes the motif in many of these accounts. Shumaker (1954, pp. 86–7) also found this to be the case in English autobiographies:

> . . . Almost without exception [subjective autobiographies] adopt as their subject some aspect of psychic development, some process of intellectual or affective becoming. . . . Few more important statements can be made about recent autobiography [through 1946] than that in one of its most common . . . forms it describes in terms of heredity and environmental causality the gradual evolution of a set of attitudes or of some other aspect or aspects of personality.

In these tales of a development of self, the authors often begin with a comment that they are concerned about the fallibility of their memories but have attempted to provide a truthful narrative. In addition, many of these autobiographers who focus on their inward lives will include a comment on why they are making their life histories public. This becomes more of a salient issue when the accounts have no particular historical use. Here the answers are varied. In an early example of a theme of development, A.J. Davis' (1857, p. 11) *The Magic Staff* includes a preface written by his wife in which she asks, "Why does Mr. Davis write his own history. . . ?", and responds that "his experience belongs to the world, inasmuch as it reveals many subtile [sic] and most important phases in the constitution of mind. . . ." Thus the autobiography can serve an instructive purpose. In contrast, Andrew Carnegie (1920, p. 1) modestly notes that the "sages" say that any man's life is interesting, while Fagan (1912, pp. 1–2) says that he wishes to tell his life story to present his unique ideas about civilization. Once these issues of the truthfulness of the account and the justification for telling one's story publicly have been raised, the authors then turn to the story of the "I" that developed over a lifetime.

In these autobiographies which have self development as the theme, the story is told of a mental or moral development. Occasionally, an author will mention a concern also with physical development of health, but I found no autobiographies that detailed the steps or stages in such a development.[4] However, in both cases of mental and moral development, the story of a self is presented chronologically—typically beginning at birth and proceeding to the time at which the author sat down to begin his or her retrospective account. It is assumed that there are causal connections between the events that are described, and thus it is important to detail one's development in this chronological manner. Others who wish to develop in a similar way may then use the autobiography as a guidebook. As Davis' wife (1857, p. i) puts it, in her introduction to his life-story, a story of a development is best told by a "systematic autobiography—beginning with his first memories and ascending step by step through every subsequent year to the present period. . . ."

In the autobiographies in which the emphasis is placed on a mental development, there often is a first period in childhood in which the author is in an
"intellectual slumber." This is followed by a significant incident sometime
during one's youth in which one is "awakened." This metaphor of awakening
implies that there is something already there which only needs prodding. Some
describe this incident as a "turning point"—a phrase, according to Webster,
which did not appear in print in the English language until 1851.[5]

For many of these awakenings, the vehicle which triggers the incident is a
book—often happened across in an accidental way by the autobiographer. For
Carnegie (1920), it was the reading of books by Darwin and Spencer. For
Mencken (1940, p. 163), the most "stupendous" event in his life was the
reading of *Huckleberry Finn*. Another trigger for the awakening may be the
influence of a mentor. Fagan (1912, p. 41) claims this about an acquaintance:

> . . . He had done me a world of good. In the short space of three weeks I had changed or been
> converted from a mere boy, perplexed with a mind full of emotional instincts, into an
> individual, with a more or less definite trade-mark, and with a certain point of view in regard
> to life and living in which I had become enthusiastically interested. . . . Every hour that
> passed added to my stock of enlightenment.

Teachers who took a special interest in the future autobiographer are also mentioned for their roles as mentors.

The goal of a mental development appears to be the formation of a coherent
philosophy of life. Such a development is discussed as a possibility for all
individuals, but it does require an active self. I found no autobiographies in
which such a development was simply attributed to outside influences such as the
process of socialization or formal education. Carnegie (1920, p. 206) describes
the attaining of this goal of mental development: "Where there had been chaos
there was now order. My mind was at rest. I had a philosophy at last." Carnegie
(1920, p. 339) also adds that the process involves absorbing mental food, "retaining what was salutary, rejecting what was deleterious. . . . "

This process of mental development is similar to that development traced in
the type of German novel known as *bildungsroman* which emerged in the mid-
eighteenth century and continued until the early twentieth century (Bruford,
1975; Cooley, 1976). In this type of novel, which includes works by Goethe,
Stendhal, and Balzac, a story is told of a male protagonist who cultivates his
"self" through a process of learning from friends, mentors, and teachers. In this
process, he acquires a "point of view in practical matters and above all a
'*Weltanschauung*' . . . or perhaps one after another" (Bruford, 1975, pp. 29–
30). Goethe also wrote an autobiography in which he discusses his self cultivation. Mazlish (1970, p. 29) has suggested that Goethe and others "merely
extended to the self notions of development concerning mankind and society"
from such writers as Saint-Simon, Comte, Condorcet, and Hegel. It might be
added that Lukes (1973, p. 67), in his book on individualism, has commented

that the notion of self development is "typically Romantic in origin" and "was most fully elaborated among the early German Romantics. . . . "

It appears that self development, then, was an aspect of the "climate of opinion" in the nineteenth century in Germany and the United States. The word "self-culture" began appearing in print in the English language in the early part of the nineteenth century and refers to the "development of one's mind or capacities through one's own efforts." One of the autobiographers made a comment about this interest: "Society at the time, from top to bottom, was absorbingly interested in personal culture and development of every description. In the year 1881, self-culture was the supreme topic in the public mind. . . . " (Fagan, 1912, p. 104).

Another type of development traced in autobiographies has been the development of character or will power—some call this a moral development. In my sample of autobiographies, I found these discussions of moral development particularly in the period from approximately 1880 until the 1930s. The use of this theme of moral development coincides with a portion of the time period during which the popular press contained many discussions of the "self-made man," e.g., articles like Teddy Roosevelt's 1900 article entitled "Character and Success" (Wyllie, 1954, p. 140). The ideas of moral development and the self-made man are closely related in their concern with will and successful action.

Current at that time, too, were interests in Herbert Spencer and the "psychology of adaptation"—a psychology that became dominant in the United States (Leahey, 1980, p. 253). Hofstadter (1955, pp. 31–32) discussed the "vogue" of Spencer in the United States in the last half of the nineteenth century, and I find evidence of this in the fact that each of the autobiographies that I analyzed from the period 1900 to approximately 1940 include a mention of Spencer. Curti (1980, p. 193) has claimed that Spencer was the first to develop the theory of evolution in relation to mental life. Spencer's interest in evolution and the "survival of the fittest" was continued by psychologists who were interested in the adaptation of individuals to their environments. Another related current of opinion was that of faculty psychology which was based on the belief that the mind was divided into separate faculties which could be strengthened through exercise of these faculties (Hofstadter, 1962, pp. 347–8). Thus effort and will were taken to be important in the development of individuals, will being viewed as "the most important of God's gifts to man" (Wyllie, 1954, p. 40). "Will power," in fact, is a term which came into use in the latter half of the nineteenth century.

In the autobiographers' discussions of will and moral development during the decades around the turn of the century, the process is described as a battle against aspects of the environment—both one's own 'negative' instincts or urges, and also possible negative influences of society. Payot (1909, pp. 42–43), in a book on the education of the will, defined negative instincts or urges as "irascibility, egotism, sensuality, and laziness. . . . " The goal of such a development appears to be the inculcation of character—a set of moral and intellectual traits.

Such a development is seen as coming from within. A metaphor used by a few of the authors stressing a moral development is that of the navigator—attempting to steer a straight course through stormy seas.

In such a development, there is an early childhood period in which the auto-biographer as a child has not yet learned self-discipline. This is treated as a foolhardy period which is illustrated by the description of childhood pranks such as playing hooky from school or minor acts of thievery. Fagan (1912, p. 5) calls it his "wilderness stage." Such descriptions of this early phase as characterized by a lack of self-discipline seem not too different from earlier religious accounts in which a stage in which "I was naturally a wild, wicked boy. . . . " precedes a religious conversion (Cartwright, 1856, p. 27).[6]

Next there is a "turning point" during which the author discovers that he (and sometimes she) can exert will and thus make progress toward goals. Brinkerhoff (1900) discusses how he learned that one can overcome both heredity and en-vironment through the use of will. And Fagan (1912, p. 11) vividly describes an incident in his childhood in which a foolhardy act almost cost him his life, but he was able to save himself through an exertion of will. "I have always looked back upon these moments as the time when my personality first emerged into real consciousness" (Fagan, 1912, p. 12). Later Fagan also overcame temptation in a confrontation with a prostitute and learned that he must use will to keep himself "unspotted" from the world.

The stories of either mental or moral development of the self imply that it is good for the self to change and develop. However, the change is not a complete metamorphosis. Rather the change is a gradual development or cultivation of something that already exists within the person. This development of self is not viewed as a natural development through which all humans must progress. In most cases, there is the triggering event sometime in childhood. From that point, then the self must actively seek this development. Because it is primarily an account of self-reliance, others (except perhaps a brief encounter with a mentor) do not play an important role in such a mental or moral development. Thus the readers find out little about wives or children. For example, Barnum (1855, p. 403) offers this comment: "I have seldom mentioned my wife and children in these pages, yet they have always been dearer to me than all things else in the wide world. . . . " Parents receive a bit more discussion, particularly mothers. But even the importance of parents is seen to be their role as a model of character and inner strength. Finally, in telling such a tale of self development, writers assume that there is almost no problem in introspecting and examining the contents of their minds. Tracing causal connections and a unilinear development is treated as a relatively unambiguous task.

Although it is impossible to precisely date the shift, there is a clear change in conceptualizations of the self in autobiographies beginning in the early twentieth century. As theorists in psychology and elsewhere began to question the pos-sibility of comprehending the self, assumptions about the self and self development began to change in autobiographical accounts. As Giddens (1979, p. 38) has noted about the effect of Freud, Marx, and Nietzsche, " . . . As radical critics of the claims of the Cartesian *cogito:* each can be seen as questioning, in a profound way, the reliability of consciousness as 'transparent to itself.' " This possibility that there may be nonrational factors which play a part in one's behavior and which are beyond one's awareness is a disquieting consideration for autobiographers.

Freud, Jung, and Ferenczi visited the United States in 1908, a visit that is believed to have played a substantial part in the psychoanalytic movement in the United States (Curti, 1980, p. 342). In my data, I find that the word "unconscious" begins to appear in the autobiographies in the 1880s. But not until after the 1930s did I begin to find clear evidence of changes in ideas about the self and self development. For those autobiographies published after 1930, I agree with Thomas Cooley's (1976, pp. 18–19) assessment that "among the altered conditions that contributed to new ways of thinking and writing about the self in America after the Civil War, perhaps none was more formative than the rise of modern psychology."

The idea that there are unconscious aspects of the self means for the autobiographer that telling a story of the self might not be as simple as once thought. Rather than tell a story of an active self which goes through a development, the story is much more of a self which is an object of scrutiny. This is the theme of the "search for the true self" that many commentators have mentioned in their attempt to understand the modern self. Although a search for the true self is not new (Cox, 1980, p. 7), this search which the autobiographers describe involves an exploration of the question of how much one is a product of social influences and human drives, instincts, or needs. It appears to be a search for understanding what uniquely belongs to one's self. The writing of an autobiography can serve as an ideal vehicle, in fact, in this search for the true self by the period of reflection that it requires. As one begins to understand more aspects of self, this is discerned as "growth." Instead of using a metaphor like the "navigator," metaphors more often refer to the process of searching for and coming to at least a partial understanding of the self, e.g., peeling away the layers, the search, the odyssey, becoming, growing, and "bursting out of the jello." This process is not envisioned as a straight path.

The concerns about comprehending the self are reflected in comments such as Anderson's (1942, p. 20) remark that "the self you seek, the true self you want at last to face, is hidden away." Or, in an autobiography which is considered a classic work, Adams (1918, p. 433) says that "the only absolute truth was the sub-conscious chaos below, which every one could feel when he sought it." Cleaver (1968, p. 15), in his book *Soul on Ice,* states this in his chapter entitled "On Becoming": "I had to find out who I am and what I want to be, what type of man I should be. . . . " Brooks (1965, p. 445) describes the process of understanding one's self in terms of "the dark caverns into which all men must descend if they are to know anything beneath the surface." And finally, in her bestseller, MacLaine (1983, p. 5) writes:

> That is what this book is about. . . . It's about the experience of getting in touch with myself in my early forties. . . . So this book is about a quest for my self—a quest which took me on a long journey that was gradually revealing and at all times simply amazing.

Freud provided a method which some viewed as a means of rationally exploring and comprehending the unconscious aspects of self. As Weber remarked in 1918, ". . . the spheres of the irrational, the only spheres that intellectualism has not yet touched, are now raised into consciousness and put under its lens" (Gerth and Mills, 1980, p. 143). And as Freud put it: "All seemingly meaningless forms of mental life are actually meaningful. . . . " (Jones, 1953, p. 366). One might be able to explore one's self by analyzing dreams, slips of the tongue or other seemingly meaningless behavior.

This new method of examining the self along with a stock of Freudian concepts was quickly picked up by many lay persons (LaPiere, 1959; Van den Berg, 1961; Crundon, 1972). Anderson (1942, p. 339) discusses the effect of Freud around 1920:

> At the time Freud had just been discovered and all the young intellectuals were busy analyzing each other and everyone they met. . . .
> . . .They psyched us. They psyched men passing in the street. It was a time when it was well for a man to be somewhat guarded in the remarks he made, what he did with his hands.

Another aspect of attempting to comprehend the self is the bringing forth of backstage information about the self in an attempt to confess and be "open." Kenneth Burke (1939, p. 391) has commented that "Freud perfected a method for being frank." One can look back to at least Rousseau to find a concern with being frank in a story of one's self. Gill (1907), concerned about such frankness, wrote an article in the *Atlantic Monthly* entitled, "The Nude in Autobiography." But it is not until after the introduction of Freud's work in the United States that there is an increasing discussion of sexuality as well as other intimacies.[7] Anderson's (1942, p. 3) opening line is this: "*If there is any value in the telling,* in this fragmentary way, of my own adventures of living, the thing to be striven for is frankness." Another apt quotation in this regard is from *The Autobiography of Malcolm X* (1965, p. 151):

> I want to say before I go on that I have never previously told anyone my sordid past in detail. I haven't done it now to sound as though I might be proud of how bad, how evil, I was.
> But people are always speculating—why am I am as I am? To understand that of any person, his whole life, from birth, must be reviewed. All of our experiences fuse into our personality. Everything that ever happened to us is an ingredient.

Thus to come to an adequate "understanding" of oneself, one must include all the details, good and bad. But it is ironic that in letting the audience of readers share these intimacies, an autobiographer cannot possibly tell all the details and must selectively choose the "sordid" events as well as the good.

This frankness about the self is, in many cases, accompanied by the use of a Freudian vocabulary after the early decades of the twentieth century. Terms like Oedipus complex, catharsis, libido, and anal retentive are used without explanation. Quinn (1972), for example, discusses the Oedipal complex, penis envy, and concerns with defecation. Angelou (1969, p. 218) describes her brother and her mother as "entangled in the Oedipal skein." Meyer (1953, p. 10) says: "Now that I look back upon my extraordinary Oedipus complex. . . . " Even in the reserved autobiography of a historian, John Nef (1973, pp. 20–21), the reader finds a detailed account of an incident in which Nef found a secret vantage point in the home of his guardian where he could watch one of the female guests taking a bath.

In addition to these concerns about comprehending the self, these autobiographies published after the first two decades of the twentieth century also manifest a concern about the management of one's life. There is less optimism about the possibility of a unilinear development of the self.[8] Discussions of will and character, for the most part, drop out as one theological writer has later lamented (Lapsley, 1967). Some autobiographers display a difficulty with seeing the story of self as a story of improvement or even any pattern at all. Hellman (1969, p. 43) chides herself when she states, "I believed I was not doing or living the way I had planned. I had planned nothing of course." Later she adds that she never figured things out. Brooks (1965, p. 641) wonders about his identity and suggests that "all these trifles [which he collected] . . . seem to me to hold my identity together in the various chances and changes of life. . . . " It is not so clear that events have meaning within an overall framework as steps in a development. These autobiographers appear more willing to admit doubts, mistakes, ambiguities, and uncertainties.

A concern about the management or control of one's life also leads some of these autobiographers to consider the role which chance or accidents play in the determination of the life course. Lundborg (1978, p. 140), a former chairman of the board at the Bank of America, emphasizes that two of the most important events of his life were due to accidents. He discusses a happening in which he was trying to decide which college to choose among those which had accepted him for admission:

> Down the street came two young friends of mine . . . and they called out, 'Hey, Louis, how about driving down to Stanford with us next week?' That did it. But for the accident of my being at that particular street corner at that particular moment, I think I would have headed east. . . . Everything that has happened to me since was certainly influenced by my being in California.

Mencken (1940) and Carnegie (1920) also stress the importance of an accidental occurrence in their lives.

The worry about the extent to which one can control one's life is accentuated as autobiographers begin to discuss more and more the social formation of the self in those works published after the mid-twentieth century. Relationships with others begins to consume a greater share of many of these autobiographies.[9] As noted earlier, autobiographers recounting their mental or moral development did not see this as an important aspect of their life-stories. But now readers do find out the names of the spouses and children and learns about aspects of their relationships. For example, Goldstein (1973) discusses his wife and his college friends. He also admits his great concern with being liked by others.

Because childhood is seen as a time in which events occurred that played a significant role in forming one's self, there is a correspondingly greater portion of the autobiographical account that is devoted to childhood. In fact, there are many such autobiographies written by adults which only discuss the authors' childhoods (Coe, 1984). This time of life is important if one accepts Freudian ideas that the personality is "set" in the first few years of life. And the role of the parents is seen as crucial in the forming of the self during this period. The relationship with the parents is scrutinized more closely, and the portrait of the parents that is painted is no longer idealized. For some, childhood also represents a time in which the child's spirit and vision had not yet been fettered by socialization, and the autobiographical account of that period is an attempt to recapture some of that "original vision."

The concerns about comprehending one's self and the extent to which one's life can be managed are reflected in a change in the way in which autobiographies are structured. First, in trying to scrutinize the self, an autobiographer does not need to adhere to a strict chronological timeline. He or she may not see causal connections between events in which one thing leads to another. Anderson (1942, p. 28) remarks, "I have no desire to put down, in chronological order, the incidental facts of [my] life but there is a loose structure, a rambling sort of house, of many rooms, occupied by many people, of which I would like to tell." It is possible to skip back and forth over time to come to some type of understanding of self.

One technique for ordering one's life that I find in autobiographies beginning in the 1960s is that of the flashback—a word which in this use originated in the movie industry in this century. This type of technique is particularly suited to telling the story of a search for self which involves delving into the past— particularly those significant first years. Anthony Quinn (1972) uses discussions with his psychoanalyst to structure his autobiography—discussions which involve flashbacks to childhood. Cleaver (1968) begins his account with a time in which he was in jail. Markers of the external world such as dates, places, and historical events do not appear to be important when one is scrutinizing one's self. The writing of the autobiography can itself be part of the search for self.

Additionally, in contrast with the earlier autobiographers who found it easier to assume that they had accurate memories, I find that in the last three decades there is more of a disregard for documentation and telling the truth in autobiographies. Moreover, some would justify a partial fictionalization of the life story because embellishments, as the argument goes, also express the individuality of the writer and this is seen as the intention of the account. It is seen as desirable to include the products of the imagination.

A decreased concern with trying to recount events accurately and truthfully means that the autobiographer can use dialogue freely with no explanation of how she or he can remember lengthy conversations word-for-word. Some contemporary autobiographers also make up characters to have a dialogue with in the book, or create fictional characters to play roles in the events which are described. For example, MacLaine (1983, p. i) begins with the comment that "some of the people who appear in this book are presented as composite characters in order to protect their privacy. . . . "

Shumaker (1954, p. 139) asserts that "personal lives, often if not generically, are adopting more of the techniques of the novel." He also claims that the novel "has frequently pretended to be autobiography" in recent years (Shumaker, 1954, p. 140). Some autobiographers are producing more artful compositions with carefully chosen titles for the chapters and more poetic titles for the whole account. But yet most autobiographies are still recognizable by the readers as autobiographies.

In sum, in these accounts in which the self is treated as an object of scrutiny, there is a notion of a self which is "buried deep inside" and which is the "real" or "true" self. But this idea of the self is a vague and ambiguous concept. On the one hand, the self is treated as that which is unique and which distinguishes one from all others. But the self also is treated as that which is unconditioned or unsocialized—some essential core that was there at birth. Here the question arises as to how much this essential core may be something common to all humans, e.g., drives, 'instincts,' or needs.

I believe that just what this core or true self is—the "I"—is actually treated as a residual concept. The focus instead is on trying to determine the process through which this self arises. This involves exploring the self—determining the extent to which one is a product of socialization, a product of childhood experience, or a product of society. The search also involves determining the extent to which one is affected by unconscious drives and motivations. As Shumaker (1954, p. 90) argues, those autobiographers interested in the subjective self have begun "to feel it their duty not only to describe how the individual life was lived, but also to explain why it was necessarily lived so and not otherwise." Andrew Carnegie (1920, p. 6) was not hesitant to say this about his mother in the early decades of this century:

> I feel her to be sacred to myself and not for others to know. None could ever really know her—I alone did that. After my father's early death she was all my own.

A later rendering of such a point of view would more likely include a discussion of such a relationship in greater detail in order to understand why this was the case. Such a careful scrutiny of the self is seen as "growth."

DISCUSSION

Peckham (1979, p. 246) has remarked that the self is a category "created by culturally conventionalized interpretation." The purpose of this study has been to explore such an idea—to determine if there are historical differences in how individuals conceive and symbolize the self. The data used are from one particular group, those who produce their life-stories for dissemination to the reading public.

To recapitulate, a theme of many of the autobiographies in the late nineteenth and early twentieth centuries was the development of self. Such a theme contrasted with earlier autobiographies in which a story of the self was only incidental to the account of some historical or social event. But in these stories which did focus on the self, a development was often traced of a mental maturation or the development of character and will power. Such a development of self was not viewed as a natural development which all human beings experience. Rather success in such an endeavor required special effort—an active self—but it did not require extraordinary "gifts" or genius (matters of heredity). Additionally, such a development of the self was not the direct result of socialization by parents, church, or school. In fact, these autobiographers tended to show concern about the negative effects of society on the self and see the development of will or a coherent philosophy of life as ways in which one could combat such negative effects.

The view of the self as active is exemplified by the metaphor of the navigator. It is not assumed that the self is a blank slate at birth. Rather, there is something already there waiting to be prodded so that it can unfold and develop. Once this self is awakened, however, it must remain active to continue this development. This is an optimistic viewpoint of the self in which, with effort and discipline, one's self can develop to a more advanced state. It also is implicit that the self is something which is knowable and which can be comprehended through direct introspection.

Since autobiographies don't all change simultaneously, it is not possible to give a specific date to changes in the conceptualization of the self. But there is a clear shift which occurs in autobiographies published in the United States in the early decades of the twentieth century. Due probably in large part to theories and concepts developed during the growth of the disciplines which study human behavior, there have been changes in ideas about the comprehension of the self, the development of the self, and the influence of the environment on the self. First, ideas about the unconscious are now evident in these life-stories and are accompanied by a doubt that it is possible to simply introspect and examine the self. It is granted that one may be influenced by unconscious motivations which are not easily dredged to the "surface." Just what the self is becomes an open question. The autobiographers begin to use metaphors of search and discovery to describe the process of looking for and trying to comprehend the self.

Second, the self is no longer viewed by many of these autobiographers as something which develops in a rather straightforward, unilinear manner. Some even begin to doubt that there is coherence in their lives. There is a greater stress on accidents, wrong decisions, ambiguities, and hesitations. Instead, a story may be told of "growth"—the steps through which the autobiographer proceeded in order to attempt to understand his or her own self. This may be a rational, self-willed exploration of the self, sometimes aided by therapists or counselors who act as "coaches."

And finally, there appears to be an increasing concession on the part of these autobiographers that a self is socially formed and (to an extent) is a product of society. The search for self includes an exploration of the extent to which one's self has been malleable, the extent to which one has been socialized and shaped by society, parents, by the educational process, by the media, and by the times.

However, there remains for many of these autobiographers a sense of freedom through their belief that individuals have at least two selves—a public or social self and a private or true self. Thus it is this social self which is malleable, which adopts various roles, which is changeable according to the expectations of various social groups, and which is used in impression management. But this malleable self is not perceived to be the real or true self. However, as pointed out earlier, what this real self is and what its relation to the social self is remains vague.

My point then is this: this data demonstrates that ways of conceiving and symbolizing the self have changed historically. Theories of the development of the self, however, seldom include this historical dimension. Most theories emphasize only one part of Mead's theory of "self" development. Mead attempted to give an account of the development of the self in an individual. On the one hand, Mead argued that the self is not something which is innate or which exists before contact with others. Rather, this self is something which develops in a dynamic process through social interaction as an individual, during childhood, learns and then incorporates the attitudes of the family and then the "generalized other" toward his or her self. On the other hand, Mead (1936) also argued that the self is constructed using cultural ideas available for manufacturing a viewpoint of the self. Not only does one learn what the attitudes are of others toward one's self, one also learns ways of thinking about and comprehending the self, and with new ways of conceptualizing the self comes, in reality, new selves. It is this second aspect of Mead's theory of the self that most modern theorists overlook.

In much of the substantial literature on the self in psychology and sociology, there is an empirical interest in individual self-concepts. For example, what traits does the individual see in herself or himself? Or what variables affect an individual's self-esteem? But this leaves aside the question of the extent to which there are, as Mills (1963) might put it, changing vocabularies of motives, concepts, and symbols which these individuals use to discuss and perceive the self and the relationship of self to environment. For not only should it be recognized that there are changing ways of conceptualizing the self, but it also should be considered whether these different ways of viewing the self may lead to differences in behavior. As Dannefer (1984, p. 107) suggests, taken-for-granted ideas such as notions of human development "may operate as self-fulfilling prophecy, producing or reinforcing the patterns it presumes." Furthermore, researchers should be cautious in their own attempts to define the self by being aware of the extent to which such a definition fits in with the current "climate of opinion."

For those researchers who are interested in the changing ideas about the self, data from autobiographies can help flesh out and refine such understandings. In autobiographies, one can find persons struggling with the questions of the effect of socialization; the importance of inherited traits, fate, accidents, drives, and unconscious motivation; and the nature of the self and self development. Zurcher (1977, p. 35) has proposed that accelerated social change has led to modifications in the self-concepts of individuals. He characterizes the changes as from an "orientation toward stability of self (self as object) to orientation toward change of self (self as process)" (Zurcher, 1977, p. 35). But he admits that his evidence is "for the most part anecdotal, illustrative, and highly selective" (Zurcher, 1977, p. 36). He relies on contemporary data from administration of the Twenty Statement Test to college students, ex-felons, and dissident priests.

Likewise, Turner (1976) has suggested that there has been a change in the "locating" of the self—from seeing one's true self displayed when performing in roles to seeing that one's true self is displayed in impulsive acts and the relation of intimate information to others. But Turner, too, relies primarily upon contemporary evidence and his own observations.

Both Zurcher and Turner surmise that a significant part of the change in the modern conception of the self has to do with dissatisfaction on the part of individuals with the social aspects of self—the extent to which one is determined and constrained by society. Both see this as a rejection by individuals of an identification with institutions. Zurcher maintains that there is a greater emphasis on the changeable self and seeing the self as process. Turner places more stress on a belief in the idea of the true self which is expressed by impulse and the revelation of intimate information about self to others.

I agree, based on my data, that there is a contemporary concern in the United States with ideas of a true self and a concern with the issue of social determinism. But I do not see these autobiographers as simply rejecting those social aspects of self and attributing value to being changeable or given to impulse. Those who chronicle their life-stories appear to have a more complex view. For example, a concern with a search for the true self involves a consideration of the relationship between self and the environment. To what extent is one a product of socialization? But also, to what extent is one affected by human drives and needs and unconscious motivations? Here there are no easy answers. I find that those who chronicle their life-histories treat impulse not as something which should be unfettered, but rather as potentially an expression of human needs and drives. It remains unclear as to whether such impulses are an expression of freedom or part of the conditions of action.

The use of autobiographical data also allows a researcher to contrast contemporary conceptualizations of the self with much earlier conceptualizations. This may help the researcher gain perspective on his or her own understandings of the self. Goffman's (1961, p. 152) comment here is pertinent:

> There is a vulgar tendency in social thought to divide the conduct of the individual into a profane and sacred part. . . . The profane part is attributed to the obligatory world of social roles; it is formal, stiff, and dead; it is exacted by society. The sacred part has to do with 'personal' matters and 'personal' relationships—with what an individual is 'really' like underneath it all when he relaxes and breaks through to those in his presence.

By examining other perspectives on the self, it is possible to see how the viewpoint of the self that Goffman describes is a sociohistorical product. Thus we must be careful not to assume that individuals always are concerned with the issue of searching for the true self or that persons have both social and private selves.

NOTES

1. According to the bibliographer, Kaplan (1961, p. v), approximately one-fourth of American autobiographies are scarce and difficult to locate. In addition, I found that most libraries select only contemporary and classic autobiographies and those memoirs with interesting supplementary information. Thus my selection of autobiographies does not include the obscure works that cannot be found in large libraries.

2. Others may wish to trace such subjective autobiographies back to Saint Augustine's *Confessions*. Shumaker (1954, p. 13) prefers to refer to these religious confessions as "directly ancestral" to subjective autobiographies because he believes that the purpose of religious confessions is only to offer oneself as an *exemplum*.

3. A concern with the self appears to be reflected in a number of words coined in the nineteenth and twentieth centuries. According to Webster, these hyphenated words appeared during that time: self-reliant (1848), self-analysis (1862), self-questioning (1862), self-identity (1866), self-reflective (1875), self-awareness (1880), self-expression (1892), self-discovery (1924), self-image (1951), and self-definition (1965).

4. Fellman and Fellman (1981) discuss the late nineteenth-century interest in physical fitness and exercise which included the belief that exercise was a way to develop the will.

5. Since it is not uncommon to find that autobiographers will mention that they have read other life histories such as Rousseau's, it may be that Rousseau's ideas about two births in a lifetime (in which the second birth involves a psychic maturation) played a part in shaping these subsequent

discussions of a significant "awakening" (Van den Berg, 1961, p. 25). Also, Weber (Gerth and Mills, 1980, p. 279) comments on the religious doctrine of rebirth and the initiation rites of youths.

6.   Rousseau may have helped to establish a convention of accounts of childhood pranks by his similar display of such deeds in his attempt to "tell all."

7.   Foucault (1978) comments on the increase of discourse about sexuality in the West, and Mills (1963) mentions the use of a sexual vocabulary of motives in the twentieth century.

8.   Wyllie (1954, p. 146) suggests that the muckrakers around the turn of the century helped tarnish the image of the self-made man by bringing to light some of the less than scrupulous methods employed by some of the better known successes. Also, Carnegie (1920) and Brooks (1965) comment that World War I created pessimism among those who had formerly believed that there was great hope for the development of both individuals and the human race.

9.   According to Wyllie (1954, p. 169), after World War I, there was less talk about improvement of character and more talk about personality and getting along with others—in the "how to win friends" style.

REFERENCES

Adams, Henry. [1906] 1918. *The Education of Henry Adams: An Autobiography*. Boston: Houghton Mifflin.
Alcott, William A. 1839. *Confessions of a School Master*. New York: Gould, Newman and Saxton.
Alden, Timothy. 1827. *An Account of Sundry Missions Performed Among the Senecas and Munsees*. New York: J. Seymour.
Anderson, Sherwood. 1942. *Sherwood Anderson's Memoirs*. Chapel Hill: University of North Carolina Press.
Angelou, Maya. 1969. *I Know Why the Caged Bird Sings*. New York: Bantam.
Bailey, Robert. 1822. *The Life and Adventures of Robert Bailey, From his Infancy up to December 1821 Interspersed with Anecdotes, and Religious and Moral Admonitions*. New York: The Author.
Ball, Charles. 1859. *Fifty Years in Chains*. New York: H. Dayton.
Barnum, Phineas T. 1855. *The Life of P.T. Barnum, Written by Himself*. New York: Redfield.
Beecher, Lyman. 1865. *Autobiography*. New York: Harper.
Brinkerhoff, Roeliff. 1900. *Recollections of a Lifetime*. Cincinnati: Robert Clarke.
Brooks, Van Wyck. 1965. *An Autobiography*. New York: E.P. Dutton.
Burroughs, Stephen. 1811. *Memoirs of the Notorious Stephen Burroughs of New Hampshire*. London: Cape.
Carnegie, Andrew. 1920. *Autobiography of Andrew Carnegie*. Boston: Houghton Mifflin.
Cartwright, Peter. 1856. *Autobiography of Peter Cartwright: The Backwoods Preacher*, edited by W.P. Strickland. New York: Phillips and Hunt.
Cleaver, Eldridge. 1968. *Soul on Ice*. New York: Delta.
Conant, James. 1970. *My Several Lives: Memoirs of a Social Inventor*. New York: Harper and Row.
Davis, Andrew Jackson. 1857. *The Magic Staff*. New York: J.S. Brown.
Davis, Rebecca Harding. 1904. *Bits of Gossip*. Westminster: Archibald Constable.
Fagan, James. 1912. *The Autobiography of an Individualist*. New York: Houghton Mifflin.
Fairbank, Calvin. 1890. *Reverend Calvin Fairbank During Slavery Times*. Chicago: R.R. McCabe.
Fox, Ebenezer. 1848. *The Adventures of Ebenezer Fox, in the Revolutionary War*. Boston: Charles Fox.
Goldstein, Andrew. 1973. *Becoming: An American Odyssey*. New York: Saturday Review Press.
Graydon, Alexander. 1811. *Memoirs of A Life, Chiefly Passed in Pennsylvania, Within the Last Sixty Years*. Harrisburg: John Wyeth.
Hellman, Lillian. 1969. *An Unfinished Woman*. New York: Bantam.

Hemingway, Ernest. 1964. *A Moveable Feast*. New York: Charles Scribner's Sons.

Jones, John Paul. 1806. *A Narrative of the Celebrated Commodore John Paul Jones*. Philadelphia: Peter K. Wagner.

Kazin, Alfred. 1951. *A Walker in the City*. New York: Harcourt, Brace, and World.

Keeler, Ralph. 1870. *Vagabond Adventures*. Boston: Field, Osgood & Co.

Kirby, Georgiana. 1887. *Years of Experience: An Autobiographical Narrative*. New York: AMS Press.

Lewisohn, Ludwig. 1922. *Up Stream: An American Chronicle*. New York: Boni and Liveright.

Lundborg, Louis. 1978. *Up to Now*. New York: W.W. Norton.

McAllister, Ward. 1890. *Society as I Have Found It*. New York: Cassell.

MacLaine, Shirley. 1983. *Out on a Limb*. Toronto: Bantam.

Malcolm X. 1965. *The Autobiography of Malcolm X*. New York: Grove.

Martin, Joseph Plumb. 1830. *A Narrative of Some of the Adventures, Dangers, and Sufferings of a Revolutionary Soldier; Interspersed with Anecdotes of Incidents that Occurred Within His Own Observation*. Hallowell, ME: Glazier, Masters & Co.

Mencken, Henry Louis. 1940. *Happy Days, 1880–1892*. New York: A.A. Knopf.

Meyer, Agnes. 1953. *Out of These Roots: The Autobiography of an American Woman*. Boston: Little, Brown.

Moody, John. 1933. *The Long Road Home*. New York: Macmillan.

Morgan, Henry. 1874. *Shadowy Hand; or Life Struggles: A Story of a Real Life*. Boston: By the Author.

Nef, John. 1973. *Search for Meaning: The Autobiography of a Nonconformist*. Washington: Public Affairs Press.

Quinn, Anthony. 1972. *The Original Sin, A Self-Portrait*. Boston: Little, Brown.

Roberts, Lemuel. 1809. *Memoirs of Captain Lemuel Roberts*. Bennington: Haswell.

Rodriguez, Richard. 1982. *Hunger of Memory: The Education of Richard Rodriguez*. New York: Bantam.

Shepard, Elihu. 1869. *Autobiography of Elihu H. Shepard, Formerly Professor of Languages in St. Louis College*. St. Louis: Knapp.

Solomon, Hannah. 1946. *Fabric of My Life*. New York: Bloch.

Steffens, Joseph Lincoln. 1931. *The Autobiography of Lincoln Steffens*. New York: Harcourt, Brace, and World.

Tanner, John. [1830] 1956. *A Narrative of the Captivity and Adventures of John Tanner*. Minneapolis: Ross and Haines.

Thomas, Ebenezer S. 1840. *Reminiscences of the Last Sixty-Five Years*. Hartford: The Author.

Vorse, Mary Marvin. 1911. *Autobiography of an Elderly Woman*. Boston: Houghton Mifflin.

Walling, George. 1887. *Recollections of a New York Chief of Police*. Montclair, NJ: Patterson Smith.

OTHER WORKS CONSULTED

Allport, Gordon. 1942. *The Use of Personal Documents in Psychological Science*. New York: Social Science Research Council.

Angell, R.C. 1945. "A Critical Review of the Development of the Personal Document Method in Sociology, 1920–1940." Pp. 177–232 in *The Use of Personal Documents in History, Anthropology and Sociology*, edited by L. Gottschalk et al. New York: Social Science Research Council.

Becker, Howard. 1966. Introduction to *The Jack-Roller*, by Clifford R. Shaw. Chicago: University of Chicago Press.

Berger, Peter. 1966. "Identity as a Problem in the Sociology of Knowledge." *European Journal of Sociology* 7: 105–115.

Bertaux, D. 1981. *Biography and Society: The Life History Approach in the Social Sciences.* Beverly Hills: Sage.

Blumer, Herbert. 1939. *Critiques of Research in the Social Sciences: I: An Appraisal of Thomas and Znaniecki's "The Polish Peasant in Europe and America."* New York: Social Science Research Council.

Briscoe, Mary Louise. 1982. *American Autobiography 1945–1980.* Madison: University of Wisconsin Press.

Bruford, Walter H. 1975. *The German Tradition of Self-Cultivation.* London: Cambridge University Press.

Burke, Kenneth. 1939. "Freud and the Analysis of Poetry." *American Journal of Sociology* 45:391–41.

Coe, Richard N. 1984. *When the Grass Was Taller: Autobiography and the Experience of Childhood.* New Haven: Yale University Press.

Cooley, Thomas. 1976. *Educated Lives: The Rise of Modern Autobiography in America.* Columbus: Ohio State University Press.

Cox, Stephen. 1980. *'The Stranger Within Thee': Concepts of the Self in Late Eighteenth-Century Literature.* Pittsburgh: University of Pittsburgh Press.

Crundon, Robert. 1972. *From Self to Society: 1919–1941.* Englewood Cliffs: Prentice-Hall.

Curti, Merle. 1980. *Human Nature in American Thought.* Madison: University of Wisconsin Press.

Dannefer, Dale. 1984. "Adult Development and Social Theory: A Paradigmatic Reappraisal." *American Sociological Review* 49: 100–116.

Delany, Paul. 1969. *British Autobiography in the Seventeenth Century.* London: Routledge and Kegan Paul.

Denzin, Norman. 1978. *The Research Act.* Chicago: Aldine.

Dollard, J. 1935. *Criteria for the Life History: With Analysis of Six Notable Documents.* New Haven: Yale University Press.

———. 1938. "The Life History in Community Studies." *American Sociological Review* 3:724–737.

Fellman, Anita and Michael Fellman. 1981. *Making Sense of Self: Medical Advice Literature in Late Nineteenth-Century America.* Philadelphia: University of Pennsylvania Press.

Foucault, Michel. 1978. *The History of Sexuality* 1. New York: Vintage.

Gerth, H. and C. Wright Mills (eds.) [1946] 1980. *From Max Weber: Essays in Sociology.* New York: Oxford University Press.

Giddens, Anthony. 1979. *Central Problems in Social Theory.* Berkeley: University of California Press.

Gill, W.A. 1907. "The Nude in Autobiography." *Atlantic Monthly* 99:71–79.

Goffman, Erving. 1961. *Encounters.* Indianapolis: Bobbs-Merrill.

Hart, Francis R. 1970. "Notes for an Anatomy of Modern Autobiography." *New Literary History* 1:485–511.

Hofstadter, Richard. 1955. *Social Darwinism in American Thought.* Boston: Beacon.

———. 1962. *Anti-Intellectualism in American Life.* New York: Vintage.

Horowitz, Irving. 1977. "Autobiography as the Presentation of Self for Social Immortality." *New Literary History* 9:173–179.

Jones, Ernest. 1953. *The Life and Work of Sigmund Freud.* New York: Basic Books.

Kaplan, Louis (ed.) 1961. *A Bibliography of American Autobiographies.* Madison: University of Wisconsin Press.

LaPiere, Richard. 1959. *The Freudian Ethic.* New York: Duell, Sloan, and Pearce.

Lapsley, James N. 1967. *The Concept of Willing: Outdated Idea or Essential Key to Man's Future?* Nashville: Abingdon Press.

Lasch, Christopher. 1978. *The Culture of Narcissism*. New York: Norton.

Leahey, Thomas H. 1980. *A History of Psychology*. Englewood Cliffs: Prentice-Hall.

Lukes, Stephen. 1973. *Individualism*. Oxford: Basil Blackwell.

Lyons, John O. 1978. *The Invention of the Self*. Carbondale: Southern Illinois University Press.

Marx, Leo. 1969. "American Studies—a Defense of an Unscientific Method." *New Literary History* 1:75–90.

Mazlish, Bruce. 1970. "Autobiography and Psycho-Analysis: Between Truth and Self-Deception." *Encounter* 35:28–37.

Mead, George Herbert. 1936. *Movements of Thought in the Nineteenth Century*. Chicago: University of Chicago Press.

Mills, C. Wright. [1940] 1963. "Situated Actions and Vocabularies of Motives." Pp. 439–452 in *Power, Politics, and People*, edited by Irving Horowitz. New York: Ballantine.

Olney, James. 1972. *Metaphors of Self: The Meaning of Autobiography*. Princeton: Princeton University Press.

————. 1980. *Autobiography: Essays Theoretical and Critical*. Princeton: Princeton University Press.

Pascal, Roy. 1960. *Design and Truth in Autobiography*. London: Routledge and Kegan Paul.

Payot, Jules. [1893] 1909. *The Education of the Will: The Theory and Practice of Self-Culture*. New York: Funk and Wagnalls.

Peckham, Morse. 1979. *Explanation and Power*. New York: Seabury Press.

Plummer, Ken. 1983. *Documents of Life*. London: George Allen & Unwin.

Rousseau, Jean-Jacques. [1781] 1953. *The Confessions*. New York: Penguin.

Sennett, Richard. 1974. *The Fall of Public Man*. New York: Vintage Books.

Shaw, Clifford. 1930. *The Jack-Roller*. Chicago: University of Chicago Press.

————. 1931. *The Natural History of a Delinquent Career*. Chicago: University of Chicago Press.

————. 1938. *Brother in Crime*. Chicago: University of Chicago Press.

Shumaker, Wayne. 1954. *English Autobiography: Its Emergence, Materials, and Form*. Berkeley: University of California Press.

Sutherland, E.H. 1937. *The Professional Thief by a Professional Thief*. Chicago: Phoenix.

Symonds, Percival. 1951. *The Ego and the Self*. New York: Appleton-Century-Crofts.

Thomas, W.I. and Florian Znaniecki. [1918–1920] 1958. *The Polish Peasant in Europe and America*. New York: Dover.

Trilling, Lionel. 1972. *Sincerity and Authenticity*. Cambridge: Harvard University Press.

Turner, Ralph. 1976. "The Real Self: From Institution to Impulse." *American Journal of Sociology* 81:989–1016.

Van den Berg, J.H. 1961. *The Changing Nature of Man*. New York: W.W. Norton.

Webb, E.J., D.T. Campbell, R.D. Schwartz, and L. Sechrest. 1966. *Unobtrusive Measures: Nonreactive Research in the Social Sciences*. Chicago: Rand McNally.

Weintraub, Karl. 1978. *The Value of the Individual: Self and Circumstance in Autobiography*. Chicago: University of Chicago Press.

Wyllie, Irvin. 1954. *The Self-Made Man in America: The Myth of Rags to Riches*. New Brunswick, NJ: Rutgers University Press.

Zurcher, Louis. 1977. *The Mutable Self*. Beverly Hills: Sage.

TREASURED POSSESSIONS IN ADULTHOOD AND OLD AGE

N. Laura Kamptner, Jean R. Kayano, and Joan L.
Peterson

INTRODUCTION

A young child's attachment to a special object is a natural and commonly occurring phenomenon. The term "treasured object" conjures up an image of Linus, who is never without his security blanket. Over the past 30 years, a fair amount of attention has been devoted to finding out what objects infants and young children become attached to, and what functions these attachments serve. There is little written, however, on the importance and function of "special" possessions after the years of early childhood.

In spite of the paucity of literature in this area, the importance of inanimate objects is nonetheless evident in an individual's life after childhood. Much of our

Current Perspectives on Aging and the Life Cycle,
Volume 3, pages 69–117.
Copyright © 1989 by JAI Press Inc.
All rights of reproduction in any form reserved.
ISBN 0-89232-739-1

adult life, for example, is spent acquiring and maintaining material possessions. Holidays, birthdays, and religious ceremonies involve the giving or exchanging of material objects. Heirlooms are passed down through generations of a family. We pick up "souvenirs" or mementos while on vacation. We may also have a "collection" of coins, dolls, stamps, or shells. Or, a "good-luck" charm may be carried around by its owner.

Recent popular press articles have also attested to the importance of certain "special" objects. These have included, for example, articles on teddy bears as a source of comfort and support (regardless of one's age), a columnists' response to a 19-year old college student's concern over sleeping with her baby blanket, a young woman's sadness and grief over losing her suitcase and its contents, and a story about the institutionalization of an elderly woman who continued to tenaciously cling to (and carry around) the few personal possessions that she had left (Beck, Cooper, Dallas, Abramson, and McCormick, 1984; Goldberg, 1984; Kellogg, 1985; Neal, 1985; Schaaf, 1985; Van Buren, 1985). However, the functions and meanings of these objects in people's lives are unclear.

The purpose of the current project was to investigate the role of inanimate objects in adulthood and old age. The specific goal was to define what objects are "cherished" or important, and the meaning or function of such objects for individuals. Toward this end, this chapter is organized in the following manner. First, the meaning and function of object possession is discussed. Second, the little existing literature on "treasured possessions" is reviewed. Taken together, these two sections will then provide a background for the last portion of this chapter, in which an exploratory study of treasured and other special possessions in adulthood and old age is presented.

OBJECT POSSESSION: THE MEANING AND FUNCTION OF OBJECTS

One way to begin examining the meanings of objects for people is to first investigate the meaning of object possession. The motivation to acquire and possess objects as a goal in itself seems to be a uniquely human phenomenon. Animals, for example, take only those objects that serve basic survival needs, and apparently have no innate tendency to hoard, collect, or acquire for the sake of acquiring alone (Beaglehole, 1932). Early and current writings suggest that there are a variety of different yet closely related "meanings" of object possession for the individual, as described below.

Objects as Sources of Social Status and Social Dominance

Objects symbolizing social status and power are found in almost all cultures, and function to elicit respect, consideration, and envy of others (Csikszentmihalyi and Rochberg-Halton, 1981). In particular, objects that are rare, expensive, or old may qualify as a status symbol by conveying to others who believe in its status that its owner is a person possessing distinctive or superior qualities.

Likewise, saving and owning property may be motivated by a desire for social dominance over others. Suttie, Ginsburg, Issacs, and Marshall (1935) suggested, for example, that the relation between an object and a person always involves at least two people and the object in question. Similarly, Hallowell (1943) viewed objects as pawns in a game, as instruments for controlling and defining the relationship between two or more people.

Finally, it has also been suggested that property may call attention to oneself, which may help one to overcome "separation anxiety," an anxiety associated with the realization of one's "separateness" or separation from loved ones (Suttie et al., 1935).

Objects as Sources of Effectance Motivation

Theories of motivation emphasize that humans are motivated to produce effects and interact competently with their environments (Brehm, 1972; deCharms, 1968; Deci, 1975; Rotter, 1966; Seligman, 1975; White, 1959). Object possession has been linked to effectance motivation, in that it enhances one's sense of mastery or control. Litwinski (1943/1944), for example, suggested that possessing objects makes people feel like masters of nature since they provide one with a margin of safety in the case of inaccurate judgements of future need. Furby (1978), one of the few contemporary researchers of possessions and possessive behavior, more recently suggested that possessions may be a primary manifestation of effectance motivation since control over the use of an object was found in her study to be the most salient characteristic that defined a possession for all ages. She speculated that the major motivational force behind object possession was causal efficacy or control over aspects of one's environment.

The role of objects in the development of a sense of competence and mastery has especially been noted during the early years of life. Infants and young children, for example, are constantly trying to learn how to bring their environment under their control (Piaget, 1936). Control over an object such as a blanket, for example, may provide a young child with a greater sense of accomplishment and competence (White, 1959). Along similar lines, Hong (1978) has speculated that for a young child, his/her blanket may function as being more controllable and manipulable than the mother, which may reinforce the infant's attachment to the object (Busch, 1974; Rudhe and Ekecrantz, 1974).

Effectance motivation has also been linked to the development of a sense of self. Seligman (1975) has suggested that infants learn to differentiate *me* from *not-me* according to which parts of the environment they can control. Infants interacting with objects may thus experience beginnings of control over the environment, which has been linked to the development of a sense of self (Lamb and Campos, 1982). Things (and their effects) may tell children who they are by expressing their intentions, which may confirm and strengthen a child's sense of self. Csikszentmihalyi and Rochberg-Halton (1981), in fact, suggest that toys are shapers of the self in childhood. Furby (1978) similarly speculates that possessions are included in one's concept of self to the degree that they are under the individual's control, and that they aid in defining one's individuality. This notion is developed more fully below.

Objects as Defining One's Self and Identity

The initial linking of possessions with "self" dates back to William James (1890), who postulated a close relationship between the self and one's possessions. He viewed possessions as important since part of the definition of self was the "material self," which included all that a person owned, including one's body and other possessions. James viewed the task of distinguishing between one's self and one's possessions as a difficult one.

From a cross-cultural perspective, Beaglehole (1932) examined possession and ownership in many different cultures and concluded that people in other cultures view personal property as somehow assimilated to the self. He suggested that there is a tendency to integrate part of the self with any object one uses or views as one's own, such that part of one's "life spirit" is united with the object. He also proposed that this was a basic psychological characteristic of humans.

Object ownership may also create a feeling of psychological "nearness" to an object, which in turn enhances the value of an object for its owner. Irwin and Gebhard (1946), for example, found that subjects preferred their own object even when presented with exactly the same (but unowned) object.

Possessions have been viewed as being important to the self-definition process at all ages. Csikszentmihalyi and Rochberg-Halton (1981), for example, suggest that objects influence what a person can do by either expanding or restricting the scope of that person's action and thoughts—and, since, what a person does is primarily an expression of what she/he is, objects have a determining effect on the development of the self. During childhood, objects of action are suggested as being the most important since children have a strong need to internalize their actions and define the limits of their selves through direct kinetic control (Csikszentmihalyi and Rochberg-Halton, 1981). In addition, experiences of adequacy and inadequacy as reflected by the material environment are very important to the cultivation of the self (Erikson, 1950). Thus, toys and tools used in games are of central importance in the development of children. During adolescence, the need for immediate, physical feedback that produces enjoyment is suggested to be the most salient, with objects helping youth establish concrete evidence for the existence of an autonomous self (Csikszentmihalyi and Rochberg-Halton, 1981). Finally, with age there may be a change from concerns about current enjoyable experiences to enduring family ties. The last three of Erikson's (1950) stages of psychosocial development, for example, reveal a shift of the structure of one's self from one's own actions to one's position in a network of enduring relationships. This structure of the self within an interpersonal network is often represented and embodied within concrete objects, which may often symbolize social ties.

Identity and maturity involve integrating past experiences and views of self into an on-going idea of self (Erikson, 1968). Objects may help a person maintain a sense of continuity of the self over time, especially as other things in the environment change (Csikszentmihalyi and Rochberg-Halton, 1981). Valued material possessions may act as signs of the self by continually cultivating it and "reminding" one of the world of meaning that they have created for themselves. They may also reconfirm and represent one's notions of one's self. For example, the tools of one's trade may help to define and remind one who s/he is as an individual. Rochberg-Halton (1984) suggests, in fact, that transactions with cherished possessions are really communicative dialogues with one's self.

Objects as Sources of Security, Solace, and Positive Affect

Possessions may also provide an individual with feelings of security and act as substitutes for needs or desires that have not been met (Neal, 1985; Rochberg-Halton, 1984; Vlosky, 1979). Objects can be counted on to provide certain pleasures, whereas one may not be able to count on other people for consistent emotional fulfillment (Furby, 1978). Stevenson (1954), a psychiatrist, tells of a 24-year-old female patient whose teddy bear had been her constant companion since she was two and one-half-years old. She would become quite distressed when it was misplaced or lost. Her personal history revealed that she was one of three grown children in a family in which all the children had suffered to some degree from the conflict between their two parents and the neurotic detachment of the mother. Another patient, a 17-year-old male, was described as having a relationship with a teddy bear that was similar to that of Aloysius, the stuffed bear from the televised series Brideshead Revisited. Personification of this object was used as a means of escape from direct personal comment and thus from direct personal relationships. The qualities of these types of objects that seem to be important in order for the object to serve in this capacity are its softness, huggableness, and "responsiveness" to one's immediate needs (Rochberg-Halton, 1984).

Objects may also help buffer feelings of fear. The need for a specific object that was soothing to one as a child may reappear at a later date when "deprivation" threatens (Winnicott, 1953). Similarly, Horton, Louy, and Coppolillo (1974) contend that adolescents who are suffering from the pains of separation from the family may be reassured and soothed in a manner similar to a young child by a blanket or stuffed animal. Beck et al. (1984) has asserted that teddy bears are especially solacing objects to people of all ages, since they act as a comforting fortress against anxiety by providing support and encouragement.

Finally, objects may also promote positive affect in the individual, either through the pleasant experience or enjoyment they provide (Furby, 1978).

Objects as Representing Interpersonal Ties

Objects may also be symbols of social integration by signifying enduring bonds, ties, and relations with others—a meaning which appears to become more salient as a person ages (Csikszentmihalyi and Rochberg-Halton, 1981). Gifts, for example, although embedded in the context of exchange, have been thought to symbolize containers for the being of the donor, who gives a part of him/herself to another person. If a gift is reciprocated, a definite tie may be established between the partners in the exchange (Csikszentmihalyi and Rochberg-Halton, 1981). In addition, objects such as heirlooms may serve as ties that bind people or family members to each other and across generations.

The Utilitarian Function of Objects

Another reason why people possess things may be to make possible certain activities and conveniences for the owner. Objects may give an individual more "freedom," and enhance their ability to effect desired outcomes in their environment (Furby, 1978), as discussed previously.

Objects as Sources of Memories

Finally, objects may be receptacles of memories of past events. In their study of special objects in the home, Csikszentmihalyi and Rochberg-Halton (1981) found that there was a strong desire to remember the good times of the past, with object meanings referring to the past increasing with age.

"TREASURED" POSSESSIONS

As shown above, object possession may serve a variety of functions, including being a source of social status, effectance motivation, defining one's self, security, solace, and positive affect, representing ties with others, serving a utilitarian function, and be sources of memories. But what about objects that are considered to be "cherished" or "treasured" by the individual? What special meanings or functions do they have for a person throughout the life cycle?

The study of "treasured" possessions started with Winnicott, a British pediatrician and psychoanalyst, who noted the possible function and developmental significance of young children's attachments to special objects. In a paper published in 1953, he characterized these objects as a child's first "not-me" possession, and said that such attachments are a part of normal, healthy emotional development. From a psychoanalytic viewpoint, he theorized that a child symbolically re-establishes union with his/her mother by stroking or cuddling a teddy bear, blanket, or other soft object. Winnicott labeled these attachments as "transitional objects," whose primary functional value was to soothe and comfort in the face of anxiety, stress, or separation from the primary caregiver, and also to aid the child in making the transition from the waking to sleeping state.

Such object attachments are said to develop at two times during the early years of life—during the second half of the first year (i.e., "primary" transitional objects), and at around two years of age (i.e., "secondary transitional objects") (Busch, Nagera, McKnight, and Pezzarossi, 1973). Primary transitional objects typically involve a crib sheet, pillow, baby blanket (Stevenson, 1954), or other soft malleable object with which the child has had a close association since birth either during sleep or feeding (Busch et al., 1973). Babies born in autumn and winter, for example, often become attached to woolen objects, while spring and summer infants tend to become attached to objects made of linen or nylon (Gaddini, 1975). The texture and odor of the object seem to be important discriminating features of these objects, and they are usually used in conjunction with some type of oral gratification behavior (e.g., thumb-sucking, using a pacifier or a bottle) (Busch et al., 1973). Secondary transitional objects usually involve a teddy bear or other soft toy, and may be used in imaginary play. According to Busch (1974), the specific way in which a "special" object is used depends on the degree of internalization of the soothing process that has occurred.

In addition to being a soother in the face of separation, other functions attributed to transitional objects include relieving loneliness, aiding in the development of toleration for frustration, and aiding in successful weaning (Winnicott, 1953). In addition, these objects are theorized to serve a critical role in the developmental process of self versus object differentiation. From a psychoanalytic viewpoint, young children are struggling to establish their individuality and to adapt to the increasing demands placed on them for independent functioning. Transitional objects are though to provide the child with a "resting place" in the face of this struggle, providing a sense of security that was previously provided by parents (Goldberg, 1984). In a sense, then, the transitional object concept is rooted in the concept of separation-individuation (i.e., the on-going differentiation of one's self from the rest of the world) (Vlosky, 1979).

Infancy and Early Childhood

Aside from its origin in the psychoanalytic literature, writings on "treasured" possessions during the early years of life appear in other theoretical camps as well. Researchers have noted that attachments to blankets, diapers, or other soft objects are a natural (but not universal) phenomenon. Studies indicate that slightly over half (54 percent) of normally developing children from middle-class backgrounds become attached to a particular treasured object during the early years of life (Sherman, 1986).

The objects to which young children develop attachments are usually soft and commonly used for comforting (Busch, 1974; Busch et al., 1973). Passman's (1976) work on toddlers' attachments to objects suggests that a critical dimension of these objects may be their suitability for clinging, since it provides for more arousal reduction. They are typically used at bedtime at night or during naptime; when the child is in a strange place or is feeling anxious, or during trips of extended duration; when the child is hurt, upset, distressed, stressed, fatigued, or bored; and during the mother's arrival and departure (Busch, 1974; Busch et al., 1973).

What is the function of these objects? The presence of an inanimate attachment object has been shown to aid a child in becoming familiar with and exploring a novel environment (Halonen and Passman, 1978; Passman and Weisberg, 1975), with the removal of the attachment object resulting in less adaptive behaviors and in more distress. Children with treasured objects have been shown to display a greater capacity to tolerate anxiety and to play alone than those without favored possessions (Passman, 1977). Children without treasured objects have been shown to have a greater need for sustained proximity to, and close physical contact with their mothers upon being reunited with them compared to those who possessed such an object (Parker, 1980). In addition, children with an attachment to a special object seem to be more likely to fall asleep easily, and are perceived as being more independent, persistent, and attentive than those without an object (Boniface and Graham, 1979; Garrison and Earls, 1982). Specifically, these objects have been suggested to function as simply "substitutes" for the absent mother (Bowlby, 1969; Passman, 1976, 1977) and may be a substitute for, or an equivalent of, certain components of mothering, especially tactile "contact" comfort (Hong, 1978).

Comparative studies reveal that attachments to objects also occur in many nonhuman species. Removal of an attachment object, whether animate or inanimate, has been shown to result in decreases in adaptive behaviors and increases in distress (Harlow, 1958; Harlow and Suomi, 1970; Mason and Kenney, 1974). These inanimate objects thus appear to serve as secure bases for a variety of species—including birds (styrofoam objects), monkeys (cloth surrogates), and human children (blankets and other objects). It has been suggested that some sense of psychological security is derived from the attachment to these objects, and that this psychological state originates within the individual child or animal and is not the result of cues emitted from the object (Rajecki, Lamb, and Obmascher, 1978). Arousal can be alleviated and play promoted if an object which provides clinging or contact-seeking behaviors is available to the child or animal (Mason, 1968).

The attachments of young children to objects are considered to be supplementary or secondary attachments rather than primary or "first" attachments, since infants without attachment figures do not become attached to inanimate objects (Ainsworth, Blehar, Waters, and Wall, 1978; Provence and Lipton, 1962). It is hypothesized that once an infant can actively gain and maintain proximity or contact with an attachment figure on his/her own behalf, she/he can then do so through a variety of specific behaviors. Such behaviors can either stop serving the attachment system or become directed toward other objects of attachment.

At least in the United States, treasured object attachment does not appear to be related to social class. In other parts of the world, however, object attachments seem to be related to cultural patterns. While Passman and Halonen (1979) found that 58 percent of 3-year-old American children were attached to blankets and 20 percent were attached to hard objects, Boniface and Graham (1979) found that only 16 percent of the 3-year-old children they studied in a London borough were attached to soft objects. Gaddini and Gaddini (1970) and Hong and Townes (1976) found object attachment (especially to blankets and pacifiers) to be more prevalent (54 percent to 61 percent) among children of upper socio-economic groups. Factors such as breastfeeding, weaning at a later age, sleeping in close proximity to the mother, and physical contact with the mothering person were demonstrated to be related to a lower frequency (5 percent to 18 percent) of object attachments among rural, lower socio-economic groups in Italy and Korea.

Finally, object attachment in childhood seems to follow a developmental course that is influenced by cognitive and personality functioning. Children with serious developmental disorders (e.g., mental retardation or multiple developmental problems in communicating and relating to other people) are much less likely to become attached to objects during early life than are normal children. Retarded children may become attached to a soft object at a later age, and use it in a manner similar to normal children (e.g., to soothe and comfort). However, children with serious social and emotional problems (e.g., autism) rarely become attached to objects, and do not seem to use them for comfort in times of distress (Sherman and Hertzig, 1983).

Middle Childhood

The literature on attachment to objects after the early childhood period is relatively sparse. However, it does appear that these early object attachments frequently persists well beyond the early childhood years. Sherman, Hertzig, Austrian, and Shapiro (1981) report that almost half of the children who initially formed such an attachment to an object were still attached to it at nine years of age. These children typically continued to use the object as a soother at bedtime, during periods of stress, or during periods of inactivity such as while watching television. No significant behavioral differences between those who had and those who had not given up their object were apparent, or between those who never had an object and those who used it after nine years of age. Furthermore, sibling use, number of siblings, birth order of the subject, gender, history of parental object use, personality, temperament, parent's marital status, and history of thumb-sucking were all found to be unrelated to object attachment in normally developing children.

Adolescence

In a study of "special" objects in the home, Csikszentmihalyi and Rochberg-Halton (1981) found that action-oriented objects with egocentric meanings, and objects that were interpreted as being associated with their developing abilities were the most salient for adolescents. The objects that children and adolescents tended to identify as being special to them were those that involved kinetic interaction—they required some kind of physical manipulation to release their "meaning" or function. Such objects included stereos, musical instruments, pets, sports equipment, vehicles, refrigerators, and stuffed animals. Stereos were viewed as offering a means to listen to their records, to create moods, and seemed to reflect adolescents' need for activity. The researchers speculated that phonograph records allowed adolescents to "try on" many different identities, provided references to their developing sense of self, modulated their emotions, provided a means of "release" or escape, and helped them express (and compensate for) negative feelings. This was viewed as being beneficial to adolescents since they usually experience a greater number of daily mood swings compared to older individuals. The ongoing experience that the special object provided in terms of activity, enjoyment, or the release it made possible was also emphasized by the youths.

Adulthood and Old Age

Only two studies to date have investigated valued possessions in adulthood and old age. First, Csikszentmihalyi and Rochberg-Halton's (1981) study of special objects in the home found that object preferences and the meanings attributed to them changed in adulthood from what they were in adolescence. During adulthood, preferences for "contemplation"-oriented objects (vs. the "action"-oriented objects of adolescence) emerged, and meanings of objects referring to the past (e.g., signifying memories) increased with age. Also, the meaning attributed to these objects tended to shift with age from what one could do with the object at the present to what one had done with it in the past. Rather than providing information mainly about the personal self and reflecting the egocentric attitudes of adolescence, objects now referred more to other people— they typically represented familial or social ties and memories that reflected "belongingness."

In adulthood, the objects in one's home that were most frequently identified as being of special significance included furniture, visual art, sculpture, books, musical instruments, and photos (Csikszentmihalyi and Rochberg-Halton, 1981). Furniture typically held important memories, and symbolized ties to family and other people. Visual art was valued for reasons of the immediate life history of the individual, such as memories and links with immediate family and other social ties.

In late adulthood, photographs, books, paintings, sculpture, plateware, silverware, furniture, television, and visual art and plates were frequently identified as being "special". Meanings typically attributed to these objects included memories, personal relationships and family (e.g., "belongingness"), continuity of self and personal identity, and mastery and control, which seemed to become more prominent with age (Csikszentmihalyi and Rochberg-Halton, 1981).

In a second study that has investigated valued possessions in adulthood and old age, Sherman and Newman (1977–78) examined what objects were considered by the elderly to be treasured possessions. They found that 81 percent of the ninety-four elderly adults they interviewed could readily identify a cherished possession. Those objects identified included photographs (20 percent), consumer items (T.V., radio, books) (14 percent), religious items (bibles, rosaries, and torahs) (12 percent), symbolic jewelry (e.g., wedding rings) (12 percent), and objects signifying personal performance (6 percent). Age seemed to make a difference as to whether or not an individual claimed to have a cherished possession, with significantly more individuals 75 years of age and older claiming not to have a cherished possession. This latter finding may have been partly due to the fact that proportionately more of these older individuals lived in nursing homes (where such objects may have been taken from them), or that the oldest old were involved in an overall disengagement process.

Mediating Variables of "Treasured" Possessions

Studies on "treasured" possessions suggest a mediating function of at least two variables: gender and personality. While some studies have found no sex differences in object attachment or object use (Sherman et al, 1981; Winnicott, 1953), others have. For example, boys have been found to be more likely to use their favored possessions in relatively active modes like carrying it, while girls tend to sit passively and stroke themselves with their possessions (Parker, 1980). Sex differences have also been found in the types of objects preferred. Csikszentmihalyi and Rochberg-Halton (1981), for example, found that males tended to name T.V.'s, stereo sets, tools, sports equipment, vehicles, and trophies significantly more often than did females. Females, by contrast, named photos, sculpture, plants, plates and glassware, and textiles significantly more frequently than did males. Also, while Sherman and Newman (1977–78) found no difference between elderly males and females in the ability to identify a cherished possession, they did find a significant difference in the types of possessions identified. Females named more jewelry and photographs, while males named primarily consumer items (T.V., etc.).

Personality factors may also mediate object attachment. Cohen and Clark (1984), for example, examined personality factors in college-aged students related to early object attachments. Using the Sixteen Personality Factor Questionnaire (Cattell, Eber, and Tatsouka, 1970), the found that tenseness (i.e., excitability, restless, impatience, and anxiety) was related to early object attachment, while being reserved (i.e., aloof, detached, rigid, precise, objective, and being a loner or unsociable) was related to the absence of an early object attachment. It has also been suggested that having had a treasured object is an indication of a kind of sensitivity and imagination in childhood that seems to facilitate forming relationships with others in adolescence and adulthood (Goldberg, 1984). Also, people who denied meanings of objects have been found by Csikszentmihalyi and Rochberg-Halton (1981) to lack a close network of human relationships.

Goldberg (1984) reports research suggesting that disturbed adolescents were more likely to have experienced parental interference with the object (e.g., their parents disposed of the object before the child was ready to part with it), and that a disproportionately high percentage of sociopaths appear to never have had such objects as children. However, Sherman (1986) asserts that attachment to a treasured object is neither essential for normal development nor a reflection of pathology. The issue may perhaps rest on whether or not early person attachments were established, with attachments to objects occurring only as secondary bonds.

SUMMARY AND PURPOSE OF STUDY

In summary, the above review suggests that personal possessions may serve a variety of functions, including being a source of social status and effectance motivation; aid one in defining one's self; provide feelings of security, solace, and positive affect; represent interpersonal ties; serve a utilitarian function; and remind one of past events. The studies reviewed here suggest that individuals do have possessions which they consider to be "treasured" or "cherished" throughout the various life stages, but the specific objects and the meanings attributed to them appear to vary with life stage, gender, and possibly personality. As reviewed above, only two studies have examined the role of special or treasured possessions in adulthood and old age. One study investigated special objects in the home, while the other examined what one's cherished possession was in old age.

The purpose of the current exploratory study was to examine the role of inanimate objects in adulthood and old age more comprehensively than has been done to date. Specifically, we were interested in finding out what objects are important to adults, and why. In addition to inquiring about "treasured" objects, we, unlike previous studies, also investigated what objects are salient under a variety of other conditions (e.g., what objects are "most important," favorite, preferred as gifts, etc.) in order to gain a broader understanding of objects and their meanings in adults' lives. In both of these cases, we were also interested in examining how the objects named and the meanings associated with them varied with age and with gender.

A second goal of this study was to examine the role of objects in life course development, and to find out how the objects named, and the meanings associated with them, varied with life stage. In other words, is continuity or discontinuity more characteristic of salient objects and their meanings as one progresses through the life course? To address this issue, we (unlike previous studies) inquired about the subjects' treasured possessions (and their meanings) during childhood and adolescence.

Finally, we were interested in what meanings are typically associated with which objects, and whether there was any connection between object possession during the early years of life and later personality attributes.

METHOD

Subjects

The subjects were 175 predominately white, middle-class adults from a small state university campus and its surrounding community in southern California. The sample included 133 females and 42 males who ranged in age from 18 to 89 years. Subjects were divided into three age groups: young adults (18- to 29 years old; n = 96), middle-aged adults (30- to 59 years old; n = 41), and older adults (60- to 89 years old; n = 38). Thirty-nine percent of the sample was currently married; 17 percent was either divorced, separated, or widowed; and 44 percent had never married.

Participation in the study was solicited through locating volunteers from undergraduate psychology courses and from the surrounding campus community, and by using the "snowball" technique of having these subjects locate other adults to participate in the study. In addition, older adults from a nearby non-residential senior center who were living independently and who were ambulatory and sufficiently well-oriented to respond to the questionnaire survey were included. The selection of the sample was thus not random, but since the nature of this study was exploratory, the primary intent was to achieve an adequate number of subjects in the three different age groups.

Instrument

A 43-item questionnaire designed to be self-administered was developed (see Appendix A). Many of the items were open-ended questions since the study was quite exploratory in nature. The questionnaire was divided into seven sections:

(1) *Background information* (age, gender, and marital status—the exact socio-economic status was not recorded since Csikszentmihalyi and Rochberg-Halton (1981) found it to be relatively uninfluential in their study of special possessions in the home), (2) *Early childhood treasured possessions* (what it was, why it was special, when and how it was used, its special characteristics or qualities, and when, how, or if it was given up), (3) *Adolescent treasured possessions* (what the subject's treasured possessions were during adolescence), (4) *Current adult treasured possessions* (what they are, why they are important or special to their owner, how they were acquired, and how and when the objects are used), (5) *"Important" and "favorite" possessions* (what other possessions are considered to be very important and favorite, and why), (6) *Other* (what objects would be rescued in case of a fire, and why; what objects are "comforting"; what "good-luck" objects are owned; what one would purchase with an extra $1000; what kinds of things are most desired to receive as gifts; what family heirlooms one has, and their meanings; what collections one has; and what types of gifts of objects are exchanged when friends move away), and (7) *Self-description* (fifteen self-report Likert-scale items on how subjects describe themselves regarding their desired degree of interpersonal affiliation, how tense they are, their perceived adaptability, perceived security, coping ability, worrisomeness, depression, anxiety, and how "warm" they viewed each of their parents as being).

Procedure

The questionnaire took approximately one hour to complete, and for most subjects it was self-administered. For several of the elderly adults for whom reading and/or writing were difficult, the questionnaire items were read to them and their responses recorded by a research assistant.

It was intriguing to note that, when subjects completed the questionnaire and returned it to the experimenter, they would (almost without exception) comment on the "effect" that the questionnaire had on them. Many remarked that while initially they hadn't thought that they had (or currently have) any "special" possessions, working through the questionnaire items "forced" or prodded them to think about some things that they had not thought about for years. They also tended to remark that having to think about what these objects meant was an "enlightening" process for them, and that, for the most part, accessing these memories had brought on feelings of significant positive affect. These subjects also commented that filling out the questionnaire encouraged them to think about things in their environment in a new way—forcing them to closely examine what they owned and why, and what was really important to them. The consistent display of interest, curiosity, and desire by the subjects to know more about what we were looking at was a genuine surprise. It is also worth noting that for the elderly who were interviewed, the interviews became exceedingly lengthy and emotionally-charged. In a sense, it seemed as if the interview session became a format for reminiscence, which is really not surprising given the nature of our questions and the issues of old age. Several of the older widows would name, for example, their wedding ring as being their most treasured objects, and would begin crying at the memories that it evoked of the deceased spouse. One elderly couple insisted on being interviewed together, and together they wept as they discussed the memories stimulated by the objects they valued.

RESULTS

The analysis of the resulting data first required the development of a classification scheme from the total inventory of objects named, and the meanings attributed to those objects. The resulting classification scheme is based in part on the classification scheme developed by Csikszentmihalyi and Rochberg-Halton (1981), but these categories were modified to fit the existing data. The resulting classification scheme for categories of objects named included twenty-three categories, and the classification scheme for the meanings attributed to the objects included ten categories. The classification schemes, along with the definition of each category, are presented in Appendix B.

Next, frequencies for the total group, for the three age groups, and for males and females were computed for both objects and meanings attributed to those objects for childhood, adolescent, and adult possessions. The percentages recorded in the following tables are the percentages based on the total number of objects (or meanings) named for each category of subjects (i.e., total group, by age, and by sex). The presentation of the data in this format was necessary because there were unequal numbers of subjects in both the age and gender groupings of subjects, and because different numbers of objects named by each grouping of subjects for each prescribed situation of object possession for which information was obtained.

TREASURED POSSESSIONS

Childhood

Eighty-three percent of the adult subjects reported having had a special object to which they were attached in early childhood. Of those objects named as being "treasured," 31.7 percent were teddy bears/stuffed animals and 34.4 percent were dolls for the total group. The frequency with which teddy bears/stuffed animals were identified as being cherished possessions decreased as the current age of the subject increased. The frequency with which dolls were mentioned, however, increased with the current age of the subject. No males were reportedly attached to dolls, but in addition to teddy bears/stuffed animals, males identified pillows, blankets, and other childhood toys as being cherished objects more frequently than females (Table 1).

The reasons given for why this object was special included the intrinsic quality

Table 1.   Childhood Treasured Possessions

	Total Group	Age of Adults			Gender	
		Young	Middle	Older	Males	Females
	(N = 175)	(N = 96)	(N = 41)	(N = 38)	(N = 42)	(N = 133)
Percentage of total objects (or meanings) named for each of the six subject groups						
Object Categories						
1. Teddy bears/ stuffed animals	31.7	38.8	25.8	14.3	38.7	28.9
2. Dolls	34.4	27.0	38.7	53.6	.0	43.8
3. Pillows/blankets	9.6	12.9	9.6	.0	16.1	7.9
4. Books	.6	1.1	.0	.0	3.2	.8
5. Clothing	2.0	3.5	.0	.0	6.4	1.7
6. Childhood toys	5.5	2.3	6.4	14.3	19.3	.8
7. Sports equipment	1.2	2.3	.0	.0	3.2	.0
8. Motor vehicles	.0	.0	.0	.0	.0	.0
9. Phonograph records/music	.0	.0	.0	.0	.0	.8
10. Photos	.6	.0	3.2	.0	.0	.8
11. Memorabilia	.6	1.1	.0	.0	.0	.0
12. Personal accomplishment	.0	.0	.0	.0	.0	.8
13. Furniture/antiques	.6	1.1	.0	.0	.0	.8
14. Dishware/ silverware	.6	1.1	.0	.0	.0	.8
15. Jewelry	1.2	1.1	.0	3.5	3.2	.0
16. Religious items	.0	.0	.0	.0	.0	.0
17. Collections	.6	.0	.0	3.5	3.2	.0
18. Small appliances	.0	.0	.0	.0	.0	.0
19. Important papers, documents, records	.0	.0	.0	.0	.0	.0
20. Money	.0	.0	.0	.0	.0	.0
21. Artwork	.0	.0	.0	.0	.0	.0
22. Tools	.0	.0	.0	.0	.0	.0
23. Personal items	9.6	7.0	16.1	10.7	6.4	10.5
Object Meanings						
1. Memories: General	6.5	11.3	.0	7.4	3.7	9.4
2. Memories: Interpersonal- familial	17.1	16.5	23.5	11.1	14.8	17.6
3. Personification	13.1	15.6	8.8	7.4	3.7	14.8
4. Cultural/Religious Association	.0	.0	.0	.0	.0	.0
5. Pleasant Experience	12.0	6.9	20.6	22.2	29.6	8.7

Table 1.   (Continued)

	Percentage of total objects (or meanings) named for each of the six subject groups					
	Total Group	Age of Adults			Gender	
		Young	Middle	Older	Males	Females
	(N = 175)	(N = 96)	(N = 41)	(N = 38)	(N = 42)	(N = 133)
6. Intrinsic Qualities of Object	18.3	19.1	17.6	14.8	11.1	19.6
7. Utilitarian Value	4.0	3.4	2.9	7.4	7.4	3.3
8. Personal Values	6.2	6.8	5.8	3.7	7.4	6.1
9. Comfort/Security	12.5	15.6	11.6	3.7	16.1	11.5
10. Other	81.0	4.3	8.8	22.2	3.7	8.7

of the object (18.3 percent) (in most cases, physical features of the objects were mentioned), and its interpersonal-familial association (17.1 percent) (most frequently, someone special gave it to them). Other reasons included personification (13.1 percent) (e.g., "It always listened to me") and comfort/security (12.5 percent). These early cherished objects were either received as gifts (40.8 percent), or the subject always remembered having it (42.2 percent). These objects were primarily used at bedtime in going to sleep (55.9 percent), and secondarily were played with or used anytime (37.1 percent). Few of these objects were given up in early childhood—most were given up in middle childhood. Interestingly, 16.5 percent of our total sample still have not given this object up!

Adolescence

The objects that adults identified as being treasured possessions during adolescence were not as clear-cut. For the total group, of the objects reported, the most frequently mentioned objects were personal items (19.1 percent), jewelry (9 percent), motor vehicles (9 percent), and clothing (8.7 percent) (Table 2). Developmental trends across adulthood showed that older subjects more frequently identified books, clothing, and personal items, while younger adults identified motor vehicles, phonograph records, and teddy bears/stuffed animals as being treasured during youth, which may reflect cohort differences. For females, personal items, jewelry, clothing, and stuffed animals were the most frequently identified as being "treasured" during this time. For males, motor vehicles were mentioned the most frequently, possibly reflecting sex-role stereotypes for males as well as adolescents' urges for autonomy and independence. Teddy bears/ stuffed animals, dolls, pillows/blankets, jewelry, and personal items were mentioned more frequently by females compared to males. Males, on the other hand, mentioned childhood toys, motor vehicles, and phonograph records/music much

Table 2.   Adolescent Treasured Possessions

	Percentage of total objects (or meanings) named for each of the six subject groups					
	Total Group	Age of Adults			Gender	
		Young	Middle	Older	Males	Females
	(N = 175)	(N = 96)	(N = 41)	(N = 38)	(N = 42)	(N = 133)
Object Categories						
1. Teddy bears/ stuffed animals	6.6	8.0	6.5	.0	1.4	8.2
2. Dolls	4.1	3.7	6.5	2.5	.0	5.5
3. Pillows/blankets	1.3	2.1	.0	.0	.0	1.8
4. Books	5.9	3.7	4.9	17.9	7.4	5.5
5. Clothing	8.7	7.4	11.4	20.5	5.9	9.6
6. Childhood toys	7.3	6.4	8.1	10.2	16.4	4.6
7. Sports equipment	3.4	4.2	3.2	.0	8.9	1.8
8. Motor vehicles	9.0	11.2	4.9	5.1	25.4	4.1
9. Phonograph records/music	6.6	8.0	3.2	5.1	10.4	5.5
10. Photos	2.8	2.1	4.9	2.5	1.4	3.2
11. Memorabilia	5.9	6.4	6.4	2.5	.0	7.7
12. Personal accomplishment	1.0	1.6	.0	.0	1.4	.9
13. Furniture/antiques	1.0	1.0	1.6	.0	.0	1.4
14. Dishware/ silverware	1.0	1.0	.0	2.5	.0	1.4
15. Jewelry	9.0	8.0	13.0	7.6	1.4	11.8
16. Religious items	.7	.5	1.6	.0	5.9	.4
17. Collections	2.4	2.1	.0	7.6	1.4	1.4
18. Small appliances	3.8	3.7	6.4	.0	.0	4.1
19. Important papers, documents, records	.0	.0	.0	.0	.0	.0
20. Money	.0	.0	.0	.0	.0	.0
21. Artwork	.0	.0	.0	.0	.0	.0
22. Tools	.0	.0	.0	.0	.0	.0
23. Personal items	19.1	18.1	16.2	25.6	11.8	21.4

more frequently than did females. Although the particular meanings ascribed to these objects were not specifically requested, crosstabulations calculated on the objects and their meanings during adulthood demonstrated that the objects named by females during adolescence held meanings that were primarily interpersonal-familial associations, while the objects named by males had meanings emphasizing the utilitarian value and expressions of personal values.

Adulthood

For the total group, the objects most frequently identified as treasured possessions were jewelry (22.2 percent), personal items (17.6 percent), and photos (11.7 percent) (Table 3). Developmental trends across adulthood showed that of the objects named, for young adults jewelry (27.8 percent), personal items (13.9 percent), motor vehicles (11.5 percent), and photos (9.8 percent) were most salient. For middle-aged adults, personal items (27.8 percent), photos (11.4 percent), dishware/silverware (11.4 percent), and jewelry (9.8 percent) were most frequently mentioned. For older adults, jewelry (28.8 percent), photos (20.0 percent), dishware/silverware (15.5 percent), and personal items (15.5 percent) were most often identified. Thus, for younger adults motor vehicles were more valued, while photos and dishware/silverware increased in importance with age. Males most frequently named personal items, motor vehicles, and photos as treasured objects, and appeared to have a greater preference than females for motor vehicles, phonograph records/music, and collections. Females most frequently identified jewelry, personal items, and photos as their treasured objects, and showed a greater preference than did males for dishware/silverware and jewelry.

Subjects were also asked both why they thought they were attached to this object, and why the object was special to them. The responses to both questions were fairly similar. Subjects felt they were attached to the object primarily because of interpersonal-familial associations (46.4 percent) and for reasons of comfort/security (18.3 percent). Developmental trends demonstrated that interpersonal-familial associations became more important with age, while meanings regarding expressions of personal values declined with age. Females mentioned comfort/security and interpersonal-familial associations more frequently than did males, while males mentioned non-interpersonal memories and expressions of personal values more frequently than did females.

As to why this object was special to them, it was primarily because of interpersonal-familial associations (46.4 percent). As before, interpersonal-familial associations increased in importance with age. Females mentioned interpersonal-familial associations and comfort/security more frequently than males, who mentioned personification, pleasant experience, and expressions of personal values more frequently than did females.

These treasured possessions were acquired primarily by receiving it from someone special (48.0 percent) and by the subject acquiring or making it themselves (30.0 percent). Fifteen percent said it was associated with a special (emotional) situation. In terms of how and when these treasured possessions were used, the most frequent responses included "look at it often" (20 percent), "wear it" (15 percent), "use it at bedtime or sleep" (13 percent), and "when upset, bored, lonely, or afraid" (10 percent).

Table 3.　Treasured Possessions in Adulthood

	Total Group	Age of Adults			Gender	
		Young	Middle	Older	Males	Females
	(N = 175)	(N = 96)	(N = 41)	(N = 38)	(N = 42)	(N = 133)

Percentage of total objects (or meanings) named for each of the six subject groups

Object Categories						
1. Teddy bears/ stuffed animals	2.5	4.9	.0	.0	2.2	2.6
2. Dolls	1.2	.8	3.2	.0	.0	1.5
3. Pillows/blankets	1.7	3.2	.0	.0	.0	2.0
4. Books	2.9	2.4	6.4	.0	4.5	2.6
5. Clothing	.8	1.6	.0	.0	.0	1.0
6. Childhood toys	.4	.8	.0	.0	2.2	.0
7. Sports equipment	1.2	1.6	.0	2.2	2.2	1.0
8. Motor vehicles	7.9	11.5	1.6	4.4	15.9	5.1
9. Phonograph records/music	1.2	2.4	.0	.0	4.5	.5
10. Photos	11.7	9.8	11.4	20.0	13.6	11.3
11. Memorabilia	6.7	8.1	6.4	4.4	6.8	6.6
12. Personal accomplishment	.0	.0	.0	.0	.0	.0
13. Furniture/antiques	4.6	.8	8.1	6.6	2.2	5.1
14. Dishware/ silverware	7.9	4.1	11.4	15.5	.0	9.7
15. Jewelry	22.2	27.8	9.8	28.8	11.3	24.6
16. Religious items	2.5	.8	3.2	6.6	2.2	2.6
17. Collections	1.2	.0	.0	6.6	6.8	.0
18. Small appliances	4.6	3.2	4.9	8.8	4.5	4.6
19. Important papers, documents, records	.0	.0	.0	.0	.0	.0
20. Money	.0	.0	.0	.0	.0	.0
21. Artwork	1.7	.0	4.9	2.2	2.2	1.5
22. Tools	.0	.0	.0	.0	.0	.0
23. Personal items	17.6	13.9	27.8	15.5	18.1	17.4
Object Meanings: (Why Attached)						
1. Memories: General	6.4	7.9	6.9	.0	12.8	4.9
2. Memories: Interpersonal- familial	46.4	42.4	40.0	72.7	38.4	48.6
3. Personification	1.3	.7	4.0	.0	7.6	1.0
4. Cultural/Religious Association	.4	.0	.0	3.0	5.1	.5
5. Pleasant Experience	7.6	8.6	6.0	6.0	5.1	7.1

Table 3. (Continued)

	Total Group	Age of Adults			Gender	
		Young	Middle	Older	Males	Females
	(N = 175)	(N = 96)	(N = 41)	(N = 38)	(N = 42)	(N = 133)

Percentage of total objects (or meanings) named for each of the six subject groups

6. Intrinsic Qualities of Object	6.2	5.7	8.0	6.0	7.6	6.5
7. Utilitarian Value	3.1	2.1	6.0	3.0	5.1	2.7
8. Personal Values	9.8	12.2	8.0	3.0	20.5	7.6
9. Comfort/Security	18.3	20.1	20.0	9.0	10.2	20.2
10. Other	.4	.0	2.0	.0	.0	.5
Object Meanings: (Why Special)						
1. Memories: General	8.1	9.7	10.0	.0	8.6	8.6
2. Memories: Interpersonal-familial	46.4	44.2	35.0	67.7	33.3	46.0
3. Personification	3.2	2.6	2.5	6.4	11.1	1.4
4. Cultural/Religious Association	.0	.0	.0	.0	.0	.0
5. Pleasant Experience	8.1	11.5	5.0	.0	13.8	7.1
6. Intrinsic Qualities of Object	8.1	5.2	12.5	12.8	5.5	9.3
7. Utilitarian Value	3.2	1.7	7.5	3.2	.0	4.3
8. Personal Values	12.9	15.0	12.5	6.4	22.2	11.5
9. Comfort/Security	8.1	7.9	12.5	3.2	2.7	10.1
10. Other	1.6	1.7	2.5	.0	2.7	1.4

OTHER VALUED POSSESSIONS

In this section, subjects were asked what their favorite possessions were, what objects they considered to be the *most* "important" (of all they owned), and what other possessions they had that were important to them.

Favorite Possessions

For the total group, the objects identified most frequently as being "favorite" possessions were personal items (19.8 percent), jewelry (14.1 percent), photos (10.2 percent), and motor vehicles (8.1 percent) (Table 4). As the age of the subject increased, motor vehicles and jewelry declined in importance, while dishware/silverware and personal items increased in importance. Males' most

Table 4.   Adults' Favorite Possessions

	Total Group	Age of Adults			Gender	
		Young	Middle	Older	Males	Females
	(N = 175)	(N = 96)	(N = 41)	(N = 38)	(N = 42)	(N = 133)
Object Categories						
1. Teddy bears/ stuffed animals	2.4	3.4	1.5	.6	1.1	2.8
2. Dolls	1.0	.6	1.5	.6	.0	1.3
3. Pillows/blankets	1.1	1.2	1.5	.0	1.1	1.1
4. Books	4.6	3.1	9.1	4.0	4.4	5.0
5. Clothing	4.2	4.7	4.5	2.0	2.8	4.7
6. Childhood toys	.9	.9	1.5	.6	1.6	.6
7. Sports equipment	1.8	2.4	.5	.6	5.6	.6
8. Motor vehicles	8.1	10.4	6.0	3.3	16.8	5.5
9. Phonograph records/music	3.1	4.3	2.0	.6	4.4	2.7
10. Photos	10.2	9.9	10.2	9.3	2.8	12.3
11. Memorabilia	3.6	4.5	2.0	2.6	.5	4.5
12. Personal accomplishment	1.0	1.5	.5	.0	.0	1.3
13. Furniture/antiques	5.6	3.8	7.0	8.0	1.6	6.7
14. Dishware/ silverware	5.3	3.6	4.0	18.0	.5	6.7
15. Jewelry	14.1	16.1	10.2	12.0	5.0	16.7
16. Religious items	1.5	1.3	2.0	1.2	1.1	1.7
17. Collections	1.4	.9	1.5	2.4	1.1	1.5
18. Small appliances	6.2	5.6	9.1	3.3	11.8	4.5
19. Important papers, documents, records	.2	.2	.0	.6	.0	.3
20. Money	1.0	1.3	1.0	.0	3.3	.3
21. Artwork	1.7	.9	1.5	4.0	2.2	1.5
22. Tools	6.4	.6	.0	1.2	2.2	.1
23. Personal items	19.8	17.4	22.3	25.3	29.2	16.9

Percentage of total objects (or meanings) named for each of the six subject groups

favorite possessions included personal items, motor vehicles, and small appliances, and they mentioned sports equipment, motor vehicles, small appliances, and personal items more frequently than did females. Females most frequently named personal items, jewelry, and photos as their favorite possessions, and mentioned photos and jewelry more often than did males.

"Most Important" Possessions

When asked what the *most* important possessions were of all they owned, subjects as a total group mentioned personal items (23.5 percent), photos (12.7

Table 5. Adults' Most Important Possessions

	Total Group	Age of Adults			Gender	
		Young	Middle	Older	Males	Females
	(N = 175)	(N = 96)	(N = 41)	(N = 38)	(N = 42)	(N = 133)
Object Categories						
1. Teddy bears/ stuffed animals	2.4	3.2	1.3	.0	.0	3.0
2. Dolls	.5	.9	.0	.0	.0	.7
3. Pillows/blankets	.8	1.3	.0	.0	.0	1.1
4. Books	4.1	3.7	3.4	.0	2.6	4.6
5. Clothing	2.3	3.2	.0	2.1	2.6	2.3
6. Childhood toys	.2	.4	.0	.0	.0	.3
7. Sports equipment	1.4	1.8	.0	2.1	4.0	.7
8. Motor vehicles	11.3	13.0	3.4	8.6	22.6	8.0
9. Phonograph records/music	4.1	5.5	2.6	.0	5.2	3.8
10. Photos	12.7	11.1	18.9	8.6	2.6	15.8
11. Memorabilia	3.2	4.6	.0	2.1	.0	4.2
12. Personal accomplishment	.2	.4	.0	.0	.0	.0
13. Furniture/anitques	3.8	2.6	6.7	4.2	.0	5.0
14. Dishware/ silverware	5.1	3.2	4.0	15.2	.0	6.5
15. Jewelry	10.7	12.0	4.0	15.2	4.0	12.7
16. Religious items	2.3	1.8	2.6	4.2	2.6	2.3
17. Collections	.8	.9	.0	2.1	1.3	1.5
18. Small appliances	6.3	4.6	13.5	.0	10.4	4.2
19. Important papers, documents, records	.5	.4	.0	2.1	.0	.7
20. Money	.5	.4	1.3	.0	1.3	.3
21. Artwork	2.1	1.8	2.6	2.1	1.3	2.3
22. Tools	.5	.4	.0	.0	.0	.0
23. Personal items	23.5	21.4	25.6	30.4	38.6	19.2
Object Meanings						
1. Memories: General	4.8	6.6	.0	2.9	8.7	3.8
2. Memories: Interpersonal- familial	31.6	31.8	19.6	52.9	17.4	34.9
3. Personification	2.4	3.6	.0	2.9	3.5	2.5
4. Cultural/Religious Association	.0	.0	.0	.0	.0	.0
5. Pleasant Experience	10.2	9.7	14.7	5.8	7.0	11.0

(continued)

Table 5.   (Continued)

	Total Group	Age of Adults			Gender	
		Young	Middle	Older	Males	Females
	(N = 175)	(N = 96)	(N = 41)	(N = 38)	(N = 42)	(N = 133)
6. Intrinsic Qualties of Object	12.0	12.3	11.4	8.8	10.5	12.3
7. Utilitarian Value	15.8	7.6	39.3	20.6	22.8	14.0
8. Personal Values	14.8	18.9	6.5	5.8	28.1	11.5
9. Comfort/Security	7.6	8.2	8.2	.0	1.7	8.9
10. Other	.3	1.0	.0	.0	.0	.8

Percentage of total objects (or meanings) named for each of the six subject groups

percent), motor vehicles (11.3 percent), and jewelry (10.7 percent) most frequently (Table 5). While younger and older adults fit this same pattern, middle-aged adults had a greater preference for small appliances and photos. Males' most important possessions were primarily personal items, motor vehicles, and small appliances, which were all mentioned more frequently compared to females. For females, personal items, photos, and jewelry were the most frequently identified; and photos, furniture, dishware/silverware, and jewelry were named more frequently compared to males.

Of the reasons that were given as to why these objects were most important, 31.6 percent were interpersonal-familial associations, 15.8 percent were because of their utilitarian value, 14.8 percent were because they expressed personal values of the subject, and 12.0 percent were because of the intrinsic qualities of the object. The relative frequency of interpersonal-familial associations increased with age, while meanings of expressions of personal values declined with age. Middle-aged adults most frequently mentioned the object's utilitarian value (39.3 percent). The primary meaning attributed to these objects by females was its interpersonal-familial associations, while males most frequently named expression of personal values and the object's utilitarian value as being the most salient.

Other Important Possessions

When asked what other possessions were important, subjects most frequently replied personal items (17.5 percent), jewelry (15.8 percent), and photos (13.4 percent) (Table 6). Personal items and furniture/antiques increased in importance with age. For males, important objects were mainly personal items, motor vehicles, and phonograph records/music, and they mentioned more often than did females sports equipment, motor vehicles, phonograph records/music, small

Table 6. Adults' Other Important Possessions

	Percentage of total objects (or meanings) named for each of the six subject groups					
	Total Group	Age of Adults			Gender	
		Young	Middle	Older	Males	Females
	(N = 175)	(N = 96)	(N = 41)	(N = 38)	(N = 42)	(N = 133)
Object Categories						
1. Teddy bears/ stuffed animals	6.8	1.2	.0	.0	.0	.8
2. Dolls	1.0	1.2	.0	.0	.0	1.2
3. Pillows/blankets	1.0	1.2	1.3	.0	.0	1.2
4. Books	3.7	3.0	6.6	2.0	.0	4.6
5. Clothing	6.1	6.6	5.3	6.0	3.7	6.6
6. Childhood toys	.6	.6	1.3	.0	.0	.8
7. Sports equipment	1.7	3.0	.0	.0	7.5	.4
8. Motor vehicles	8.6	9.1	10.6	4.0	22.6	5.4
9. Phonograph records/music	4.8	4.2	5.3	6.0	11.3	3.3
10. Photos	13.4	12.1	17.3	12.0	5.6	14.9
11. Memorabilia	5.1	8.4	.0	2.0	.0	6.2
12. Personal accomplishment	1.0	1.2	.0	2.0	.0	1.2
13. Furniture/antiques	5.4	4.2	5.3	10.0	.0	6.6
14. Dishware/ silverware	4.1	3.6	4.0	6.0	.0	4.9
15. Jewelry	15.8	17.6	13.3	14.0	3.7	18.2
16. Religious items	1.2	1.2	1.3	2.0	.0	1.6
17. Collections	1.0	.6	1.3	2.0	1.8	.8
18. Small appliances	4.1	4.2	6.6	.0	9.4	2.9
19. Important papers, documents, records	.0	.0	.0	.0	.0	1.2
20. Money	.6	.6	1.3	.0	1.8	.4
21. Artwork	1.2	1.2	1.3	2.0	1.8	1.2
22. Tools	.3	.0	.0	2.0	1.8	.0
23. Personal items	17.5	14.5	17.3	28.0	28.3	14.9
Object Meanings						
1. Memories: General	7.3	5.3	2.5	1.9	4.2	8.0
2. Memories: Interpersonal- familial	37.2	38.7	24.0	53.8	8.4	44.5
3. Personification	.8	1.7	.0	.0	2.1	.3
4. Cultural/Religious Association	.8	1.1	1.2	.0	.0	.9
5. Pleasant Experience	18.0	17.3	30.3	15.3	25.3	17.1

(*continued*)

Table 6.   (Continued)

	Total Group	Age of Adults			Gender	
		Young	*Middle*	*Older*	*Males*	*Females*
	(N = 175)	*(N = 96)*	*(N = 41)*	*(N = 38)*	*(N = 42)*	*(N = 133)*
6. Intrinsic Qualities of Object	13.0	6.5	25.3	13.4	23.9	10.2
7. Utilitarian Value	8.7	6.5	13.9	3.8	16.9	5.8
8. Personal Values	9.9	18.4	1.2	.0	14.1	8.7
9. Comfort/Security	3.2	4.1	1.2	5.7	4.2	2.9
10. Other	.8	.0	.0	5.7	.0	3.9

Percentage of total objects (or meanings) named for each of the six subject groups

appliances, and personal items. For females, the most frequently identified objects were jewelry, personal items, and photos, and they mentioned photos and jewelry more frequently than did males.

Reasons given for why these objects were important included interpersonal-familial associations (37.2 percent), pleasant experiences (13.0 percent), and intrinsic qualities of the object (13.0 percent). Interpersonal-familial associations increased in importance with age, while reasons of expressions of personal values declined with age. For middle-aged adults, reasons of pleasant experience, intrinsic qualities of the object, and its utilitarian value were more important than for the other two age groups. Females mainly ascribed meanings of interpersonal-familial associations to these objects, while the most salient meanings for males were the pleasant experience it afforded and its own intrinsic qualities.

OBJECTS AND OTHER CIRCUMSTANCES

The purpose of this section was to analyze how objects and the meanings attributed to them might change depending on the specific circumstances.

Objects Rescued in Case of a Fire

Subjects were asked which five inanimate objects they would rescue in case of a fire, and why. For the total group, subjects most frequently mentioned photos (16.0 percent), personal items (15.2 percent), clothing (11.2 percent), and jewelry (9.0 percent) (Table 7). Photos, jewelry, and important documents increased in importance with age, while clothing declined in importance. Males named personal items, clothing, and sports equipment most often, and identified sports equipment, small appliances, and important documents more frequently than did

Table 7.   Objects Rescued in Case of Fire

	Total Group	Age of Adults			Gender	
		Young	*Middle*	*Older*	*Males*	*Females*
	(N = 175)	*(N = 96)*	*(N = 41)*	*(N = 38)*	*(N = 42)*	*(N = 133)*

Percentage of total objects (or meanings) named for each of the six subject groups

Object Categories						
1. Teddy bears/ stuffed animals	2.1	3.4	.5	.0	.0	2.6
2. Dolls	1.5	1.4	1.7	.7	.0	1.9
3. Pillows/blankets	1.7	1.9	2.3	.0	1.3	1.7
4. Books	3.5	3.2	5.3	2.3	4.6	3.2
5. Clothing	11.2	13.7	8.2	7.9	13.4	10.8
6. Childhood toys	.0	.2	.0	.0	.6	.0
7. Sports equipment	.9	1.4	.0	.0	12.6	.5
8. Motor vehicles	.9	1.7	.0	.7	1.3	.8
9. Phonograph records/music	3.5	4.7	2.3	1.4	4.0	3.3
10. Photos	16.0	14.4	18.3	18.2	12.0	16.9
11. Memorabilia	4.1	5.7	1.7	2.3	1.3	4.8
12. Personal accomplishment	1.1	1.4	1.0	.0	1.3	1.0
13. Furniture/antiques	3.4	3.4	4.1	2.3	1.3	3.9
14. Dishware/ silverware	3.9	3.7	3.4	5.5	.0	4.9
15. Jewelry	9.0	9.2	6.8	11.9	3.3	10.5
16. Religious items	2.5	1.7	2.9	4.6	1.3	2.8
17. Collections	1.4	.0	2.3	1.4	2.0	1.2
18. Small appliances	6.0	7.4	4.1	4.6	12.0	4.6
19. Important papers, documents, records	6.9	4.7	8.2	12.6	11.4	6.4
20. Money	2.1	1.9	2.3	2.3	4.0	1.6
21. Artwork	1.8	.9	4.1	2.3	2.0	1.9
22. Tools	.2	.0	.0	.0	.6	.1
23. Personal items	15.2	12.7	18.9	17.5	18.8	13.7
Object Meanings						
1. Memories: General	3.5	2.6	6.7	.0	8.0	2.6
2. Memories: Interpersonal- familial	28.1	26.8	28.3	33.3	14.0	30.7
3. Personification	.7	1.0	.0	.0	.0	.8
4. Cultural/Religious Association	.3	.0	.0	5.5	2.0	.0
5. Pleasant Experience	7.4	6.9	10.8	.0	6.0	7.8

(continued)

Table 7.   (Continued)

	Percentage of total objects (or meanings) named for each of the six subject groups				
	Total Group	Age of Adults			Gender
		Young	Middle	Older	Males	Females
	(N = 175)	(N = 96)	(N = 41)	(N = 38)	(N = 42)	(N = 133)
6. Intrinsic Qualities of Object	31.7	27.9	40.5	38.8	50.0	28.1
7. Utilitarian Value	16.3	19.3	9.4	16.6	8.0	18.4
8. Personal Values	6.4	9.6	.0	.0	4.0	7.0
9. Comfort/Security	2.1	1.6	2.7	5.5	4.0	1.6
10. Other	3.2	3.7	1.3	.0	4.0	2.6

females. Females mentioned photos and personal items most frequently, and name photos and jewelry more frequently than did males.

For the total group, the reasons given as to why these objects would be rescued included interpersonal-familial associations (28.1 percent), intrinsic qualities of the object (31.7 percent), and the object's utilitarian value (16.3 percent). Interpersonal-familial associations increased with age, while "pleasant experience" and "expressions of personal values" declined with the age of the subject. Males mentioned the intrinsic quality of the object as the most salient meaning, while for females the most salient meaning was interpersonal-familial associations.

Heirlooms

Subjects were next asked whether or not they had any family heirlooms, and if they did, what they were and what they signified. Of the objects named, the most frequently mentioned ones for the total group were dishware/silverware (25.9 percent), jewelry (24.0 percent), furniture/antiques (15.4 percent), and personal items (11.1 percent) (Table 8). The frequency with which dishware/silverware was mentioned increased with age, while jewelry and photos declined with age. Items named most frequently by females were dishware/silverware and jewelry, while jewelry and personal items were identified the most often by males.

The primary significance of these heirlooms was, not surprisingly, interpersonal-familial associations (28.1 percent) and the intrinsic qualities of the objects (31.7 percent) (e.g., respondents simply liked the object for its inherent physical characteristics). The utilitarian value of the heirloom was more frequently cited as the age of the subject increased, while the pleasant experience and the intrinsic qualities of the object declined with age. This may likely have been due to the nature of objects older people had as heirlooms (i.e., dishware/silverware). Males mentioned pleasant experience more often than did females, who mentioned the intrinsic qualities of the object more than did males.

Table 8. Family Heirlooms

	Total Group	Age of Adults			Gender	
		Young	*Middle*	*Older*	*Males*	*Females*
	(N = 175)	*(N = 96)*	*(N = 41)*	*(N = 38)*	*(N = 42)*	*(N = 133)*

Percentage of total objects (or meanings) named for each of the six subject groups

Object Categories						
1. Teddy bears/ stuffed animals	.0	.0	.0	.0	.0	.0
2. Dolls	1.8	1.2	2.4	2.3	.0	2.3
3. Pillows/blankets	1.2	1.2	.0	2.3	.0	1.5
4. Books	.0	.0	.0	.0	.0	.0
5. Clothing	1.2	1.2	.0	2.3	.0	1.5
6. Childhood toys	.0	.0	.0	.0	.0	.0
7. Sports equipment	.0	.0	.0	.0	.0	.0
8. Motor vehicles	.0	.0	.0	.0	.0	.0
9. Phonograph records/music	.6	1.2	.0	.0	.0	.7
10. Photos	8.0	11.5	7.3	2.3	11.7	6.9
11. Memorabilia	.0	.0	.0	.0	.0	.0
12. Personal accomplishment	.0	.0	.0	.0	.0	.0
13. Furniture/antiques	15.4	15.3	17.1	13.9	11.7	17.1
14. Dishware/ silverware	25.9	21.7	21.9	37.2	11.7	29.5
15. Jewelry	24.0	28.2	21.9	18.6	23.4	24.0
16. Religious items	2.4	1.2	4.8	2.3	2.9	2.3
17. Collections	1.2	1.2	2.4	.0	5.8	.0
18. Small appliances	3.6	2.4	4.2	4.6	8.8	2.3
19. Important papers, documents, records	.6	.0	2.4	.0	2.9	.0
20. Money	.0	.0	.0	.0	.0	.0
21. Artwork	2.4	2.4	2.4	2.3	5.8	1.5
22. Tools	.0	.0	.0	.0	.0	.0
23. Personal items	11.1	10.2	12.2	11.6	14.7	10.1
Object Meanings						
1. Memories: General	3.5	2.6	6.7	.0	8.0	2.6
2. Memories: Interpersonal- familial	28.1	26.8	28.3	33.3	14.0	30.7
3. Personification	.7	1.0	.0	.0	.0	.8
4. Cultural/Religious Association	.3	.0	.0	5.5	2.0	.0
5. Pleasant Experience	7.4	6.9	10.8	.0	6.0	7.8

(*continued*)

Table 8.   (Continued)

	Total Group	Age of Adults			Gender	
		Young	*Middle*	*Older*	*Males*	*Females*
	(N = 175)	*(N = 96)*	*(N = 41)*	*(N = 38)*	*(N = 42)*	*(N = 133)*
6. Intrinsic Qualities of Object	31.7	27.9	40.5	38.8	50.0	28.1
7. Utilitarian Value	16.3	19.3	9.4	16.6	8.0	18.4
8. Personal Values	6.4	9.6	.0	.0	4.0	7.0
9. Comfort/Security	2.1	1.6	2.7	5.5	4.0	1.6
10. Other	3.2	3.7	1.3	.0	4.0	2.6

*Percentage of total objects (or meanings) named for each of the six subject groups*

"Comforting" Objects

When asked what objects provide the most comfort when one is alone, afraid, lonely, or upset, subjects most frequently mentioned personal items (19.6 percent), small appliances (16.1 percent), books (14.3 percent), and phonograph records/music (11.3 percent) (Table 9). Pillows/blankets and photos were also mentioned (8.3 percent each). Developmental trends across adulthood showed that books, small appliances, and personal items were most comforting to the older adults, while phonograph records/music, pillows/blankets, and small appliances were the most comforting objects for young adults. For middle-aged adults, books, photos, religious items, and personal items were the most comforting. For males, small appliances were the most frequently mentioned, along with personal items. Females named personal items and books as being the most comforting, and mentioned soft "cuddlies" (e.g., stuffed animals, pillows/blankets), dolls, and books more frequently than did males.

How $1000 Would Be Spent

When asked what they would spend an extra $1000 on, subjects as a total group responded that they would give it to others (for a specific reason) (20.6 percent), spend it on sports activity or equipment (20.1 percent), spend it on a vacation (18.4 percent), buy clothes or jewelry (17.8 percent), purchase miscellaneous objects (11.7 percent), or put it into an investment, school tuition, or a savings account (7.8 percent).

Objects Preferred to Receive as Gifts

When asked what they like to receive as gifts, subjects most frequently mentioned personal items (38.8 percent), clothing (19.1 percent), and jewelry (14.0

Table 9.   Objects That Provide Comfort

Object Categories	Total Group	Age of Adults			Gender	
		Young	Middle	Older	Males	Females
	(N = 175)	(N = 96)	(N = 41)	(N = 38)	(N = 42)	(N = 133)
1. Teddy bears/ stuffed animals	4.7	8.0	.0	.0	.0	5.7
2. Dolls	1.7	2.0	.0	3.1	.0	2.1
3. Pillows/blankets	8.3	13.0	2.7	.0	.0	10.0
4. Books	14.3	7.0	24.3	25.0	10.3	15.0
5. Clothing	.5	1.0	.0	.0	.0	.7
6. Childhood toys	.0	.0	.0	.0	.0	.0
7. Sports equipment	.0	.0	.0	.0	.0	.0
8. Motor vehicles	.0	.0	.0	.0	.0	.0
9. Phonograph records/music	11.3	16.0	8.1	.0	10.3	11.4
10. Photos	8.3	8.0	18.1	6.2	6.8	8.5
11. Memorabilia	3.4	6.0	.0	.0	.0	4.2
12. Personal accomplishment	.0	.0	.0	.0	.0	.0
13. Furniture/antiques	2.9	4.0	.0	3.1	.0	3.5
14. Dishware/ silverware	.5	1.0	.0	.0	.0	.7
15. Jewelry	1.0	1.0	.0	3.1	.0	1.4
16. Religious items	6.5	7.0	18.1	.0	3.4	7.0
17. Collections	.0	.0	.0	.0	.0	.0
18. Small appliances	16.1	13.0	9.0	37.5	41.3	10.7
19. Important papers, documents, records	.0	.0	.0	.0	.0	.0
20. Money	.0	.0	.0	.0	.0	.0
21. Artwork	.0	.0	2.7	.0	.0	.7
22. Tools	.0	.0	.0	.0	.0	.0
23. Personal items	19.6	13.0	35.1	21.8	27.5	17.8

Percentage of total objects (or meanings) named for each of the six subject groups

percent) (Table 10). Dishware/silverware and artwork became more salient with age, while jewelry became less important with age. Males mentioned personal items, clothing, and money the most frequently, while females named personal items, clothing, and jewelry the most often.

Collections

Subjects were also asked whether they have (or had) a collection. For the total group, of those objects mentioned, stamps, coins, and shells were mentioned

Table 10.   Objects Preferred to Receive as Gifts

	Total Group	Age of Adults			Gender	
		Young	Middle	Older	Males	Females
	(N = 175)	(N = 96)	(N = 41)	(N = 38)	(N = 42)	(N = 133)
Object Categories						
1. Teddy bears/ stuffed animals	.5	.9	.0	.0	.0	.7
2. Dolls	.0	.0	.0	.0	.0	.0
3. Pillows/blankets	.0	.0	.0	.0	.0	.0
4. Books	6.7	4.6	10.5	10.0	3.6	7.6
5. Clothing	19.1	18.9	17.6	18.0	15.8	20.1
6. Childhood toys	.8	.0	3.5	.0	3.6	.0
7. Sports equipment	2.2	3.2	1.1	.0	8.5	.3
8. Motor vehicles	.2	.0	1.1	.0	.0	.3
9. Phonograph records/music	2.2	2.7	1.1	2.0	3.6	1.8
10. Photos	1.6	1.3	1.1	4.0	.0	2.2
11. Memorabilia	1.9	1.3	3.5	2.0	3.6	1.4
12. Personal accomplishment	.0	.0	.0	.0	.0	.0
13. Furniture/antiques	.2	.4	.0	2.0	.0	.3
14. Dishware/ silverware	1.1	.9	1.1	14.0	.0	1.4
15. Jewelry	14.0	16.6	8.2	.0	2.4	17.6
16. Religious items	.2	.4	.0	2.0	.0	.3
17. Collections	.2	.0	.0	.0	1.2	.0
18. Small appliances	.5	.0	2.2	.0	1.2	.3
19. Important papers, documents, records	.0	.0	.0	.0	.0	.0
20. Money	7.0	8.7	5.8	2.0	14.6	4.7
21. Artwork	1.4	.0	2.2	26.0	.0	1.8
22. Tools	.8	.9	.0	.0	3.6	.0
23. Personal items	38.8	38.7	40.0	36.0	37.8	38.4

most frequently (25.1 percent), followed by personal items (24.7 percent). Dishware/silverware was more frequently mentioned by older adults, while books were more salient to the middle-aged adults. Females collected stuffed animals and dolls (while males did not at all); males reported books and stamps, coins, and shells as collections more frequently than females (Table 11).

"Goodbye" Gifts

Subjects were also asked whether they gave or received a gift when they or a good friend moved away. Of those items mentioned, the most frequently identi-

Table 11.   Collections

Object Categories	Total Group (N = 175)	Age of Adults			Gender	
		Young (N = 96)	Middle (N = 41)	Older (N = 38)	Males (N = 42)	Females (N = 133)
1. Teddy bears/ stuffed animals	4.1	6.1	1.5	.0	.0	5.3
2. Dolls	6.1	6.8	6.2	2.8	.0	7.9
3. Pillows/blankets	.0	.0	.0	.0	.0	.0
4. Books	6.1	6.8	10.9	2.8	10.7	4.8
5. Clothing	1.6	1.3	1.5	2.8	.0	2.1
6. Childhood toys	.8	1.3	.0	.0	1.7	.5
7. Sports equipment	1.2	2.0	.0	.0	5.3	.0
8. Motor vehicles	.8	1.3	.0	.0	3.5	.0
9. Phonograph records/music	3.7	5.4	1.5	.0	5.3	3.2
10. Photos	.0	2.7	1.5	2.8	1.7	2.6
11. Memorabilia	7.4	9.5	3.0	2.8	10.7	6.3
12. Personal accomplishment	.0	.6	.0	.0	.0	.5
13. Furniture/antiques	2.0	2.0	3.0	.0	.0	2.6
14. Dishware/ silverware	7.4	5.4	6.0	17.1	.0	9.6
15. Jewelry	3.2	2.7	3.0	5.7	.0	4.2
16. Religious items	.0	.0	.0	.0	.0	.0
17. Collections	25.1	23.1	25.0	31.4	39.2	20.7
18. Small appliances	1.2	.0	3.0	2.8	.0	1.6
19. Important papers, documents, records	.0	.0	.0	.0	.0	.0
20. Money	.0	.0	.0	.0	.0	.0
21. Artwork	1.6	.6	4.6	.0	1.7	1.6
22. Tools	.0	.0	.0	.0	.0	.0
23. Personal items	24.7	21.7	28.1	28.6	19.6	26.1

Percentage of total objects (or meanings) named for each of the six subject groups

fied items included personal items (30.4 percent), memorabilia (13.4 percent), jewelry (12.2 percent), and photos (12.2 percent) (Table 12). Young adults were most likely to exchange personal items, memorabilia, and jewelry; middle-aged adults personal items, photos, books, dishware/silverware, and jewelry; and older adults personal items, dishware/silverware, and photos. Males mentioned photos and personal items the most often, while females named personal items and jewelry the most frequently.

Table 12.   Objects Exchanged as "Good-Bye" Gifts

	Total Group (N = 175)	Age of Adults			Gender	
		Young (N = 96)	Middle (N = 41)	Older (N = 38)	Males (N = 42)	Females (N = 133)
Object Categories						
1. Teddy bears/ stuffed animals	2.4	4.4	.0	.0	.0	2.9
2. Dolls	.0	.0	.0	.0	.0	.0
3. Pillows/blankets	.0	.0	.0	.0	.0	.0
4. Books	4.8	.0	13.6	6.6	.0	5.9
5. Clothing	2.4	2.2	4.5	.0	.0	2.9
6. Childhood toys	.0	.0	.0	.0	.0	.0
7. Sports equipment	.0	.0	.0	.0	.0	.0
8. Motor vehicles	.0	.0	.0	.0	.0	.0
9. Phonograph records/music	2.4	4.4	.0	.0	13.3	.0
10. Photos	12.2	8.8	18.2	13.2	26.6	8.9
11. Memorabilia	13.4	24.4	.0	.0	13.3	13.4
12. Personal accomplishment	.0	.0	.0	.0	.0	.0
13. Furniture/antiques	3.6	.0	4.5	13.2	.0	4.5
14. Dishware/ silverware	7.3	.0	13.6	20.0	.0	8.9
15. Jewelry	14.6	20.0	13.6	.0	3.3	14.9
16. Religious items	1.2	2.2	.0	.0	6.6	.0
17. Collections	1.2	.0	.0	6.6	6.6	.0
18. Small appliances	.0	.0	.0	.0	.0	.0
19. Important papers, documents, records	.0	.0	.0	.0	.0	.0
20. Money	.0	.0	.0	.0	.0	.0
21. Artwork	3.6	4.4	4.5	.0	.0	4.5
22. Tools	.0	.0	.0	.0	.0	.0
23. Personal items	30.4	28.8	27.2	40.0	20.0	32.8

Percentage of total objects (or meanings) named for each of the six subject groups

Good-luck Objects

Finally, subjects were asked whether or not they had any "good-luck" objects, and if so, what they were. There were surprisingly few responses—two subjects had a rabbit's foot; sixteen had a piece of jewelry; and fifteen had a piece of clothing that was used as a good-luck object.

MEANINGS ASSOCIATED WITH SPECIFIC OBJECTS

Another area of inquiry in this study focused on whether there were different meanings and referents associated with different possessions. In order to determine the frequencies of meanings attributed to specific objects, cross-tabulations were calculated.

The results suggested that teddy bears/stuffed animals were most frequently associated with interpersonal-familial associations. Dolls were either associated with interpersonal-familial associations, personification, or were valued for their intrinsic qualities. Pillows/blankets were liked for their comfort/security qualities. Books seemed to provide one with a pleasant experience and they expressed one's personal values. The meanings most frequently attributed to clothing were utilitarian value, intrinsic qualities, and (surprisingly) interpersonal-familial associations. Childhood toys and sports equipment were liked primarily for their pleasant experience. Motor vehicles were linked mainly with utilitarian value and as expressions of the owner's personal values. Phonograph records/music were enjoyed primarily for the pleasant experience they offered, and also as expressions of personal values. Photos were associated overwhelmingly with interpersonal-familial associations, as were memorabilia, furniture/antiques, dishware/silverware, jewelry, religious items, and artwork. Small appliances were liked primarily for their pleasant experience, intrinsic qualities, and their utilitarian value. The meanings attributed to personal items were mainly interpersonal-familial associations and their intrinsic qualities. The meaning most frequently associated with important papers and documents was its utilitarian value. Collections were most frequently associated with pleasant experiences and intrinsic qualities. Money provided for pleasant experiences and was also valued for utilitarian reasons. Objects representing personal accomplishment were valued primarily for their pleasant experience. Finally, meanings associated with tools were mainly that they expressed one's personal values.

The total percentages of how often the different meaning categories were mentioned were as follows: interpersonal-familial associations (38.4 percent), intrinsic qualities of the object (16.0 percent), pleasant experience (11.2 percent), utilitarian value (10.9 percent), personal values (9.2 percent), comfort/security (6.2 percent), non-interpersonal memories (4.7 percent), personification (2.4 percent), and cultural/religious association (.7 percent).

PERSONALITY AND EARLY TREASURED POSSESSIONS

Finally, we were curious to see whether there was a relationship between those who did report having had a treasured object in early childhood and those who did not, and their corresponding current "personality" attributes. Chi squares were calculated on the two groups, but there were significant differences on only two of the fifteen personality-related items. Those who reported having had a treasured object in early childhood were significantly more outgoing (χ^2 = 18.009; p = .006), and reported a significantly greater desire for social affiliation (χ^2 = 13.352; p = .037) compared to those who did not have such an object. There were the usual problems of interpreting these data, however, especially since the questions regarding childhood possessions were retrospective and fairly general, and the "personality" items were self-report. Although speculative, these findings are consistent with Cohen and Clark's (1984) finding that lack of early object attachment was related to later reservedness and social detachment, and Csikszentimihalyi and Rochberg-Halton's (1981) finding that lack of treasured possessions was correlated with a lack of close, interpersonal ties.

DISCUSSION

This exploratory study was designed to examine what objects are important to adults and why, and their role within the larger framework of life course development. The results demonstrated that objects continue to be significant to individuals throughout the life span, and not just during the early years of life, primarily because of the meanings that they hold for their owner. The objects identified by the adult subjects in this study and the meanings attributed to them were in part mediated by age and life stage, cohort, gender, and the prescribed circumstance. However, in spite of these varied influences, there were several common themes among the meanings attributed to the objects.

Objects and Their Meanings in Adulthood and Old Age

In general, the objects most often identified as being "treasured" in adulthood were jewelry, personal items, and photographs. Developmental trends across adulthood showed that jewelry was mentioned the most often by younger and older adults, while personal items were more salient for middle-aged adults. Because the meanings associated with these objects were primarily interpersonal-familial associations, it appears that what adults hold most dear is based to a large extent on the object's association with interpersonal ties.

For objects not listed as "treasured" but which were important or favorite objects, personal items were frequently named and they became more salient as the age of the subject increased. Jewelry and motor vehicles were valued more by the younger than the older adults, while dishware/silverware increased in importance with age. As with the treasured objects, interpersonal-familial associations increased in importance with age, while meanings of expressions of personal values declined with age.

The same two trends of increasing interpersonal-familial associations and decreasing expressions of personal values were noted in the objects named under the prescribed circumstances. Objects most often identified as being rescued in case of a fire were photos, personal items, and clothing. The reasons given as to why these objects would be rescued were primarily because they signified inter-personal-familial associations, because of the object's intrinsic quality, and because it served a utilitarian function. As above, the frequency with which the meaning "expression of personal values" was given declined with age. Objects most frequently identified as heirlooms (dishware/silverware, jewelry, fur-niture/antiques, and personal items) also held meanings that were primarily interpersonal-familial ties, which became more salient with age. Objects that provided comfort included personal items, small appliances, books, and music. (These items may function to comfort one by creating positive affect or pleasant feelings of enjoyment, altering one's mood. T.V. in particular may be a source of "social" contact with the outside world, helping to rid one of loneliness or distract one from his/her current emotional state). Objects most frequently de-sired as gifts included personal items, clothing, and jewelry. The meanings associated with these objects, according to the cross-tabulations, were primarily interpersonal-familial associations and the intrinsic qualities of the object. It appeared from the crosstabulations that collections offered individuals enjoyment and pleasant experiences, and that they were valued because of their intrinsic qualities and interpersonal-familial ties. Finally, items identified as being ex-changed as gifts all held meanings that were primarily interpersonal-familial associations.

One other issue deserves mention—the effect of cohort on object preferences. Some objects were more salient for one age group than for others, which seemed to indicate the impact of the historical period of time in which one grew up on the objects that were mentioned. This appeared to be most evident with such objects as cars, stereo equipment, and phonograph records which were more salient for the younger, compared to the older generations.

In summary, three developmental trends were apparent in these objects. First, the frequency with which the meaning "expression of personal values" was mentioned declined with age. This may in part be due to individuals having a more clearly defined and internalized sense of self, and no longer feeling such a need to "project" their beliefs or attitudes into objects in the environment as much as adolescents and young adults, who are more apt to be facing the developmental task of consolidating a personal identity (Erikson, 1968). This finding may also be due to greater interiority, whereby an individual is proposed to become more introspective and self-reflective with age (Rosen & Neugarten, 1964). Second, meanings signifying interpersonal-familial ties were the most predominant meaning attributed to these objects, and it generally became more salient with age. Third, there seemed to be an underlying theme of objects and their meanings acting as "referents" to the self—to both its development and its maintenance. These latter two trends are discussed in more detail below.

Interpersonal Relations

Why were object meanings signifying interpersonal-familial ties mentioned so frequently? The salience of interpersonal relations in an individual's life is well established. Whether they exist within a friendship or familial framework, close interpersonal relations provide individuals with a variety of assets, including emotional support and social resources. As an emotional support, interpersonal relationships may provide one with a sense of security based on affection, mutual trust, and support (Henderson, 1982 Henderson et al., 1978; Weiss, 1982); emotional comfort (Henderson, 1982); a sense of reliable alliance (Weiss, 1974); reassurance of one's self-worth (Henderson, 1982; Weiss, 1974); and help one to cope with environmental stressors. As a social resource, interpersonal relations provide opportunities for social integration through companionship and social activity (Henderson, 1982), and provide help and guidance, social support, feelings of positive affect, and affirmation about who one is as an individual (Kahn & Antonucci, 1980). Erikson (1969), in fact, states that establishing intimacy in interpersonal relations is a major developmental task of adulthood. The existence of at least one confidant has been found to be a critical factor in adjustment and high morale during adulthood (Lowenthal and Haven, 1968).

This importance of interpersonal relations was repeatedly expressed in the meanings attributed to the objects in the current study. The results of the current study demonstrated that interpersonal-familial associations were frequently attributed to the objects during early childhood, and that during adolescence, especially for females, interpersonal-familial ties were the most salient meaning attributed to objects. In adulthood and old age, interpersonal-familial ties were found to become even more important, especially as age increased. The object meaning categories "personification" (where the object takes on the qualities of a person) and "comfort/security" also had undertones that were interpersonal in nature. The theme of interpersonal-familial associations as a meaning of object possession is also implied in several of the early and current theories of object possession reviewed earlier, such as in: 1) the social dominance and social status theory (especially in regard to the notion of separation anxiety), 2) the theory that objects offer comfort, security, and solace (i.e., objects act as substitutes that provide comfort in place of human comfort), 3) the theory that objects represent ties and define relationships with others, and 4) the theory that during the early years of life, treasured objects function primarily as mother-substitutes.

Interpersonal-familial associations were mentioned more often for both older and younger adults compared to the middle-aged groups. For middle-aged adults, although interpersonal-familial ties were still important, the utilitarian value or pleasant experience offered by the object appeared to be more salient. It may be that the establishment and maintenance of intimate relations is a larger or more salient issue for younger adults, who may be dealing with the establishment of intimate relationships (Erikson, 1968), and older adults, who may be dealing with loss or other lifestyle changes. Conversely, middle-aged adults may more likely be settled in intimate networks, making other object meanings more salient.

In late adulthood, close interpersonal ties appear to become more salient than at any other age period. Why? Research has found that subjective well-being and family-life satisfaction are strongly related to life satisfaction as a whole (Medley, 1976). Friendships may compensate for the loss of other social roles and may thus increase in importance in late adulthood (Hess, 1972). Close kin relationships have been found to be more common in older than in middle-aged adults (Dickens and Perlman, 1981)—adult children are typically the main providers of emotional support for their parents, especially their widowed parents (Lopata, 1973). Older people turn to their families first when in need, and are most likely to name a family member as the person closest to them (Lopata, 1975; Lowenthal and Robinson, 1977; Troll and Bengtson, 1982).

Several other characteristics of the elderly may contribute to the increased salience of interpersonal ties with age. Stevens-Long (1984) has suggested that as instrumental roles become less urgent, the expressive ones may become more salient. A change in the perception of time also appears to occur with age, such that older people experience a sense of immediacy, with the elemental things of life such as intimacy, children, plants, nature, and human touching assuming a greater significance in their lives (Butler and Lewis, 1983). Finally, one's social group is suggested to function to maintain an individual's sense of self and personal identity during later life (Myerhoff, 1978).

Why are close, interpersonal ties so important in a person's life? Several theoretical views offer some suggestions. First, ethological theory suggests that human are biologically programmed for the predisposition toward establishing and maintaining social relationships (Bowlby, 1969). Second, learning theory suggests that individuals grow up with the expectation that the meaning in life is primarily derived from close relationships (Marris, 1982). Third, from a psychosocial perspective, individuals derive affect, affirmation of self, and aid from close relations (Kahn and Antonucci, 1980). (Although this view purports that interpersonal relations are necessary and important to mental well-being, it does not actually explain why intimate relations are important). Finally, existential psychology views the primary task of humans as searching for, and establishing meaning. According to Fromm (1955), the basic passions of humans are rooted in the need to find a relatedness to other people and to nature. Newman and Newman (1979) suggest that as our boundaries of self become increasingly defined we inevitably confront feelings of separation and isolation, and thus our desire for affiliation with others increases. This last viewpoint in particular seemed to be consistent with a major theme found in this study, which was the role of objects in the development and maintenance (and expression) of one's sense of self and personal identity.

Sense of Self and Personal Identity

Rochberg-Halton (1984) commented that even before an infant is born, its parents have created an environment of clothing, toys, and furnishings that will begin the socialization process. The self arises in this environment, which constantly addresses it and reminds it of who it is and what it is to become. Later, throughout the varied life stages, objects influence what a person can do, either by expanding or restricting the scope of that person's actions and thoughts. Thus, objects have a determining effect on the development of the self since what a person does is primarily an expression of what she/he is (Csikszentmihalyi and Rochberg-Halton, 1981).

The role of objects in the development of self (particularly during childhood and youth) was evident in several of the theories of object function described earlier, including the development of effectance motivation, learning self-object differentiation, and using objects to express one's personal values or autonomy. All of these kinds of activities that involve objects function to give individuals additional feedback regarding their "self," thus aiding in the development of a personal identity.

In adulthood and old age, however, there seemed to be a movement from an emphasis on the development of self to the maintenance of self or identity, with objects assisting in this task by providing continuity with one's past, embodying past memories or interpersonal ties, providing a utilitarian function, and expressing one's personal values. Perceiving continuity of self over time is one of the major components of the development and maintenance of a sense of self and identity (Erikson, 1968). Rochberg-Halton (1984) suggests that the things we value act as signs of the self that are necessary for its continued cultivation.

This function of objects (i.e., maintaining one's sense of self) appears particularly evident in late adulthood, when people may increasingly think of death. This may in turn stimulate more frequent reflections on the past, especially on matters of self and personal identity (Unruh, 1983). Butler and Lewis (1983) suggest that in late adulthood there may be an increasing emotional investment in the things surrounding the elderly's daily lives, especially in regard to homes, pets, familiar objects, heirlooms, keepsakes, photo albums, scrapbooks, and old letters. Such objects provide a sense of continuity, aid the memory, and provide comfort, security, and satisfaction.

Sherman and Newman (1977–78) speculate that some kinds of cherished possessions may serve as adaptive objects of reminiscence in the life review process, which allows elderly persons to make a positive adjustment to, and come to terms with, old age. Cherished possessions may enhance older people's ability to reminisce in the life review process, identified by Butler (1963) as being an adaptive and constructive process that assists older adults in coming to know themselves better and communicate to others who they were (and are).

Memories, often embodied in objects, may function to maintain the elderly's sense of self by integrating the various patterns around which the self is constructed at different points in time. Objects may also help older adults preserve their identities over time, especially if such objects are passed down as holders of family history—some objects often represent the last symbolic remnants of who and what an elderly person once was (Unruh, 1983).

Thus, the self of older adults tends to be structured around networks of past and present relationships, which are often embodied in concrete objects. Depriving an older person of such objects may precipitate the destruction of his/her self (Csikszentmihalyi and Rochberg-Halton, 1981; Sherman and Newman, 1977–78). Object possession may serve as a link with life and "being" for an individual, assisting in the validation of one's identity, self, and being. Frankl (1955), for example, noted in the concentration camps of WWII that when all possessions were taken away, persons had nothing with which to form an external link with their former life. Along similar lines, Goffman (1961) notes that institutionalization causes certain possessions to take on significant value, with personal possessions (such as clothing, cosmetics, scrapbooks, photograph albums, mementos, souvenirs, and jewelry) forming a person's "identity kit," allowing one to present their usual self to others and which also serves as an embodiment of one's personal history. The frequent removal of such personal possessions that occurs upon institutional admission may cause ego-debasement since it takes away an individual's representations or symbols of self and autonomy.

Thus, the role of objects in one's personal history is apparent. Objects may serve as reminders of past events by stimulating one's recall of, and embodying memories that are both non-interpersonal as well as interpersonal-familial in nature. Objects may also cultivate, maintain, reaffirm, and validate one's sense of self, and provide continuity of one's past self with one's present self. As Erikson (1968) contends, identity is a developmental task which is interwoven throughout the stages of the life cycle.

Gender Differences

In general, females attributed more meanings to objects that were interpersonal-familial associations more frequently than did males. Males tended to attribute meanings to objects that were primarily non-interpersonal memories, pers nifications, expressions of personal values, intrinsic quality of the object, and pleasant experience more often than did females. These findings are similar to those noted by Csikszentmihalyi and Rochberg-Halton (1981), who noted that their male subjects' valued possessions tended to represent professional roles more often than did females, who were more focused on interpersonal and familial ties. (Their study showed adult females to be more similar to the grandparents in their study, whereas men tend to represent the children in their study).

Why does this difference exist? Csikszentmihalyi and Rochberg-Halton (1981) suggest that socialization (at least in American culture) may be partly responsible. The sex-role socialization of males and females may thus lead them to value different experiences and teach them to choose different types of belonging to objectify these experiences. Gilligan (1982) suggests that because of their socialization, women and men may experience the world in different ways—i.e., throughout childhood, separation and autonomy are seen as basic to masculinity, whereas attachment and empathy are seen as basic to femininity. By adolescence, each sex is viewed as having a different interpersonal orientation and a different range of social experiences.

In Gilligan's analysis of emotional development, relationships are at the center of women's experience of life, with their identity being defined in the context of relationships (i.e., through intimacy and affiliation). Other studies attest to affiliation being a primary concern of females (Dickens and Perlman, 1981; Douvan and Adelson, 1966; Dusek and Flaherty, 1981; Hodgson and Fischer, 1979). In male development, however, separation and autonomy define the self (Gilligan, 1982), since society endorses separation and autonomy as masculine. Male adolescents' identity development, for example, revolves around the issues of gaining autonomy, assertiveness, and independence (Dusek and Flaherty, 1981). Matteson (1975) suggests that this process in males seems to reflect the cultural expectation of autonomy and personality differentiation. This self-definition may also affect later intimacy. Dickens and Perlman (1981) argue that cultural norms in North America prohibit men from developing intimate friendships. Traditional male roles, such as pressure to compete, homophobia, and aversion to vulnerability and being "open," create emotional barriers to intimacy. Also, in Levinson et al.'s (1978) study of male adult development, relationships with others were subordinated to other pursuits. All the men in his study were fairly distant from others, and few of them had any close friends.

LIFE COURSE THEMES

In addition to the above influences, the age and life stage of an individual appeared to influence the importance and meaning of objects for individuals. Differences in maturation, cognitive level, and developmental (psychosocial) tasks may together make each of the stages in life course development qualitatively unique from one another. In early childhood, for example, the perceptual features of objects and materials in the environment are the most salient, and the child's lack of internalized coping ability in the face of struggling with separation-individuation issues may in part explain why soft objects and their "contact comfort" qualities are the most predominant during this time. In adolescence, a period marked by high physical activity and the struggle toward autonomy and a personal identity, objects that facilitate the development or enhancement of these features predominate. However, the effects of gender-role socialization are already apparent, with females favoring objects that signify interpersonal ties, and males preferring objects that represent autonomy and action. Finally, adulthood and old age are characterized by an increased awareness and concern with one's interpersonal ties and the maintenance of one's sense of self, and an increasing awareness of one's mortality.

In a sense, then, the study of the meaning and function of objects for people is the study of human psychosocial development in a microcosm. The themes of attachment and separation, of identity and sense of self, of seeking security and mastery over aspects of one's environment are all evident in the study of objects and their meanings, as well as being primary themes of human life course development. These themes are very similar to those outlined by Erikson (1950, 1968) in his stage theory of psychosocial development, whereby each life stage is characterized by a certain developmental issue. Similarly, Neugarten (1968) suggests that the major themes of life course development (identity, intimacy, mastery, reformulating life goals, etc.) are psychological preoccupations that are continually worked and reworked throughout the life course.

The study of the function of objects in individuals' lives sheds new light on objects and their significance to people. On a general level, this exploratory study complements and adds to current knowledge of human psychological development. On a more individual or personal level, this study found that objects trace or outline the patterns and continuity of one's life course, with objects functioning as silent markers or reminders for an individual of their personal life history. Finally, the findings presented here provide some preliminary groundwork for future research in unraveling the complex relationship between people and objects.

REFERENCES

Ainsworth, M.D.S, Blehar, M.C., Waters, E., and Wall, S. 1978. *Patterns of Attachment.* Hillsdale, NJ: Erlbaum.

Beaglehole, E. 1932. *Property: A Study in Social Psychology.* New York: MacMillan.

Beck, M., Cooper, N., Dallas, R., Abramson, P. and McCormick, J. "Looking for Mr. Good Bear." *Newsweek,* December 24, 1984, 66–71.

Boniface, D. and Graham P. 1979. The Three-year Old and His Attachment to a Special Soft Object. *Journal of Child Psychology and Psychiatry* 20:217–224.

Bowlby, J. 1969. *Attachment and Loss: Attachment.* New York: Basic.

Brehm, J.W. 1972. *Response to Loss of Freedom: A Theory of Psychological Reactance.* Morristown, NJ: General Learning Press.

Busch, F. 1974. Dimensions of the First Transitional Object. *Psychoanalytic Study of the Child* 15:243–260.

Busch, F., Nagera, H., McKnight, J., and Pezzarossi, G. 1973. Primary Transitional Objects. *Journal of the American Academy of Child Psychiatry* 12:193–214.

Butler, R. 1963. The Life Review: An Interpretation of Reminiscence in the Aged. *Psychiatry* 26:65–75.

Butler, R. and Lewis, M. 1983. *Aging and Mental Health.* New York: Mosby.

Cattell, R., Eber, H., and Tatsouka, M. 1970. *Handbook for the Sixteen Personality Factor Questionnaire*. Champaign, IL: Institute for Personality and Ability Testing.

Cohen, K.N. and Clark, J.A. 1984. Transitional Object Attachments in Early Childhood and Personality Characteristics in Later Life. *Journal of Personality and Social Psychology* 46:106–111.

Csikszentimihalyi, M. and Rochberg-Halton, E. 1981. *The Meaning of Things: Domestic Symbols and the Self*. Cambridge: Cambridge University Press.

deCharms, R. 1968. *Personal Causation: The Internal Affective Determinants of Behavior*. New York: Academic Press.

Deci, E.L. 1975. *Intrinsic Motivation*. New York: Plenum.

Dickens, W.J. and Perlman, D. 1981. Friendship over the Life Cycle. In *Developing Personal Relationships*, edited by S. Duck and R. Gilmour. London: Academic Press.

Douvan, E. and Adelson, J. 1966. *The Adolescent Experience*. New York: Wiley.

Dusek, J.B. and Flaherty, J.F. 1981. The Development of the Self-concept during the Adolescent Years. *Monographs of the Society for Research in Child Development* 46, Serial No. 191.

Erikson, E.H. 1950. *Childhood and Society*. New York: Norton.

Erikson, E.H. 1968. *Identity: Youth and Crisis*. New York: Norton.

Erikson, E.H. 1969. Identity and the Life Cycle. *Psychological Issues* 1:18–171.

Frankl, V. 1955. *The Doctor and the Soul*. New York: Bantam.

Fromm, E. 1955. *The Sane Society*. New York: Rinehart.

Furby, L. 1978. Possessions in Humans: An Exploratory Study of its Meaning and Motivation. *Journal of Social Behavior and Personality* 6, 49–65.

Gaddini, R. and Gaddini, E. 1970. Transitional Objects and the Process of Individuation: A Study in Three Different Social Groups. *Journal of the American Academy of Child Psychiatry* 9:347–365.

Garrison, W. and Earls, F. 1982. Attachment to a Special Object at the Age of Three Years: Behavior and Temperament Characteristics. *Child Psychiatry and Human Development* 12:131–141.

Gilligan, C. 1982. *In a Different Voice*. Cambridge, MA: Harvard University Press.

Goffman, E. 1961. *Asylums*. New York: Anchor.

Goldberg, J. 1984. Teddy Bears as Soul Soothers. *Psychology Today* 19:71.

Hallowell, A. 1943. The Nature and Function of Property as a Social Institution. *Journal of Legal and Political Sociology* 1:115–138.

Halonen, J.S. and Passman, R.H. 1978. Pacifiers' Effects upon Play and Separation from the Mother for the One-year-old in a Novel Environment. *Infant Behavior and Development* 1: 70–78.

Harlow, H.F. 1958. The Nature of Love. *American Psychologist* 13:673–685.

Harlow, H.F., and Suomi, S.J. 1970. Nature of Love-simplified. *American Psychologist* 25:161–168.

Henderson, S. 1982. The Significance of Social Relationships in the Etiology of Neurosis. In *The Place of Attachment in Human Behavior*, edited by C.M. Parkes and J. Stevenson-Hinde. New York: Basic.

Henderson, S., Byrne, D.G., Duncan-Jones, P., Adcock, S., Scott, R., and Steele, G. 1978. Social Bonds in the Epidemiology of Neurosis: A Preliminary Communication. *British Journal of Psychiatry* 132:463–466.

Hess, B. 1972. Friendship. In *Aging and Society*, edited by Riley and Foner. New York: Sage.

Hodgson, J.W. and Fischer, J.L. 1979. Sex Differences in Identity and Intimacy Development in College Youth. *Journal of Youth and Adolescence* 8:37–80.

Hong, K. 1978. The Transitional Phenomena. *Psychoanalytic Study of the Child*, 3:47–49.

Hong, K.M. and Townes, B.D. 1976. Infants' Attachment to Inanimate Objects: A Cross-cultural Study. *Journal of Child Psychiatry* 15:49–61.

Horton, P.C., Louy, J.W., and Coppolillo, H.P. 1974. Personality and Transitional Relatedness. *Archives of General Psychiatry* 30:618–622.

Irwin, F.W. and Gebhard, M.E. 1946. Studies in Object Preferences: The Effect of Ownership and Other Social Influences. *American Journal of Psychology:* 59, 633–651.

James, W. 1890. *The Principles of Psychology*. New York: Dover.

Kahn, R.L. and Antonucci, T.C. 1980. Convoys over the Life Course: Attachment, Roles, and Social Support. In *Life-span Development and Behavior,* edited by P.B. Baltes. New York: Academic Press.

Kellogg, M. 1985. My Lost Luggage. *Glamour,* 83, (4):204.

Lamb, M.E. and Campos, J.J. 1982. *Development in Infancy*. New York: Random House.

Levinson, D., Darrow, C., Klein, B., Levinson, M., and McKee, B. 1978. *Seasons of a Man's Life*. New York: Knopf.

Litwinski, L. 1943/1944. Is there an Instinct of Possession? *British Journal of Psychology* 33:28–39.

Lopata, H.Z. 1973. *Widowhood in an American City*. Cambridge, MA: Schenkman.

Lopata, H.Z. 1975. Widowhood: Societal Factors in Life-span Disruptions and Alternatives. In *Life-span Developmental Psychology: Normative Life Crises,* edited by N. Datan and L. Ginsburg. New York: Academic Press.

Lowenthal, M.F. and Haven, C. 1978. Interaction and Adaptation: Intimacy as a Critical Variable. *American Sociological Review* 33:20–30.

Lowenthal, M.F. and Robinson, B. 1977. Social Networks and Isolation. In *Handbook of Aging and the Social Sciences,* edited by R.H. Binstock and E. Shanas. New York: Van Nostrand.

Marris, P. 1982. Attachment and Society. In *The Place of Attachment in Human Behavior,* edited by C.M. Parkes and J. Stevenson-Hinde. New York: Basic.

Mason, W.A. 1968. Early Social Deprivation in the Nonhuman Primates: Implications for Human Behavior. In *Environmental Influences,* edited by D. Glass. New York: Rockefeller University.

Mason, W.A. and Kenney, M.D. 1974. Redirection of Filial Attachments in Rhesus Monkeys: Dogs as Mother Surrogates. *Science* 183:1209–1211.

Matteson, D.R. 1975. *Adolescence Today: Sex Roles and the Search for Identity*. Homewood, IL: Dorsey Press.

Medley, M.L. 1976. Satisfaction with Life among Persons Sixty-five Years and Older: A Causal Model. *Journal of Gerontology* 31:448–455.

Myerhoff, B. 1978. *Life's Career: Aging*. Beverly Hills, CA: Sage.

Neal, P. My Grandmother, the Bag Lady. *Newsweek,* February 11, 1985, 14.

Neugarten, B. 1968. Adult Personality: Toward a Psychology of the Life Cycle. In *Middle Age and Aging,* edited by B. Neugarten. Chicago: University of Chicago Press.

Newman, B.M. and Newman, P.R. 1979. *An Introduction to the Psychology of Adolescence*. Homewood, IL: Dorsey Press.

Parker, C. 1980. Mother-infant Interactions and Infants' Use of Transitional Objects. *DAI*, 40 (10-B):5038–5039.

Passman, R.H., 1976. Arousal Reducing Properties of Attachment Objects: Testing the Functional Limits of the Security Blanket Relate to the Mother. *Developmental Psychology* 12:448–449.

Passman, R.H. 1977. Providing Attachment Objects to Facilitate Learning and Reduce Distress: Effects of Mothers and Security Blankets. *Developmental Psychology* 13, 26–28.

Passman, R.H. and Halonen, J. 1979. A Developmental Survey of Young Children's Attachments to Inanimate Objects. *Journal of Genetic Psychology* 143:165–178.

Passman, R.H. and Weisberg, P. 1975. Mothers and Blankets as Agents for Promoting Play and Exploration by Young Children in a Novel Environment: The Effects of Social and Nonsocial Attachment Objects. *Developmental Psychology* 11:170–177.

Piaget, J. 1936. *The Origins of Intelligence in Children*. New York: International Universities Press.

Provence, S. and Lipton, R. 1962. *Infants in Institutions*. New York: International Universities Press.

Rajecki, D., Lamb, M., and Obmascher, P. 1978. Toward a General Theory of Infantile Attachment:

A Comprehensive Review of Aspects of the Social Bond. *Behavioral and Brain Sciences,* 1:417–436.

Rochberg-Halton, E. 1984. Object Relations, Role Models, and Cultivation of the Self. *Environment and Behavior* 16:335–368.

Rosen, J.L. and Neugarten, B.L. 1964. Ego Functions in the Middle and Later Years: A Thematic Apperception Study. In *Personality in Middle and Later Life,* edited by B.L. Neugarten and Associates. New York: Atherton.

Rotter, J.B. 1966. Generalized Expectancies for Internal versus External Control of Reinforcment. *Psychological Monographs* 80:1–28.

Rudhe, L. and Ekecrantz, L. 1974. Transitional Phenomena. *Acta Psychiat. Schandanavia,* 50: 381–400.

Schaaf, M. Remembering When. 1985. *Los Angeles Times Magazine* 52, December 8.

Seligman, M. 1975. *Helplessness.* San Francisco: Freeman.

Sherman, M. 1986. Importance of Treasured Objects in Young Children. *Medical Aspects of Human Sexuality* 20:47–60.

Sherman, M. and Hertzig, M. 1983. Treasured Object Use: A Cognitive and Developmental Marker. *Journal of the American Academy of Child Psychiatry* 22:541–544.

Sherman, M., Hertzig, M., Austrian, R., and Shapiro, T. 1981. Treasured Objects in School-aged Children. *Pediatrics* 68:379–386.

Sherman, E. and Newman, E. (1977–78). The Meaning of Cherished Personal Possessions for the Elderly. *Journal of Aging and Human Development* 8:181–192.

Stevens-Long, J. 1984. *Adult Life.* Palo Alto, CA: Mayfield.

Stevenson, O. 1954. The First Treasured Possession: A Study of the Part Played by Specially Loved Objects and Toys in the Lives of Certain Children. *Psychoanalytic Study of the Child* 9:199–217.

Suttie, I., Ginsberg, M., Isaacs, S., and Marshall, T. 1935. A Symposium on Property and Possessiveness. *British Journal of Medical Psychology* 15:51–83.

Troll, L.E. and Bengston, V. 1982. Intergenerational Relations throughout the Life-span. In *Handbook of Developmental Psychology,* edited by B.B. Wolman. Englewood Cliffs, NJ: Prentice-Hall.

Unruh, D. 1983. Death and Personal History: Strategies of Identity Preservation. *Social Problems* 30:340–351.

Van Buren, A. Don't Lose Sleep over Security Blanket. 1985, *Los Angeles Times,* September 13, 18.

Vlosky, M. 1979. Adults' Transitional Objects. *DAI,* 41 (2-B):703.

Weiss, R.S. 1974. The Provisions of Social Relationships. In *Doing unto Others,* edited by Z. Rubin. Englewood Cliffs, NJ: Prentice-Hall.

Weiss, R.S. 1982. Attachment in Adult Life. In *The Place of Attachment in Human Behavior,* edited by C.M. Parkes and J. Stevenson-Hinde. New York: Basic.

White, R.W. 1959. Motivation Reconsidered: The Concept of Competence. *Psychological Review* 66:297–333.

Winnicott, D.W. 1953. Transitional Objects and Transitional Phenomena: A Study of the First Not-me Possession. *International Journal of Psycho-Analysis* 34:89–97.

APPENDIX A. TREASURED OBJECTS QUESTIONNAIRE

Instructions:

This study is designed to explore what personal possessions are important to adults, and why. Please think carefully about each question, and answer each as thoroughly as possible. Do not omit any question. Your frank and candid responses to each item will be most appreciated. If you have any questions, or if any of the items are unclear to you, please ask the experimenter. Thank you for participating in this project.

I.   Background Information:
1.   Your gender (circle one): male female
2.   Your age:
3.   Your marital status:

II. Childhood Treasured Objects:
1.   When you were a young child (1 to 6 years old), did you have a special toy or object that was your favorite and that was more important to you than the others? If so, what was it?
2.   If yes, please describe it.
3.   Do you remember specifically receiving the toy or object?
4.   In what situations, places, or times did you most often have it with you?
5.   Until approximately what age did you have this object? Do you remember any specific incident in which it was given up?
6.   Where there any special qualities about the object that stand out in your mind?
7.   Why do you think you became attached to it?

III. Treasured Objects in Adolescence:
1.   What were your treasured possessions as an adolescent (between 12 to 18 years of age)?

IV. Current Treasured Possessions:
1.   Do you currently have a treasured object(s) or possession(s)—one that is very special or important to you? If so, what is it (are they)?
2.   Please describe the object(s).
3.   How was the object(s) acquired?
4.   Why do think you are ''attached'' to it?
5.   Why is it special to you? What does it mean to you?
6.   How and when are you most likely to ''use'' it?

V.   Other Special Possessions:
1.   What other possessions do you have that are very important to you?
2.   Why are the above possessions important to you?
3.   If you were asked to list your five favorite possessions, what would they be?
4.   Of all the things you own, which are the most important to you, and why?

VI. Miscellaneous Possessions:
1.   If your house caught on fire, what five inanimate objects would you rescue, and why?

2. When you are all alone, anxious or upset, are there certain objects that are comforting to you?
3. Do you have any "good luck" objects? If so, what are they?
4. If you had $1000 to spend, what would you buy?
5. What kinds of things do you like to receive as gifts?
6. Do you have any family heirlooms? If so, what are they? What "significance" do they have in you family?
7. Did you (or do you now) have any collections? If so, what are they?
8. When you have moved away from a good friend (or a good friend has moved away from you), have you given or received an object or gift from that friend? If so, what was it?

APPENDIX B. CODING CATEGORIES AND DEFINITIONS: OBJECTS

1. *Teddy bears/stuffed animals*
2. *Dolls*
3. *Pillows/blankets*
4. *Books*
5. *Clothing* (clothing, hats, purses, etc.)
6. *Childhood toys* (bikes, weapons, wagons, soldiers, dress-up items, etc.)
7. *Sports equipment*
8. *Motor vehicles* (cars, motorcycles, boats, planes)
9. *Phonograph records/music* (records, tapes, music, musical instruments)
10. *Photos* (photos, portraits)
11. *Memorabilia* (diaries, journals, yearbooks, pennants, souvenirs)
12. *Personal accomplishment* (awards, trophies, etc.)
13. *Furniture/antiques*
14. *Dishware/silverware* (dishware, silverware, china, glassware)
15. *Jewelry* (jewelry, watches, rings)
16. *Religious items* (Bible, torah, rosary)
17. *Collections* (stamps, coins, shells)
18. *Small appliances* (stereo, camera, television, typewriter, etc.)
19. *Important papers/documents* (documents, records)
20. *Money*
21. *Artwork*
22. *Tools*
23. *Personal items* (jewelry box, perfume, wallet, flowers, trinkets, etc.)

CODING CATEGORIES AND DEFINITIONS: MEANINGS

1. *Memories: General* (non-interpersonal: recollection of a specific occasion, place, or event)

2. *Memories: Interpersonal-familial* (interpersonal-familial associations; sentimental associations; reminds them of someone special; was given by or belonged to someone special; represents attachments to others)
3. *Personification* (object described as having the qualities of a person; represents "someone to talk to"; was described as a "companion" or something to care for; is reliable and always there)
4. *Cultural/religious associations* (ethnic, cultural, or religious association)
5. *Pleasant experience* (object brings enjoyment; positive feelings associated with it; object provides for feelings of "release" or "escape")
6. *Intrinsic qualities of object* (physical characteristics and/or a description of the object is given; including "handmade", "one of a kind", mention of its "worth", or comments about its design, style, color, etc.)
7. *Utilitarian value* (object provides convenience; saves time; is used for learning or guidance)
8. *Personal value* (object is an expression of one's personal values; is an embodiment of an ideal; represents a desired accomplishment; is an expression of self; represents freedom or independence; is something that was always wanted)
9. *Comfort/security* (object provides comfort and/or a sense of security)
10. *Other*

BEGINNING WITH LIFE HISTORIES:
INTERVIEWING IN THE FAMILIES OF WELSH STEELWORKERS

W.R. Bytheway

INTRODUCTION

At a seminar held in Oslo in February 1986, John Modell presented a paper that was something of a revelation to me, brought up as I was in the more statistical traditions of social science. In his paper, Modell observed that:

> the evidence-gathering interventions by economists, political scientists, geographers, demographers, sociologists are even more apparent (than those of anthropologists). We historians can do none of these, and so we must almost inevitably apply frameworks for assessing and inferring from evidence that violate social science assumptions . . . 'New' historians call

Current Perspectives on Aging and the Life Cycle,
Volume 3, pages 119–126.
Copyright © 1989 by JAI Press Inc.
All rights of reproduction in any form reserved.
ISBN 0-89232-739-1

upon themselves to write plausibly about unintentional, systemic phenomena of which only highly imperfect records exist.

He concluded his paper:

> There is more than merely a perverse pleasure to be had in using data in ways quite athwart their creator's purposes: such inferences are inherently less subject than others to the kinds of bias that enter when one accepts the limitations their creators have imposed on data.

This encounter led me to perceive the potential for re-examining my own research material: to analyze it in order to consider the questions it was intended to address, but also and quite separately to 're-visit' it, like a future historian might, in order to see what else there may be to learn from it.

Thus in this paper I want to put to one side the purpose of my recent research and the resulting findings. It is sufficient to say that in 1984, I interviewed 108 men who were aged 55 or over when in 1980 they took the offer of redundancy from the large Abbey Steelworks in Port Talbot in South Wales. The objective was to investigate the ways in which redundancy might have become an early retirement (Bytheway, 1986). In 1985 and 1986 I returned to distribute to the 108 a copy of a report on the 1984 study and to enquire further into broader aspects of family life (Bytheway, 1987).

The primary objectives in neither of these two studies included the collection of life histories. The fact is, however, that I was given a number of detailed biographical accounts which I value not just because some are so exotic—for example, the man who worked in the German coal mines as a First World War prisoner-of-war—but also because they amplify both the significance of the redundancy experience and family relations in later life.

Fieldwork

In the first interview I held a proforma which imposed a certain structure upon the course of the interview, and this began with a question about what brought them to Port Talbot or what they did after the War (whichever was the more recent). This triggered off a number of detailed blow-by-blow accounts and in such circumstances I tended eventually to intervene in order to move on to more immediate questions of redundancy—I was aware in most interview situations that time was limited and I did want to work through to the end of the proforma. Others were more reticent and some only started to become expansive towards the end of the interview. At that point I was asking them to compare their experience of retirement and that of their fathers and this was a second opportunity to reminisce—rather more freely since by then I was less anxious about the possibility of failing to complete the interview.

In the 1985/1986 interviewing, I dispensed with proformas and adopted a less structured approach. The reason why I inserted the word 'beginning' into the title of this paper is that a question that is oriented to the interviewee's life history changes rather significantly the relationship between interviewer and interviewee. This kind of question tends to free the exchange from the expectation that the interviewer has specific questions to ask, and grants the interviewee the freedom to construct a selective autobiography and the interviewer the freedom to interrupt with questions that are then prompted. As Graham (1984, pp. 118–9) has observed: "Story-telling offers advantages over traditional interviewing, more effectively safeguarding the rights of informants to participate as subjects as well as objects in the construction of sociological knowledge." On occasions a steelworker would ask what brought me, the interviewer, to Port Talbot or where I was brought up and I, like them, constructed ways of answering this kind of question.

Perhaps I should say that this approach has still tended to meet the formal requirements of the research project. Through the repeated experience of interviewing in one family after another, certain significant commonalities and contrasts emerged which formed the basis of a series of carefully worded and generalizable conclusions*.

I attended the Oslo seminar in the middle of the second period of fieldwork and, following Modell's lead, I decided that I should re-examine my data in order to relate it to the life course approach to the study of social change. I felt that there was much in the life histories that I was being offered which could be directly referenced to the dramatic recent history of heavy industry in this part of Wales.

What I wish to do in this paper, however, is discuss the experience of doing the interviewing from what I think of as a more phenomenological perspective. One of the strengths of sociology is its power to lead one to see the familiar from significantly different perspectives. Unfortunately, however, it seems to have failed to perform the same service for what are the all-too-familiar elements of its own practices. This may reflect Hammond's observation that "because most research methods texts, like research monographs, deal exclusively with the context of justification, matters related to the context of discovery are seldom found in print." (1964, p. 3).

Participation

I have come to think of these interviews as episodes of participant observation. The trouble of course is that participant observers are usually expected to participate at most at only a menial or marginal level. When they have been forced to take an initiative, or indeed 'the lead', then the experience tends to be seen as a problem or even as a failure, rather than as an opportunity. In this sense inter-

*Accounts of these will be published in due course: contact the author for details.

viewing researchers are not expected to interpret the experience of the exchanges that they have instigated in this way—their research objective is to extract not to observe. As Galtung (1967, p. 150) remarked, surveys "tear individuals from their social context."

The problem, of course, is that what is extracted—answers to questions—is at best second-hand information, regarding the potential for which the first hand has variable and arguably selective access. It is much more than the problem of recall, or the interviewee's task of constructing a biography. It is also the problem of representing lived-through experiences by spoken words. It is true, of course, that the interviewer, through training and experience, can become particularly adept at extracting accounts that inform and enlighten, and likewise that many people prove to be articulate and perceptive interviewees, but what tends to be forgotten in planning and subsequently in interpreting this choice material is the fact that the interviewer is actually 'there': there is first-hand information to be recorded that documents the immediate social context of the interview.

Institutions

At this point, I would like to put to one side the idea of 'interviewer as participant observer' and introduce a second alternative perspective upon standard empirical practices: the idea of 'household as social institution'. It is interesting to note that the U.K. Office of Populations, Censuses and Surveys has recently reviewed the definition of a household and went to considerable lengths to examine variations in shared living arrangements, but there was no reference to the potentially problematic relationship of people to places (Todd and Griffiths, 1986). What I would suggest is that the concept of 'household', like a number of other concepts, seems to be too central in the organization of a broad range of empirical research, for it to be accorded any kind of critical attention or indeed simple curiosity. Certainly it is surrounded by much critical work on such matters as: family life, domestic labor, child rearing practices, health care, the household economy and so on, and perhaps because of this few seem to have sat down to discuss the thing itself: the household, its membership, its points of entry and exit, its boundaries, its internal history and on-going organization, and its relations with its neighbors and other households with which it does business. It tends to be defined unproblematically and instrumentally in order to facilitate the doing of research on other issues.

As soon as the household is seen to occupy a place of substantive sociological interest, then being in 'its dwelling place' gives the researcher first-hand insight into its way of life. Moreover because the researcher—at least in my case— arrives unannounced, it is, within certain limits, a fairly random and uncontrived glimpse into the life of the household.

I appreciate that all kinds of problems spring to mind: the participant observers, for example, traditionally spends extensive periods of time within the observed setting, documenting the activities of pre-occupied people. The researcher who knocks on the door with a view to extracting life histories or answers to questions about other aspects of 'ordinary life' is usually under pressure to minimize the time spent in the household because of the demands of a tight schedule, and in this regard I was not exceptional. The fact remains, however, that when I am spending time attempting to glimpse ordinary life in times past and places elsewhere through the interviewee's spoken word, I am actually in a position to see it in the time present. To add substance to this, consider two questions that I faced in the course of fieldwork:

> What do I make of the woman who insists that she has a busy weekly schedule of all kinds of formal and informal activities outside the home, but who was in the first time I called, and who gave no indication of a need to be elsewhere during the hour or so during which she answered my questions?

> What do I make of the experience, in interviewing a woman, 75 or so and living on her own, of another woman walking straight into her house, warming her hands on the fire, listening to us but not being introduced or joining in, walking out and then returning twenty minutes later cursing that 'it' hadn't opened yet?

What I think is that both instances constitute relevant first-hand information about the realities of ordinary household life. The first woman presented the account of someone who is free of ties to the household, who is busy and active in a range of social activities, but who also proved to be someone who was easy to locate by being at home. She contrasts with the widowed steelworker who was out on eight out of ten occasions when I called and whose neighbor came to know me as a consequence. The second woman gave me a full account of her way of life that indicated it would be entirely inappropriate to describe her as 'living alone'. Several friends, as I observed in the course of the interview, had open access to her home and she likewise was a 'member' of a number of other households.

Biographies

Returning now to the life course and the biographies that can be constructed in the course of an interview, one interesting aspect of an inquiry into such matters as redundancy in 1980 or ill-health over the last five years, is that the biography can and perhaps should merge into the present. In effect the line of inquiry begins with a question that takes the interviewee back into the past and then works its way back (forward?) to the present. Ideally this can mean a continuing account wherein what happened last year is related to what happened last month, this in turn is related to what happened last week, and that in turn to what happened yesterday, and what is due to happen today and tomorrow. There is then a real merging of the past and the present so that one can understand the historical antecedents of the household observed.

In the reality of interviewing practice, however, there appears to be a curious gap between the biography of the past and the circumstances of the present. In part this may be explained by the strategy that we adopt in formulating questions and, more generally, by the fact that the English language makes a coarse distinction between the past and the present: 'And what did you do then?' leading on eventually to 'And what do you do now?'. The important point is that in practice it is not easy to establish a sequence of events or changes that leads up to the present. This is a familiar and potentially intractable problem for researchers (as in a national census) who have to design—indeed to word—a tightly structured questionnaire for others to administer, but even in the unstructured open-ended interview one runs the risk of the answer 'Nothing happened then' ambiguously implying 'So that is how I am now'.

Quite apart from language, however, a further reason for the gap is the common tendency for people to maintain a clear and marked distinction between 'then' and 'now'. From a gerontological point of view this is fascinating since it raises all sorts of questions about the experience of aging and the relationships between those of different generations, but from a life course or life histories perspective it is potentially very damaging methodologically. Quite apart from the fact that it fosters an undifferentiated and undated popular past, it divorces the two primary sources of data that are obtained from the interview: the second-hand account of the past and the first-hand glimpse of the present. Moreover it adds confusion to the longitudinal element of the study. My interviews of 1984 and of 1985 and 1986, for example, both tend to represent the 'same' present each possessing the same past.

My objectives in interviewing within these families have been essentially exploratory. Among other objectives I have been exploring the processes of inquiry by which we can learn about the ways in which life has been and is being lived. There was one improbable incident which highlighted the importance of seeing interviews as being bedded within the on-going life. One morning I attempted to obtain an interview from a steelworker who had been repeatedly refusing on the grounds that every day he and his wife were having to visit his sick aunt in the hospital. It seemed that they were spending most of each day out of the house and they had neither time nor energy to spare. On this occasion I was politely refused once again but as before was given some doorstep information about the aunt's progress. An hour later, I was interviewing another steelworker who happened to be a Borough Councillor and, in the course of this, he received a phone call from the first seeking his assistance in having an inspector check on the house of the sick aunt.

Sitting on the sofa while the Councillor handled the request, I turned my mind to two conflicting questions: one was what kind of inspector was being requested, and the other was what kind of ethical issues for research were involved. I resolved neither but what the experience brought home to me was the neglected truth that interviews take up real time in the course of continuing real lives. On the one hand, there is the instrumental problem of minimizing the inconvenience caused to the family and maximizing the information obtained. On the other there is the phenomenological problem of interpreting and acting within a real and dynamic setting in a way such that lessons are learned to be passed on to others.

Less dramatically, there were occasional references made to events of the previous day or week: 'Funny you should say that, but only last Thursday . . . '. This kind of response implies a popular assumption that life histories, or questions about life experiences, should be located well and truly in the past. More recent events that are ambiguously still part of the present, are seen as surprising coincidences. Frequently I suspected that the interviewee had difficulty relating a question beginning: 'Have you ever . . .' to the events of the previous week. It was when I realized that the hesitant answer 'once or twice a year' in reply to a 'how often' question related to a period of only eighteen months, that I came to realize that most households are in a state of flux most of the time wherein specific events do not occur at a perceivable 'rate'. Histories are being collected at points within processes that are difficult to locate within the chronology of change. It is hardly surprising then if both interviewer and interviewee prefer to divorce the life history from present circumstance.

CONCLUSION

It is my view that empirical studies that focus upon the life course must actively seek to prevent this divorce, and that recording first-hand observations in the course of receiving personal life histories is the most effective means of doing so.

REFERENCES

Bytheway, W.R. 1986. ''Redundancy and the Older Worker'' in *Redundancy, Lay-offs and Plant Closures,* edited by R.M. Lee. London: Croom Helm.

Bytheway, W.R. 1987. *Informal Care Systems,* Report to Joseph Rowntree Memorial Trust, University College of Swansea.

Galtung, J. 1967. *Theory and Methods of Social Research.* London: Allen & Unwin.

Graham, H. 1984. 'Story-telling' in *Social Researching,* edited by C. Bell and H. Roberts. London: Routledge and Kegan Paul.

Hammond, P.E. 1964. *Sociologists at Work.* New York: Basic Books.

Modell, J. February 1986. 'Do New Historical Questions Merely Permit, or, Instead, Actually Demand New Sources?' Seminar on Life Course, Family and Work, Oslo.

Todd, J. and Griffiths, D. 1986. *Changing the Definition of a Household,* Office of Population, Censuses and Surveys. London: Her Majesty's Stationery Office.

LIFE NARRATIVES:

A STRUCTURAL MODEL FOR THE STUDY OF BLACK WOMEN'S CULTURE

Beverly J. Robinson

INTRODUCTION

Life narratives fall within the province of studies traditionally associated with life history, oral history and oral narratives. The latter terms are often interchangeable and their usage is academically determined by one's disciplinary training. For instance, there are literally hundreds of life history accounts emanating from anthropology and the effect they have made on social science research will be discussed later. Oral history, a term developed just after World War II, has primarily been the purview of historians documenting the lives of men (Titon 1980, p. 281). The oldest of the three, oral narrative, is a more generalized phrase not restricted to a specific discipline, but of particular concern to more

Current Perspectives on Aging and the Life Cycle,
Volume 3, pages 127–140.
Copyright © 1989 by JAI Press Inc.
All rights of reproduction in any form reserved.
ISBN 0-89232-739-1

than 6,000 extant narratives of enslaved and emancipated Africans in the United States from 1703 to the last one published in 1944.[1] Each of the above terms are concerned with written accounts of peoples' lives as a result of oral interview techniques, and rely heavily on capturing the spoken word. Other terms, such as stories, autobiography and biography, have provided important life capturings, but often their construction has depended on the audience as opposed to the spontaneity of the individual simply wanting to share aspects of her (or his) life.

What is recorded as an historical event or "fact" can be perceived as the product of the prevailing cultural values of any society—any culture. However, those activities and/or events that are not considered important by the chroniclers of an era (i.e., family structure, social support, copying mechanism, health care, practices and attitudes, social networks, etc.), coexist with experiences which are closely related to the lives of women. This portion of history is not only ignored, but the recorded past, if any, is lost due to an assumed lack of importance. For African American women, this is clearly borne out by Darling (1987, p. 48), where she states that "until recently, Black women have either been invisible in scholarly analyses of the American past or they have served as a foil for the actions of others. . . Despite their invisibility, generations of Black women have lived a 'secret' life, and validated their own visions and images of themselves with remarkable creativity and resilience." The technique of the life narrative interview provides an opportunity to recapture such events—at least from those persons as far back as the late nineteenth century.

We must begin to question Black women of this period about their lives; learn how widespread certain cultural beliefs and practices were (are?); and uncover the extent to which they identified with the assumed homogeneity of a "female culture." These and other historical aspects can be explored to give us a sense of women's past and a sense of personal continuity. This is particularly essential to studies devoted to African American women in lieu of their multi-faceted roles during any one life period in American history. Before presenting examples of a life narrative study, the significance of an individual's story through the use of oral interviews will be given as a way of understanding the implications of life narratives to social and cultural sciences.

The Development of a Folkloric Perspective

The scholarly development of life history studies is primarily a result of social science research. The research has been an integral part of anthropology's history and since the latter part of the 1960s, the work of Langness stands out as one of the more notable investigations. His work is primarily devoted to anthropological history and use of life histories, though he does include methodological constructs and outlines the importance of life account studies. Based on the earlier works of Kluckhohn and Aberle, Langness (1965, pp. 4–5) defines life history as "an extensive record of a person's life as it is reported by the person himself or by others or by both, and whether it is written or in interviews or both.'' Another source which has involved itself with not only life histories but many personal documents to supplement these histories (particularly during the 1930s and '40s) is the Social Science Research Council. The contributors to the Council's work were primarily sociologists and psychologists and much of the social science guidelines for life history research was developed by these two schools of thought.[2]

Other disciplines, such as folklore and folklife, have devoted studies to women utilizing the life history model based on oral tradition. The results were primarily biographical statements (similar to those of the historian), but basically derived from research examining specific genres; e.g., narratives, folk art, games, songs, beliefs. As late as the 1960s, these statements were described as ''thumbnail sketches of varying size and scope'' (Goldstein 1964, pp. 121–126). Since the development of folklife and social science programs on state and national levels (the Folklife Program at the Smithsonian Institution began in January, 1967 and the American Folklife Center at the Library of Congress in 1976), scholars of these disciplines have developed research and compilation tools when interviewing informants; or, in this instance, recording life narratives of women—specifically African American women. However, these tools (the tape recorder, various types of probing questions and other elements of the interview process), are interwoven with methodology. On the whole, as Mandelbaum (1973, pp. 177–182) maintains, ''the study of lives for purposes of social science has been more advocated than practiced. . . [Even] longitudinal studies, notably those conducted in the Institute of Human Development at the University of California, Berkeley, have yielded many significant observations of growth and social development, but these have yet to be placed in their social and cultural context.''

The concept of using life narratives proposes an interesting challenge to the social science researcher. Not so much as an addendum to a specific research model, but possibly from what Alan Dundes has repeatedly described as eliciting important emotional and attitudinal information which the ethnographer may not have otherwise captured. Though native statements are subject to bias and perhaps even more distortion, ''the distortion itself can be invaluable raw data. . . What an individual distorts can tell the sensitive ethnographer a great deal about how that individual thinks about [oneself] and about different aspects of [one's] culture'' (Dundes 1968, p. 440). More importantly, life narratives can serve as catalysts encouraging the elicitation of political, social and cultural information as it relates to the informant telling her own story.

Life narrative studies on African American women encourage contributors to define themselves. Since a single aspect of their life is not the prerequisite of the interview, a woman is given an open-ended opportunity to tell her story where no previous basis of analysis exists. On one hand, the results allow a woman to define herself via the oral interview for an ethnocentric record. While on another, a context is provided for understanding several genres arising out of the life narrative interview. Historically, such institutions as family, religion, health and socialization have been individually researched, but not interwoven in an overall experience of the individual. This method allows for the expansion of our previous notions of the cultural practices that characterize these institutions. For instance, research on Mrs. Phyllis Carter, an elderly Southern Black woman in her 90's from Brookfield, Georgia provides the opportunity to discover these genres. Older women, such as Mrs. Carter, in small Southern towns represent repositories of time past, the transmitters of culture, the administers of health practices, and portrayals of folk-way survivals in contemporary society. In addition, these Southern women represent a merging of primary and secondary institutions because of their traditional role in childbearing, keeping the family together, church, midwife activities, and womanhood training.

One of the major difficulties with much of the literature on older Blacks—specifically women and their families—is absence of an appreciation for cultural diversity in the analysis. The heterogeneity of African American culture is not fully represented in scholarly literature. Complexities further abound when identifying literature on women in small (specifically rural) communities. This is an essential, yet grossly neglected area of Black studies.[3] Rather, traditional research on African American women has relied on descriptive analysis often concerned with the single predicative interests of the researcher; for example, matriarchal, sexual relationships and survey studies. Hence, there have been continuous distortions resulting from what Jacquelyne Jackson (Bell, Parker and Guy-Sheftall 1979: xxvii–xxviii) cites as "inadequate conceptualizations of questions confronting historians investigating Black women," which has also effected creative literature. These "continued distortions influence the patterns and behavior of Black women in and out of fiction. And to be sure, growing accounts of sophisticated studies—i.e., suicide, sex, matriarchal, and others—are indicative of the incessant trend to shape and reshape the ways of Black womanhood. (Ibid)"

In many cases, social science research on African American women has been conducted from a myopic view or not included in some of the more prominent works devoted to African American life-styles. For example, Staples' 1973 work, *Black Women in America,* concentrates on descriptive analyses of life experiences. John Hope Franklin (1947), a noted historian, confined his discussion to referencing a few women with little or no details. The historical research of Gerda Lerner, to expand further on Jackson's previous citation, "fall[s] within the genre of hasty and inefficient work developed quickly to capitalize on 'blackness'." *Deep Down in the Jungle* and *Positively Black,* Abrahams' (1963, 1970) early cultural landmark studies, left the reader with the task of analyzing African American women's silence; wherein their presence is limited to butt-of-the-joke references in male verbal play exercises within the narrative traditions of "toasts," "playing the dozens" and "oral poetry." The examples are not merely restricted to male writers, but the majority of the accredited (i.e., "known") studies on or about African American women have not been traditionally documented by the women themselves.[4]

Any analysis should be from a periscopic or a holistic multidisciplinary approach. Anthropological, folkloric, gerontological, historical, psychological, and sociological considerations are *all* necessary for an accurate assessment of Black women. The requisite history and records of African American women have been such that their roles have demanded a multi-faceted world involvement socially, domestically, culturally, and in many cases, politically.

Writings appearing in the late 1970s and early '80s by African Americans began taking steps to correct some of the past inequities. One outstanding work has been Kathryn Morgan's *Children of Strangers* (1980) and a subsequent in-progress work on Edith Wilson,[5] which is modeled after Morgan's. There are also notable efforts made by John L. Gwaltney (1980) and Theodore Kennedy's *You Gotta Deal With It* (1980); though these latter two contributions used informant pseudonyms which are not a common feature of life narrative studies. Morgan's work has been particularly successful because it recounts narrative experiences learned from her childhood and reinforced throughout her adult years. Her work is written from a woman's point of view with the cultural and intellectual understanding of a scholar who is African American. As previously mentioned, the unexpurgated interview is indeed ethnocentric. Yet it represents an overview of those aspects in a woman's life as she relates them in recreating her experiences. Further, it is a way of establishing and understanding the interrelations of her environment. The research on Phyllis Carter exemplifies this perspective for it draws heavily on the original interview texts to depict the breadth of Mrs. Carter's recorded knowledge. The unexpurgated interview elicits personality traits, wherein information is personally shared. Thereby major emotional and attitudinal data can be gleaned as illustrated in Mrs. Carter's advice to younger people:

> Bad company will get you away from this world. I ain't joking and you ain't got to do nothing to them. . . they'll kill you for your own money and rob you. Be particular. When you see a rough crowd, win off and you'll always hold your bones. Hold your whole bones. But if you don't do it, you can't. If I didn't know what I was talking about, I wouldn't of said it.[6]

Additionally, by virtue of her lifestyle, Phyllis Carter has contributed to several dimensions of the Black female experiences in America and she offers an indelible narration toward understanding chapters in African American history. Like other older women in small Southern communities, she provides observations of clan interaction, collective conscience, folk medicine and sources of African survivals.

Documenting the world of Mrs. Phyllis Carter is a contribution to a concern for what may be termed "reputational sampling." A reputational sample is drawn from participants (or leaders) who are identified by the community as having been involved with the growth of the community "when enumeration of

the entire group would be impossible. . . these kinds of samples require strong
assumptions or considerable knowledge of population and subgroup selected. . .
[and] are appropriate when probability samples are not possible, practicable or
desirable" (Anderson 1976, pp. 14–15 and Miller 1969).

The initial interview with Mrs. Phyllis Carter began when she was ninety-
seven years old. She attributes her long life and many accomplishments to living
a "good life." The good life to Mrs. Carter is defined as not mistreating anyone,
always being able to return to places one has lived or visited, and above all,
treating people the way you want to be treated and not the way you want them to
treat you. To Mrs. Carter's way of thinking, longevity and earthly success are
spiritual gifts given to those who obey God's laws: "The Lord just give me them
gifts and I hold them and I don't forget them." She has been active in numerous
projects over the years and her vitality is reminiscent of the belief that age is not
in numbers, but a quality of the mind.[7] She is popularly known as "Aunt (Añt)
Phyllis," but she also bears the name of "Auntie" to young Blacks, "Granny"
to Whites, or in the past she was affectionately called "Lil' Bit" by one of the
four husbands she has outlived. Commensurate with time, she has developed a
prophetic and honorable *reputation* throughout South Central Georgia to various
parts of Florida as evidenced by conversations with numerous residents within
the area who consistently referred to the importance of her presence in various
aspects of the community. There are several reasons why she has developed a
reputation and is deemed important by her community. They include the fact that
"she still totes a pistol, cooks, quilts, and lives alone." Then, too, it might be all
of these reasons.

Among rural Southern Blacks, according to Plumpp (1972, pp. 27–28), older
Blacks were respected and admired leaders of the community. Plumpp and others
(Davie 1949, pp. 206–209 and Frazier 1949, pp. 114–124) further write that
"women in the South were primarily homemakers." They took pride in "how
well their children behaved and looked, clean and well-fed, how well they kept
house and cooked, how well their flower beds and garden looked, [or] how many
eggs their hen laid a day. These women were the doctors and nurses of rural
communities." For Phyllis Carter, conversations praise her as "one of the finest
midwives in the area;" "a griot" (better known in Western terminology as an
"oral historian"); "a woman who never forgets;" or, "an advisor/confidante."
The praise is not a reticent reflection of her life being an exceptional one; rather,
one which is more representative of the attributes and personage her community
is proud of and respects.

Field Research Model

The research was part of a South Georgia fieldwork project for the Library of
Congress Folklife Center. The project marked my introduction to the State of
Georgia. Upon arrival, the first goal was to identify the African American

churches and social organizations in the area and any women's based groups (e.g., PTA, 4–H and homemakers' groups) in the immediate and surrounding areas. This early personal identification (which is extremely important when seeking out persons to be part of your research) was supplemented with meeting the various dignitaries and lay people. It was a result of informally asking various people about persons whom they felt were important to their community, which then led to Mrs. Carter. People would refer to her as "old granny," "Auntie," "Doc" and other terms of endearment that were indeed paying respect to one person–Phyllis Carter. Her more popular name among African Americans in the area, and the one which made her smile, was Aunt [Añt] Phyllis.

The initial meetings (3) with Mrs. Carter were introductory ones. They were primarily designed to see if (a) she was comfortable with me (I personally shared narratives related to my family and academic background); (b) developing a conversational friendship; (c) identifying her verbal and nonverbal communication style; and (d) developing a basic respect wherein she would openly share some of her life narratives and feel assured that I would collect and present them in a prideful manner. The overall research is a result of several extensive interviews during the summer of 1977 in South Central Georgia with follow-up contacts through 1979, and is comprised of four basic sections: (1) The Formative Years, (2) The Dispatching of Dreams, (3) Collected Beliefs, and (4) The Later Years. However, only two sections are included in this paper.

"The Formative Years" includes basic background information on Phyllis Carter, the matrilineal and patrilineal history of her family, and her marriages. Aunt Phyllis was born on April 15, 1880 in Wilmington, North Carolina. This was during post-reconstruction when hundreds of Blacks were being lynched and burned at the stake throughout the South. Though Emancipation and Reconstruction offered changes, many adherents of the "old ways"—particularly in the South—refused to abide with national laws, and places like Wilmington "gave many a poor colored person a tough way to go."[8] The racial climate in Wilmington had become like a keg of dynamite about to be ignited over disfranchisement regulations, miscegenation and outright segregation. Many Blacks, including Aunt Phyllis' family were experiencing grave social and economic hardships. The situation was almost intolerable and motivated her father to flee with his family to Willacoochie, Georgia.

I remember when we left there, way out yonder in Wilmington, North Carolina. We run away from there one night. Well my Daddy just kept saying to her [mother], 'Fannie, I just can't stay here and make a living for these children. We just got to get away from here. I done went to Willacoochie and I done bought ten acres of ground and there's two houses on it and let's us leave here tonight.' And we packed things up and put them on a flat car. The man came and picked it up and brought it on into Willacoochie. That's the way we had to get away to come out her to Georgia from there—yonder in North Carolina. I was in my elevens—I remember that good. When we come to Georgia it was different. We could have anything we wanted to eat and cook like we wanted. You growed it yourself [and] got something for it. When you was working for them you didn't get nothing.

Aunt Phyllis' recollections of activities during nineteenth century America provides important historical insight. She was exposed to family memories of slavery, and her recounted text attests to the institutionalization of human bondage. Many of the oldest incidents were often those of hardships for African Americans. They were times plagued with women being sexually abused; men whipped for being sick in the field and therefore unable to work (and some died); or, what it was like in a later era earning twenty-five cents a day picking cotton, pulling fodder, or carrying peanuts and tobacco.

The formative years also include the traditional extended family construct which was a part of her childhood as opposed to one which developed in her later years. She was reared in a household with immediate parents and two grandmothers. In the Black community, the traditional primary agents of socialization for preadolescent and adolescent girls are her immediate and extended family. Although she may spend more time with her nuclear family, vital role models are also represented by aunts, great-aunts, uncles, grandparents, cousins, "make believe" kin and others. It is within this environment that the early patterns of learning occur.

Occasionally, the child may become confused about to whom she should relate and by whom she should be supervised as the authority figure. In the case of a disagreement between the parent and grandparent over the child's behavior, the child is prone to take sides with the one she favors. Although some parents may feel that the grandparents' ideas about child rearing are somewhat antiquated, studies have proven that few parents are able to dismiss grandparental advice as simply being old-fashioned (Ladner 1971, p. 72). An example of this is when Aunt Phyllis recalled how once there was a difference of opinion when she was told to secure eggs hidden under the house flooring where she encountered a snake which her father later killed:

> I remember one day it was raining. Mama said, 'Phyllis, I want you to go under that house and get all them eggs. . . ' The house was low. I got near about to them nest eggs and honey it was running over with eggs. A snake helt his head up . . . and grandma was there, she say 'go under there.' I say 'I ain't going to let that snake bite me. I know that's a snake 'cause I done seen my Daddy killed them.' And he was a rattlesnake. They killed him, but had to rip up the boards in the middle of the house—my Daddy did—took them up and he shot that snake, he did. Drug him on through the house 'cause he couldn't go down in there. And one of my grandmas say, 'If she's at my house, she'd have to get them eggs else I'd beat her to death.' My other grandma say, 'well, I'm sho glad she found that.' Mama say, 'I ain't had no eggs out from under there in two or three days.' But honey that nest was full. It sho was full.

American child rearing practices have relied on two socializing agencies: One, the home; and the other, school. As Aunt Phyllis recalled her early years, school was to be a quite different experience. The schoolhouse was located at the home of a Black teacher who lived "across the creek." Eight of the Baker children attended school together. For Aunt Phyllis it was a traumatic experience that did not last very long:

I went to school three days in my life. When I did start to school it was a wide place to cross on the creek. People didn't have no bridges then. You had to walk on foot logs. I started school on a Monday morning and I went to Wednesday evening. I seen so many children got to fighting and was pushed off of that big log. I reckon it was about twenty-five or thirty got drowned where they were fighting. This one here got to fighting and push him over there and it was a big creek. Oh it was wider than here from them pecan trees cross yonder [approximately one-half mile]. Me and my sisters and brothers honey we got on them foot logs and had us a stick and got cross. Looked back and we didn't see a child no where. First house us come to was named Miss Betty Tumor. I told her she better go see about her children and everybody left the horses in the field to go see about them children. All the Baker children got saved.

So I ain't go back [to school] no more—I was scared to. No honey they'd drown some of them on them foot logs. When they tried to make me go back I just went to fighting. Teacher she grabbed to pull me in [to join the other children and to hold me]. She stuck my head between her legs. And what she done that for. I say, 'you bet not hit me no more . . . I'm gonna bring this plug out [bite her leg].' Brought it out too. She had to go to the hospital and tried to have that stuck back on, but she couldn't. She didn't try that no more. So that expelled me from school for good and you'd a done the same!

The attempted disciplinary action of her teacher did not daunt Aunt Phyllis' commitment to remove herself from any further fear that developed as a result of the drownings, not necessarily from a classroom perspective, but from a life outlook which she has never forgotten. The experience may have also influenced the reasons why she never learned to read or write.

Since school proved less than desirable, Aunt Phyllis spent considerable time at home learning a lot from the matriarchal lineage of her family. Learning adult responsibilities is a part of the early training of Black children. The females are encouraged to be independent rather than passive individuals because many will assume family and economic responsibilities (Iscoe 1964, pp. 451–460). Quilting was an alternative economic source when Aunt Phyllis was young. She would always collect the scraps from sewing and devoted long hours to making quilts:

I worked more quilts than I'll ever have. When I was eleven years old I'd go to my grandma and she'd be making some and she'd hand them to me. She'd say, 'now I want you to learn how to make quilts.' I made them. Sometimes she'd cut out what she wanted in a paper [quilt design]. She'd say, 'I want you to make it just like this.' If I had long stitches in there she'd draw them out—make me go back over it. I thank her for learning me.

How well a child learns adult responsibilities is seen as sources of strength for survival and independence—particularly among females. This type of strength does not embrace dominance as some researchers would suggest, but supports the research of those who distinguish between strong and dominant African American women (Ladner 1971; Hill 1972; Staples 1973).

After the family moved to Georgia, she became quite proficient at quilting. She not only adopted new designs but adhered to traditional quilt patterns she had learned in her earlier years. Today, her quilting is still done by hand. She only makes large (or block) patch quilts which she sells for $20 to $35 each, and she continues to make two tops—a design for each side—in accordance with her childhood training. The major innovative change is in the padding. Aunt Phyllis no longer uses batting but has discovered that rag padding makes a quilt "sleep warmer."

Quilting and leaving North Carolina were not the only memorable events which occurred during her earlier years. Another very significant occurrence was the day she "got religion." Religion has held and continues to hold a very prominent place in Black life. Since her father was not only a farmer but a preacher, it is not surprising that religion was important in their home, and over the years she:

> ain't lost nere speck of it. I prayed to get it. Everybody had revival you know, and everybody had got off the seat but me. And that Saturday morning my sisters and them they worried me so much I went in the woods and there where I got it aside of a stump. I looked at my feets, they looked new. Looked at my hands and I went to the house and told Ma. She say, 'well you got it.' I say, 'I ain't going lose it' and I ain't lost it.

Because religion has played a vital part in Aunt Phyllis' spiritual well-being, it was something she tried hard to pass on to her children regardless of the rebellious attitudes she encountered in a few of her spouses. In her own terms, her husbands consisted of "two good ones and two devils. I put up with them until the Lord took them." Indeed she remained steadfast to her religion and reared ten children.

Her first husband was particularly attentive to the other children while she recuperated after a birth. Normally one or two women (more specifically those of the immediate family of the wife) would assume responsibility over the other children until the mother was able to resume her usual domestic duties. This generally took about four weeks. Aunt Phyllis' first husband would not only tend to their children, but he would give her at least a month before he would want her to cook for him. Such caring and understanding remained with her after his death. Although she was to be married three more times, each husband brought a special meaning into her life. Her last husband, Arthur Carter, was with her during the height of her career as a midwife. The idea of being a midwife was manifested through her dreams.

The dream phenomenon (The Dispatching of Dreams) is not an uncommon aspect of African American culture and is most closely associated with religion of the unknown. This phenomenon is captured mostly in novelistic and documentary writings particularly in the folkloric works of Hurston and the brief mentioning in W.E.B. DuBois' book, *The Souls of Black Folk*.

Aunt Phyllis did not become a midwife until she was well into her thirties. Yet the recurrence of the midwife dream and other images when she was a youth (which later became realities) had convinced her parents to refer to her as a "God sent child." To Aunt Phyllis, dreams can be viewed as signs, and signs can only be properly interpreted if one has been conditioned in childhood. Because of the frequency in which dream images appeared and materialized, her dreams can be categorized into three different areas: (1) predictional, (2) curative, and (3) vocational. The latter, vocational, gave emphasis to her career as a midwife.

Aunt Phyllis midwifed for both Blacks and Whites. Just how many children she delivered in almost forty years remains a rough statistic of approximately three to four thousand. The first time she delivered (better known as "catching," to have "caught," or "slapping") a baby, Aunt Phyllis was not nervous:

> No, I didn't get nervous. I didn't get nervous when I dreamt it! Told my Mama about it 'fore I was grown. The Lord showed it. When he show you a thing in a vision, you follow it. It don't come to you to surprise you. The first woman I ever did wait on her husband say, 'if you scared now don't go.' I say, 'I ain't scared. Where I go the Lord follows me.' Things I can't do, I ask the Lord to show me. He'll show you 'cause he got all the power over us. We ain't got a bit of power.

This vocation brought her together with one of the most important processes in life and one that has several orderly processes within itself. She sees birth as both a beginning and an end: (1) It is the beginning of a child's life as well as the end of pregnancy; (2) She sees the first child in the family as bringing an end to the preoccupation of a husband and wife with each other, but the beginning of a family; and (3) It is the end of an adult's period of being solely a learner and the beginning of his or her primary function as a teacher.

CONCLUSION

The dynamic energies of Phyllis Carter have given her a prominent position in her community. She has become a mirror of cultural virtues and memorable events encountered in her lifetime. For social and cultural scientists, she contributes to several genres of interest. They include: narrative, legend, history, material folklife, belief,[9] drama, medicine, and religion. These genres are notably marked by domestic art and oral tradition, the oldest records of human cultural history. Her contributions are glimpses of her life and the lives of others with whom she has interacted over the years; indeed, life narratives. Mrs. Carter's inability to read or write does not shadow the much needed concern for literacy in today's society. Rather, it takes its place in historical time and circumstances, and contributes to a prolific kind of knowledge—a survivalistic one.

Not restricted to, but specifically for, social scientists, life narratives become an important viable tool. It is one which moves away from myopic analysis to a periscopic view; a vision which calls for a better understanding of what people want researchers to know about themselves and how such special knowledge is interwoven into the rest of their world and society as a whole. The academic wealth of the collected texts from informants demonstrates how "the printed

page is richer for the inclusion of the human voice in the historical [life] record"
(Allen and Montell 1981, p. 107). Life narratives are useful for building a
foundation for social science research, not just on African American women, but
including other women of African heritage and women of similar cultural backgrounds. Life narratives, as a research model, constitute the type of homework
which draws one into what my grandmother repeatedly described as "the construction gang rather than the wrecking crew."

As a structural model, life narratives require:

a. Going into and learning as much about a community chosen for its traditional respect for
   elders.
b. Who do the people in the community consider important? (reputational sampling)
c. Developing conversational relationships with these "important" people. A researcher
   often becomes a friend as a prerequisite to collecting data.
d. Learning to listen for an over-all glimpse of a person's life while understanding different
   areas (or genres) which may unfold in the narration. This method also allows informant
   to tell her story her way.
e. Analysis of the narrative (data) for a multi-dimensional approach that integrates various
   aspects of her life into a holistic scheme as she opts to tell it.

Though this paper is specifically concerned with African American women,
researchers have an obligation to represent people the way they want to be seen
rather than how we want to portray them.

ACKNOWLEDGMENT

The term, "life narratives," was coined in 1980 and a preliminary paper was presented at
the 1983 California Folklore Society Meeting in San Francisco, California. My deepest
appreciation goes to Mary W. Walker (California State University, Long Beach) and Dr.
Vickie Mays (University of California, Los Angeles) for their suggestions while developing this paper.

NOTES

1. Based on her 1946 dissertation, Starlings' research on slave narratives is undoubtedly one of
the finest. She is a literary scholar with a profound respect for historical analysis.
2. See Allport; Blumer; Gottschalk, Kluckhohn, and Angell for pioneering examples contribut-
ing to the Social Science Research Council.
3. See Darling's article, both of Davis' work, and DuBois' "The Damnation of Women," in
Darkwater.
4. Many women social scientists, too, are not exempt from the inability to place their collected
data on Black women into a contextual framework. More than likely this is due to the researcher's
traditional training which overlooks the multi-faceted roles and cultural innuendos used to describe
the interviewee's life experiences. For example, "I Am Annie Mae" is a very provocative study of a
seventy-year-old Black woman from Texas (Annie Mae Hunt) collected by Ruthe Winnegarten. Mrs.
Hunt's grandmother was just a child when her mother was sold during slavery. The girl wept

profusely at the loss of her mother. The white mistress to whom the youngster was enslaved reassured the girl she would be cared for, and she was. Mrs. Hunt concluded this very moving experience with the phrase, "She took care of her." My suspicion is that Annie Mae Hunt opted to use a Black language style—Ebonics—by offering the interviewer a double-entendre expression within her closing remark which has at least two possible definitions: a) the white mistress ensured her welfare; or, b) the white mistress fixed (i.e., deliberately altered) the grandmother's attitude by never letting [according to the text] the child hear from her mother as well as brother or sisters again. The wording, "She took care of her," carried an intellectual, physical and spiritual experience which is definitively left to a reader's discretion. In order to understand more fully the root meaning of the above phrase and the overall narrative, a contextual cultural analysis is imperative.

5. The author, Dr. Robinson, is currently completing a book, *"Jemima, Eliza and Edith . . . You Have Nothing to be Ashamed of. . ." The Life Narrative of Edith Wilson.*

6. All quoted statements are from interviews with Mrs. Carter. They appear in the book *Aunt [Añt] Phyllis* by B.J. Robinson. The original tapes documenting these interviews are housed at the American Folklife Center of the Library of Congress—Washington, D.C.

7. Collected from Mrs. Katie V. Greene of Berkeley, California.

8. For further discussion of this era, see DuBois, ed.; Edmonds; Meier, August and Elliott; Logan.

9. Because the term "superstition" often carries a pejorative connotation, the term "belief" is used which includes superstition.

REFERENCES

Abrahams, Roger D. 1963. *Deep Down in the Jungle.* Chicago: Aldine Publishing Company.

Abrahams, Roger D. 1970. *Positively Black.* Englewood Cliffs, New Jersey: Prentice-Hall.

Allen, Barbara and Lynwood Montell. 1981. *From Memory to History: Using Oral Sources in Local Historical Research.* Nashville, Tennessee: The American Association for State and Local History.

Allport, Gordon. 1942. *The Use of Personal Documents in Psychological Science.* New York: Social Science Research Council.

Anderson, Ernest Frederick. 1976. "The Development of Leadership and Organization Building in the Black Community of Los Angeles from 1900 through World War II" (dissertation, School of Social Work): University of Southern California.

Bell, Roseann P., Bettye J. Parker and Beverly Amy-Sheftall. 1979. *Sturdy Black Bridges.* New York: Anchor Books.

Blumer, H. 1939. *An Appraisal of Thomas and Znaniecke's The Polish Peasant in Europe and America: Critiques of Research in the Social Sciences.* New York: Social Science Research Council.

Conklin, Nancy Faires, Brenda McCallum and Marcia Wade. 1983. *The Culture of Southern Black Women: Approaches and Materials.* Birmingham: University of Alabama Press.

Darling, Marsha J. 1987. "The Disinherited As Source: Rural Black Women's Memories." *Michigan Quarterly Review* 26:48.

Davie, Maurice. 1949. *Negroes in American Society.* New York: McGraw-Hill Book Co., Inc.

Davis, Angela. 1981. "Reflections on the Black Woman's Role in the Community of Slaves." Pp. 1–14 in *The Black Scholar 12.*

Davis, Angela. 1981. *Women, Race and Class.* New York: Random House.

DuBois, W.E.B., ed. 1969. *Atlantic University Publications* (A Collection of Proceedings of the Atlanta University Conferences from 1893–1913). New York: Arno.

DuBois, W.E.B. 1920. *Darkwater.* New York: Harcourt, Brace and Howe.

DuBois, W.E.B. 1964 (c. 1903). *The Souls of Black Folk.* New York: Arno.

Dundes, Alan. 1968. "The Native Speaks for Himself." P. 440 in *Every Man His Way: Readings in Cultural Anthropology,* edited by Alan Dundes. Englewood Cliffs, New Jersey: Prentice-Hall.

Edmonds, Helen G. 1951. *The Negro and Fusion Politics in North Carolina, 1894–1901.* North Carolina: Chapel Hill.

Franklin, John H. 1947. *From Slavery to Freedom: History of Negro America.* New York: A. Knopf.

Frazier, E. Franklin. 1949. *The Negro Family in the United States.* Chicago: University of Chicago Press.

Goldstein, Kenneth S. 1964. *A Guide for Field Workers in Folklore.* Hatboro, Pennsylvania: The American Folklore Society-Folklore Associates, Inc.

Gottschalk, Louis, C. Kluckhohn and Robert Angell. 1945. *The Use of Personal Documents in History, Anthropology, and Sociology.* New York: Social Science Research Council Bulletin.

Gwaltney, John Langston. 1980. *Drylongso: A Self Portrait of Black America.* New York: Vintage Books.

Hill, Robert. 1972. *The Strength of Black Families.* New York: Emerson-Hall.

Hurston, Zora Neale. 1970. *Mules and Men.* New York: Harper and Row.

Hurston, Zora Neale. 1978. *Their Eyes Were Watching God.* Chicago: University of Illinois Press.

Iscoe, Ira. 1964. "The Developmental Approach." Pp. 451–460 in *Handbook of Marriage and the Family,* edited by H.T. Christensen. Chicago: Rand McNally.

Jackson, J.J. 1975. "Aged Negroes and Their Cultural Departures from Stereotypes of Rural Urban Differences," in *Aging Black Woman,* edited by J.J. Jackson. Washington, D.C.: National Caucus on the Black Aged.

Jones, Bessie and Bess Lomax Hawes. 1972. *Step It Down.* New York and London: Harper and Row.

Kennedy, Theodore R. 1980. *You Gotta Deal With It: Black Family Relations in a Southern Community.* New York: Oxford University Press.

Ladner, Joyce. 1971. *Tomorrow's Tomorrow.* Garden City, New York: Doubleday.

Langness, L.L. 1965. *The Life History in Anthropological Science.* New York: Holt, Rinehart and Winston.

Lerner, Gerder, ed. 1973. *Black Women in White America: A Documentary History.* New York: Vintage Books.

Logan, Frenise. 1964. *The Negro in North Carolina, 1876–1894.* North Carolina: Chapel Hill.

Mandelbaum, D.G. 1978. "The Study of Life History: Gandhi." *Current Anthropology* 14:178.

Meier, August and Elliott. 1966. *From Plantation to Ghetto.* New York: Hill and Wang.

Miller, Delbert C. 1969. *Handbook of Research.* New York: David McKay Co., Inc.

Morgan, Kathryn L. 1980. *Children of Strangers: The Stories of a Black Family.* Philadelphia: Temple University Press.

Noble, Jeanne. 1978. *Beautiful, Also, Are the Souls of My Black Sisters: A History of the Black Woman in America.* Englewood Cliffs, New Jersey: Prentice-Hall, Inc.

Plumpp, Sterling. 1972. *Black Rituals.* Chicago: Third World Press.

Robinson, B.J. 1988 (c. 1982). *Aunt [Añt] Phyllis.* Oakland: Regent Press.

Staples, Robert. 1973. *The Black Woman in America.* Chicago: Nelson-Hall Publishers.

Starling, Marion Wilson, 1981. *The Slave Narrative: Its Place in American History.* Boston: G.K. Hall and Co.

Titon, J.T. 1980. "The Life Story." *Journal of American Folklore.* 93:281.

Westkott, Marcia. 1979. "Feminist Criticism of the Social Sciences." *Harvard Educational Review* 49:422–430.

Winegarten, Ruthe, ed. 1983. *I Am Annie Mae: An Extraordinary Woman in Her Own Words.* Texas: Garden Press.

DISCOVERING THE WORLD OF TWENTIETH CENTURY TRADE UNION WAITRESSES IN THE WEST

A NASCENT ANALYSIS OF WORKING CLASS WOMEN'S MEANINGS OF SELF AND WORK.

Gail S. Livings

INTRODUCTION

INTERVIEWER: Would you consider yourself as being a union activist?
SWEET: That's all I know. My home, my union, and my church.
INTERVIEWER: How would you define a union activist?
SWEET: Well, one who really is with heart and soul with the union, and tries to sell the idea to others. A union activist is someone who is getting out and battling a bit for a cause they believe in. I am not a battler, I like to be a persuader.

Current Perspectives on Aging and the Life Cycle,
Volume 3, pages 141–174.
Copyright © 1989 by JAI Press Inc.
All rights of reproduction in any form reserved.
ISBN 0-89232-739-1

INTERVIEWER: What would you say is the most exciting part of your life?
SWEET: My work within the unions!

(Sweet, 1978, pp. 44,47)

The views being expressed by waitress union leader, Gertrude Sweet, indicate the significant impact of her work for her sense of self. Oral history interviews with several former leaders in the Waitresses Unions across the western United States reveal similar perceptions of the meaning of their lifelong work as trade unionists. In connection with the recent resurgence of feminist scholarship on working class women, this paper attempts to expand efforts to understand the formation of consciousness among wage-earning women in a period of industrialization and expansion of women's paid employment. Moreover, this paper contributes to an examination of working class women's resolutions of the conflict between societal attitudes toward women (particularly married women) engaging in paid labor and the economic necessity of their own lives.

From 1890 until the 1960s various conceptions of respectable womanhood (whether based on Victorian edicts or the Cult of Domesticity) have excluded wage-earning women. Yet wage-earning women themselves did not acquiesce in this judgment. Nor did they passively accept the harsh conditions which they confronted in their working environment. Their response to traditional social values and changing family and work relations was varied and complex. In her own analysis of working class women's consciousness, Sarah Eisenstein (1983, pp. 5–6) views this cultural domain as:

an arena of conflict, uneasy accomodation and especially, of 'negotiated response' on the part of working women to both dominant and critical feminist, or working class ideologies, in light of their own distinct experience. Working women's attitudes did not simply reflect either the dominant Victorian or alternative ideologies. Rather, they grew out of the women's active response both to new conditions of work and working-class life, and to ideas which were available to them.

One important method for discovering the "negotiated responses" of working class women is through the collection and use of oral histories. Sherna Gluck (1982, pp. 110–11) describes the theoretical importance of insights revealed through working women's oral testimony as:

help(ing) to re-cast the question of how we evaluate social change. The effect of an experience might have reverberations over many years. When we examine the effect of such experiences without this kind of time reference, without reference to the personal arena, and without differentiating among the various kinds of actors, our judgments tend to be superficial and distorted. As a result, we are likely to perpetuate the depiction of women as passive victims, rather than complex actors demonstrating resilience and strength.

Eisenstein (1983, pp. 9–10) argues that this approach avoids the simplifications of two popular historical treatments. The first, which focuses on the survival of the working-class family unit (and its centrality for women), tends to view working-class culture as fully separate, and working-class women as unaffected by dominant social ideologies.[1] An opposing, equally one-sided emphasis is found in the many studies of nineteenth-century Victorian ideology of "woman's sphere" which imply that this dominant mode of thought was the one operative cultural force affecting women of all classes.[2] Unlike these historical approaches, the perspective put forth by Eisenstein, Gluck and others permits a more complex analysis of the relationships between dominant social ideology and subordinate group experience and consciousness in the situation of working-class women.

Moreover, this approach intends to avoid the tendency to reduce the social factors determining working-class women's response to industrialization (and the subsequent rise of the service sector)[3] to a single dimension of what were multifaceted relationships. One such interpretation of working-class behavior in its encounter with industrial capitalism locates sources of radicalism or social cohesion exclusively within pre-industrial immigrant culture and community life.[4] Yet, as Eisenstein (Ibid) points out this approach neglects entirely the socializing function of employment outside the home for the emergence of a distinct sense of identity and ability to act together among working class women. Alternative interpretations treat capitalist work settings as the catalyst of female emancipation from traditional forms of subjugation. However, these interpretations differ as to whether the emancipation that takes place is one which produces a revolutionary class consciousness based on the exploitative nature of the work setting; or, a positive individualism and assimilation of modern work habits that accompany modernization. In contrast to these polarized conceptions, the present study attempts to adhere to Eisenstein's analytic model by specifying the domains of conflict and accommodation between these social forces that influence the social construction of their meanings of self and work.

Eisenstein's dialectical perspective, which informs this study, is closely tied to the examination and critical use of classical social theories of consciousness. This analysis incorporates concepts like the Marxist notion of dynamic structural conditions which generate a socialization of consciousness, with Mannheim's concept of "self-discovery of the social group" that derives from its confrontation with external definitions of group characteristics, and the notion of a "negotiated response" or social construction of meaning on the part of a subordinate class in terms of dominant and radical ideologies.[5]

In this framework, this paper concentrates on the ways in which working women experience and understand their lived experiences; specifically, this analysis attempts to understand the ways in which working-class women dealt with their participation in the labor force throughout the twentieth-century in the United States. As such, the periods from the turn of the century to the end of the First World War, the Great Depression, and World War II, all proved to be critical moments in the development of women's participation in the labor force.

The period from the turn of the century to post-World War II also marked the continued growth of permanent union organization among women workers in a variety of occupations.

This expansion of union activity is important in defining the period in which to investigate working women's consciousness. Union activity in the first place, indicates significant identification with work, at least its salience in one's life. Second, women involved in organizational activity are more likely to have occasion to develop and articulate their ideas, and to answer challenges to their position. More importantly, they are also more likely to leave some record of their consciousness. Yet, while middle- or upper-class women who worked had at least partial ideological reasons for doing so (often reflective of their organizational ties with the women's movement), working-class women often had no choice. So, how did the working-class woman deal with the conflicts between the traditional woman's role and the necessity to engage in paid labor outside the home?

Fortunately, various historical resources have been discovered which yield some answers to this question. Among these historical artifacts are some records left by working women themselves—autobiographies. letters, diaries, speeches, and articles. Most of these were produced by trade union women, and often by the most active among these, or by women who were in at least some contact with union or reform organizations. They were certainly not typical or representative of all working women. Organized women were probably the most likely to be able to react sharply to their situation and to formulate new ideas and orientations. They were not likely, though, to be totally dissimilar in reaction from other women. If their ideas are not strictly typical or representative, they still may be seen as indicative of the direction of development of women's ideas.

The extent to which working women accepted the dominant value system will be determined by analyzing their responses to attacks on working women grounded in that system, and by evaluating the degree to which they felt that the work situation had special implications or presented special problems to them as women, and by looking at their complaints and descriptions of their working conditions. By examining both their complaints and what these women valued about participation in trade unions, their meanings of self and work are revealed in their own words. And finally, by scrutinizing more diffuse expressions about working women's lives, it is possible to detect how new elements of consciousness were informed by their everyday lived experiences.

For all these reasons, this paper chooses to focus on the words and worlds of labor leaders in waitresses unions in the western United States since the turn of the century. Discussion of the attitudes and activities of union waitresses from 1900 to the advent of World War II relies on the scant evidence available from the few secondary sources currently available to researchers.[6] However, for the primary purposes of this paper, the majority of the analysis focuses on oral history testimony from waitress union leaders regarding the period from the

1930s to the present. While there were at least five oral history interviews with women leaders of the Hotel Employees and Restaurant Employees International Union (hereafter known as H.E.R.E.) archived in various locations throughout the midwest and western United States, the majority of the oral history interviews were (and continue to be) collected by the author from 1983 to the present. Previously archived oral histories appear to represent the first generation of women leaders in the Waitresses Unions when most of them were independent of the other male-dominated craft unions in the hotel and restaurant industry during the first half of the twentieth century. These "fellow" craft unions included the Waiters Union, the Cooks Union, the Bartenders Union, the Miscellaneous Workers Union, and the Hotel Service Workers Union (excluding the maids). The recently conducted interviews with female leaders of the H.E.R.E. represent the second generation of waitress union activists. Before proceeding to examine the oral testimony of waitress union leaders, the next section will examine some of the evidence of societal views regarding women who worked as waitresses. From these expressions of the dominant ideology regarding this female occupation, one gets a sense of the struggle women entering this profession faced in terms of their sense of self.

"WOMEN WHO WAIT": THE SOCIAL CONSTRUCTION OF WAITRESS WORK IN THE EARLY 1900s.

In 1910, about 8 million women were listed as wage-earners by the U.S. Census. Nevertheless, working women were the objects of much disapproval and censure. Disapproval of women going to work, and attacks on women for doing so, generally had three basic components. In the first place, women working was seen as inconsistent with the Victorian ideal of maintaining a home and raising children. Secondly, it was argued that women did not really need to work, that they did so only to earn "pin money." Connected to the second argument was the third contention, that women competed unfairly in the labor market, bringing down wages and lowering conditions in the shop. Numerous feminist historians have presented ample evidence to refute, or at least qualify, all of these arguments. With specific regard to waitresses, it will be shown that each of these arguments was without any merit in their case. Nevertheless, societal disapproval was difficult to reconcile with their working class reality and tended to make some women ashamed of working at all, and often defensive about their situation.

As a lingering tenet of Victorian moral standards for women, the aspect of employment which was most distressing was the degree to which it exposed them to improper sexual advances or to personal abuse or insult. Waitresses, because of their contact with customers, were particularly susceptible to uninvited approaches. As a consequence, some waitresses even expressed contempt for the tipping system because of what they saw as the connection between the personal relationship it implied with a customer and insulting, unwanted advances:

> girls have to live on tips . . . you have to put up with it or starve. The majority of girls . . . if they could get a good living would be glad to do without tips. . . Girls in restaurants have greater temptations than most girls. Advances are always made, especially in certain districts. A great number go wrong because of so many advances (Consumers League of New York, 1916, p. 36).

Throughout her description of the working experiences of waitresses in turn of the century Chicago, Frances Donovan notes the pervasiveness of sexual connotations in the social interactions between waitresses and their customers, or even their co-workers. In a description of the relationship between the waitresses and the male employees, Donovan (1920, p. 80) notes:

> In the kitchen the same atmosphere prevailed. A man could not cut a piece of meat for a girl without making a filthy joke about it or making a suggestive movement towards her. There was never any open violation of the proprieties but always the suggestive talk and behavior.

And with regards to the customers, Donovan (1920, p. 68) recalls getting this advice:

> 'You can't get along in any kind of restaurant,' said a girl to me, 'unless you jolly the customer.' And certainly the customers look for the jolly.

As a result of their concern over the decadent character of waitress work, the Consumers League of New York (1916, p. 7) was moved to make this observation:

> The moral danger of the work is largely confined to waitresses. Because of their position, they are peculiarly exposed to the attentions of men customers. For this very reason, the Baltimore Vice Commission recommended that only older and more experienced women be employed in this capacity, while in Norway the law sets a minimum age limit for waitresses in public places.

As a consequence of the prevalent social devaluation of waitress work, Donovan (1920, p. 130–31) offered the following incident to demonstrate the effect of the low status of their work on their self-esteem:

> The work of the waitress does not rank very high in the occupational scale. The waitress herself is ashamed of her job, and tries to conceal from her friends that she is a waitress. 'What's the use of letting everybody know that you are a hasher?' she will say.

As Eisenstein (1983, p. 68) points out in her discussion of the impact of Victorian philosophy on social values, working outside the home meant that a woman was unprotected, subject to close social and even physical contact with strange men, and to the orders of men who were not her father or husband.

Hence, service work in a restaurant left a woman open to improper advances, and put her in a position of serving the personal needs of men with whom she did not have properly sanctioned relationships. This preoccupation with the danger of falling into impropriety and sexual irregularity constituted the most threatening aspect of the working world for women during this period. The traditional Victorian guarantees of a ''lady's'' respectability had been the secluded and well-ordered nature of her social contacts. In fact, the *appearance* of respectability was almost as important as one's actual adherence to conventional morality, and women going out to earn a living were carefully advised to guard it.[18] Consequently, some occupations—like waitressing—were out of the question because they made it difficult to conform to conventional moral standards.

Yet, by the 1920s the relaxing of Victorian morality and the advent of Prohibition improved the pre-World War I social status of waitress work. In addition, the increasing numbers of native-born Anglo women (vs. immigrant women of color) working in the restaurant industry upgraded the societal value placed on waitress work. Furthermore, the economic crisis and labor movement of the 1930s continued to relax the social stigma attached to women working—even for married women if they could show ''just cause,'' not a difficult claim to make during the Depression. Specifically, Susan Ware (1982, p. 21) argues that:

Contrary to widely held views, women workers were not devastated by the Depression. Although American society was often blatantly hostile to the idea of women working in the 1930s, women fared better than men during the Depression in many cases. Moreover, women workers benefited from the new federal relief programs and the dramatic growth of unions in the 1930s.

Her position is supported by the work of Alice Kessler-Harris, Ruth Milkman, and others who stress the sexual division of labor and highly sex-segregated nature of the occupational structure during this period. As an ongoing process of sex-typing jobs, women fared better because the occupations in which they were concentrated (secondary labor market; e.g., clerical, trade, and service occupations) contracted less than those in which men were concentrated (primary labor market; e.g., manufacturing) (Milkman, 1980). However, significant sex differences did exist in the degree of successes gained through the New Deal legislation and the labor movement for particular occupations—namely, domestic service, agricultural work, and professional service (like waitressing). The oral history interviews with waitress union leaders provides occupationally-specific evidence of these claims for the differential effects of the events of the 1930s on working women. However, before moving on to examine the testimony of these women, a brief look at the origins of union activity among waitresses will be presented in the following section.

EARLY TRADE UNION ACTIVITY AMONG WAITRESSES IN THE WEST: 1900–50.

The U.S. labor movement was just beginning when union records show the first local union of workers in the catering trade was formed in Chicago in 1866 (Josephson, 1956). It was called the "Bartenders and Waiters Union, Chicago," and later, Local 57, when it affiliated with the city's trade assembly, and the National Labor Union (one of the early attempts at a federation of trade unions). According to Josephson (1956, p. 23), its members were all Germans who had absorbed unionist lessons in Europe before immigrating to the U.S. Generally, nineteenth century America had few "great hotels" that could serve several hundred lodgers and diners like the old Tremont House in Boston or the Astor House in New York. The surge in this business came during the "railroad age" after the Civil War. The railroad companies helped build hotels that primarily housed men who worked for the railroad companies. In the central quarters of major cities, large restaurants and hotels were being erected to meet the needs of a growing clientele of newly-flourishing businessmen in bludgeoning industries like coal, railroads, steel, banking, etc. At the same time, the Knights of Labor (KOL) had emerged as an important national labor union with a membership of 700,000. In 1886 Chicago Local 57 affiliated with the KOL as District Assembly 7474, KOL, with a membership of mainly waiters and "beer-slingers." Not soon afterwards, local unions were forming in major cities around the country and affiliating with the Knights as among the first workers to give their support to the KOL. However, the 1886 Chicago Haymarket incident, the subsequent decline of the KOL, and Samuel Gompers' efforts to establish the American Federation of Labor, pushed the various local unions into affiliation with the newly-formed American Federation of Labor (hereafter referred to as A.F.L.). Thus, throughout the next few years, numerous waiters and bartenders local unions were forming in the major cities and affiliating with the A.F.L. Not long after this development, the strongest of these locals in New York, Boston, Chicago, and St. Louis, began working toward establishing a national union for the catering trade laborers. On April 24, 1891, an application for a national charter under the name—Waiters and Bartenders National Union—was submitted to A.F.L. President Gompers and was approved. With the addition of hotel employees unions and Canadian locals, the name was changed to a lengthier version in April 1898: The Hotel and Restaurant Employees International Alliance and Bartenders International League.

According to Josephson (1956, p. 35), it was at this time that the International Union first made an attempt to organize woman workers. At the tenth annual convention held in St. Louis in 1901 the first woman delegate, Sister Bertye (Bertha) Greene representing Waitresses Local Union 249 of St. Louis, made her appearance (Josephson, 1956, p. 36). It was noted during the convention that year earlier in Seattle, Washington, fifty waitresses had banded together under the leadership of Alice Lord to form the first waitresses' local union, No. 240. Then in 1903, the International received an encouraging report from the Chicago Waitress Union: "Though but a few months old, we have over 2,000 members and are growing at the rate of 50 to 75 new members at every meeting night" (Josephson, 1956, p. 54). At that time in Chicago, the waiters of the new Local 336 began sporadic strikes which soon involved the six other locals, including the cooks, bartenders, "colored waiters," waitresses and miscellaneous. According to Josephson's (1956, p. 54) report:

> The colored waiters particularly were reported to be in a state of extreme discontent. Another highly militant group, the waitresses, was led by Elizabeth Maloney, one of the most famous women labor leaders of the time in Chicago. Her girls were embittered (according to testimony she gave before a Federal Commission not long afterward) at being paid fixed wages of $3.00 a week for a seven-day week, and fed with wretched food that turned many of them sick).

In 1911, Elizabeth Maloney was elected as vice-president-at-large on the Executive Board representing women workers. Testifying before the U.S. Industrial Commission in Washington in 1914, Maloney declared that, prior to the passage of an Illinois ten-hour law for women, they had been working as waitresses twelve to fourteen hours a day for seven days a week, for wages of $3.50 to $5.00 a week and small tips. In the testimony, she stated:

> I investigated the case of girls where they were poisoned on food that spoiled, and I went into quite a thorough investigation of that and found out from those girls that they received food that was in a decomposed state. These girls were poisoned through eating spoiled roast pork. I went to the room of the girls because they were in bed from those attacks of ptomaine poisoning and found on their window sills boxes of crackers and bottles of milk. I said, 'What does this mean? Don't you get your board at the hotel?' 'Yes, but it is not fit to eat half the time and whatever tips I get I use to buy things to eat (Josephson, 1956, pp. 91–92).

Reflecting the overall resurgence of the labor movement, the pre World War I period saw some of the greatest gains in the membership of the Hotel and Restaurant Employees and Bartenders International Union (HREBIU). According to Josephson,(1956, p. 93), in Chicago a powerful movement to organize women workers, led by Vice-President Elizabeth Maloney, and supported by the Women's Trade Union League, brought an influx of waitresses to the International's Local 484. By January 1914, numerous downtown lunchrooms and restaurants, including one chain system with twenty units, were being picketed by waitresses who demanded $8.00 a week as fixed wages and one day of rest. Though union shop agreements were won in number of cases, the Chicago Restaurant Keepers Association intervened and forced their repudiation. Thus the waitresses' strikes dragged on throughout the winter season in Chicago until May. Court injunctions followed, and numerous arrests were instituted by the

Chicago police. In moving testimony given before a Federal Commission at Washington, Maloney told of:

> . . . two-hundred pound cops using their brute strength upon frail girls, twisting their arms, bruising their bodies . . . In taking the girls into the police wagons they treated them very badly, though they gave no resistance . . . (Josephson, 1956, pp. 102–3).

The International's waitresses' local in Chicago, No. 484 was still small, having only 600 members in 1914; yet as Josephson noted, "these women showed no less courage than any of the thousands of brawny men who marched out on strike up and down the land in the turbulent year that brought the opening of World War I (1956, pp. 102–3). Gradually the wage level for women workers was improved and the six-day week won, such that the strike action of waitresses local 484 was instrumental in setting the wage standard for thousands of non-union people working the restaurants of the Loop District.

Receiving its first charter in 1902 and after much fluctuation in organization, Waitresses Union Local No. 249 of St. Louis finally received their charter from the International in 1910 and began trying to improve the $6.60 for an eight-hour, six-days per week, wage. By 1915, they were strong enough to demand and get increases which raised the wage scale to $12.50 a week for eight-hour days. However, where WW-I ended, St. Louis hotels and restaurants tried to reduce wages to pre-war levels and a general strike and lockout followed which lasted from October 1920 to February 21, 1921 (Rubin and Obermeier, 1943, p. 168). According to Rubin and Obermeier (Ibid), in the beginning of the strike, the St. Louis police (then under a corrupt city government) handled the waiters' union pickets with the utmost brutality, so that only the line of women pickets was left to carry on the strike. Even though the "chivalrous" policemen did not generally treat the women as roughly, nevertheless many waitresses were arrested for alleged disorderly conduct (Josephson, 1956, p. 137). As Kitty Amsler, veteran secretary of Local 249, related:

> A goodly number of the pickets had little to eat. That year, almost from the time the strike started in late autumn, the city was swept by snow and sleet. Day after day passersby saw the waitresses in their threadbare coats, sloshing along. We had to beg for money from all sorts of unions and other organizations (Josephson, 1956, p. 137).

It was reported that International Secretary Sullivan sent some sums, but the defense funds had been greatly reduced by numerous strikes elsewhere. But some other St. Louis unions asserted themselves to help the strikers, the AFL's streetcar workers' local giving one of the memorable displays of solidarity. One of their officers had seen how the waitresses continued their picketing during a blizzard, when the trolley-cars themselves were stalled. Soon he came from his union meeting to their headquarters and said, "Get the girls some shoes and something to eat,'' and delivered a check for $1,000. After a four month endurance test the waitresses returned to work in the restaurants in February 1921 at the same wages they had earned before the strike. When International President Hugo Ernst reviewed the case in 1923, he concluded that the strike should have been avoided, that the leaders had overestimated their own organizations and underestimated the Restaurant Association. Local 20, formerly with a membership of nearly 500, was soon down to 75; the cooks were virtually wiped out; and the waitresses' local 249 lost the majority of its members (Josephson, 1956, p. 138). These unions were completely shut out of the St. Louis hotels for the next 17 years—which speaks for the effectiveness of the anti-labor drive in that period.

As a result of this situation, president Ernst went on a national tour of the union locals to assess the situation for himself. After the debacle in St. Louis, Ernst's speeches carried the following message:

> . . . there was no cooperation between the waiters and waitresses, such as had existed in San Francisco. They seem to have inaugurated a system of catch-as-catch-can, and when the girls organize a house the boys try to take it away from them, and vice-versa. The consequence is that no one has anything (Josephson, 1956, p. 141).

Moreover, he lamented the lack of organization of women and minority workers who were being hired instead of union workers in St. Louis, Cincinnati, Washington, D.C., and New York. A prime example of this situation produced the 1925 strike by Local 30 in San Francisco against three leading hotels. In the midst of wage negotiations, the Palace Hotel suddenly discharged 30 union waiters and replaced them with 40 non-union waitresses. Ernst noted that much resentment existed during this time between waiters and waitresses whose local unions had developed slowly over the years. The women workers were given lower wages and expected to work fewer tables than the men. At the same time, the men feared and resented their periodic replacement by women, especially non-union women.

Later at the 1927 national convention in Portland, several women leaders were in noticeable attendance. Even though International Vice-President Elizabeth Maloney was not present due to ill health, Kitty Donnelly of Cleveland, Alice Lord of Seattle, and Kitty Amsler of St. Louis were there to speak on behalf of a resolution to establish equal rights and equal pay for women workers (Josephson, 1956, p. 159). However, union action on this measure of importance to the women members was deferred for further study. The adoption of a policy of equal wage rates for women was to wait ten years more, until the New Deal had come for labor in 1937.

After World War I and several years of decline, the HREBU finally seemed to have become one of America's major labor unions. In 1935–36 it exerted its economic power by conducting strikes and organizing drives in 322 cities of 39

states and several provinces of Canada. Josephson (1956, p. 225) describes this
period in the history of the HREBU in the following way:

> The old-time barmen . . . would have rubbed their eyes at the number of women officers
> present, such as Gertrude Sweet of Portland, Pauline Newman of Seattle, Ida Peterson of
> Bellingham, Washington, Bea Tumber and Mae Stoneman of Los Angeles, Kitty Amsler of
> St. Louis, Theresa Peterson of Kansas City, and Myra Wolfgang of Detroit.

According to Pitrone (1980), Myra Wolfgang gave a speech at the 1938
convention that was instrumental in shaping the sentiments of the delegates on
the issue of the industrial unionists. The Detroit waitresses had been involved in
a number of "sit-in" or "sit-down" strikes, their victory, as Wolfgang reported,
having been gained "mainly because of the fine support given by pickets of the
United Automobile Workers, CIO" (Ibid). The same year, 1937, that the Detroit
strike was going on, 5,000 hotel workers in St. Louis, headed by Kitty Amsler of
the waitress local went out on strike. According to Josephson (1956, p. 226),
within a few days the HREBU managed to regain all the ground that had been
lost seventeen years before in the disastrous lockout struggle of 1920.

During the post World War II period, a new "wave of strikes," like that of
1936–37 swept over the nation. Prices had been greatly inflated, the dollar had
fallen to about $.58 of its 1939 purchasing power, and wages had lagged far
behind (Josephson, 1956, p. 301). Strikes for wages, working conditions, or
union shop were large and numerous, but during this period of inflation employers were less resistant to pay raises and union demands were largely being
met. The hotel and restaurant trade, partially restricted by World War II, under-
went rapid growth in response to the changing eating patterns of the time. A
breakdown of membership statistics of the 1946–47 organizing campaign in six
cities—Chicago, Pittsburgh, Detroit, Cleveland, St. Louis, and Los Angeles—
shows the heaviest influx of new members came from (1) waitresses; (2) cooks
and kitchen assistants and; (3) hotel and service workers; thus from the groups
that had not been vigorously sought in previous organizing drives (Josephson,
1956, p. 308). After the "great push" of 1946, membership tended to become
stabilized at a level of 400,000 to 415,000. But in the years preceding the Korean
War, many powerful unions which had risen rapidly, failed to hold their ground
in the midst of adverse labor legislation at the state and Federal level (Josephson,
1956, p. 332). As a result of the great influx of women workers into the union
during the post World War II period, their numbers now constituted 45 percent of
the labor force (Ibid). Josephson (1956, p. 333) offered the following comments
on this situation:

> Fortunately, the problem of educating women workers in trade union principles was met by
> the exertions of outstanding women union officers. . . . The contribution of various women
> leaders in different sections of the country to the success of the union would require an
> extended study in itself.

TWENTIETH CENTURY UNION WAITRESSES IN THE WEST: THE WORDS OF WORKING WOMEN

Getting Into Waitress Work—

Who were these Western working class women that decided to wait tables at a time when societal values denigrated the moral character of waitresses? How did these young women reconcile the need to earn a living under circumstances of little choice with their belief in the work ethic, Christian morals, and Victorian standards of womanhood? To begin to understand the attitudes and behavior of young waitresses in the early 1900s requires an examination of their family backgrounds. For this first generation of union waitresses, their decision to enter the labor force during the late 1920s-early 1930s was clearly one of economic necessity. For nearly all of these women of first generation immigrant families from Northern Europe, the Great Depression meant that the traditional ''bread-winner'' role played by their fathers was no longer viable under these economic circumstances. In at least two cases, their fathers, having finally resorted to working in the mines, were killed in mining accidents (Metro, 1978, p. 12; Randazza, 1986). For most of the others, their fathers' valiant attempts at farming in the Midwest were abandoned during the Dust Bowl years. For all of these women (with the exception of Myra Wolfgang of Detroit, a second generation Russian Jewish immigrant), the result was to join the massive migration of this period westward in hopes of better economic conditions. Some of these young women undertook this migration with their families, others with their new husbands, and others made the trek alone.

After her mother left her father because ''she just couldn't take it (his ''stinginess'') anymore,'' Jackie Walsh (1980, p. 32) describes how she and her husband decided to move from Kansas to San Francisco in 1936:

KENDALL: Why did you move to San Francisco?
WALSH: It had always been a desire of mine to come to the West Coast. I guess the opportunity arose . . .
KENDALL: You're the one that initiated the move then?
WALSH: Oh yes.
KENDALL: And your husband went along, didn't object?
WALSH: No.
KENDALL: This was the middle of the Depression. Did you feel he could get a job out there?
WALSH: When we came to California he did a variety of things.

Once these young women, most of whom were married, had settled in the Western United States, the continuing economic hardship that had characterized most of their lives remained and required that they return to the labor market. As teenagers working to supplement the family wage, most of these young women had gained their first experiences working in a restaurant, soda fountain, or hotel. Insofar as the economic survival of their families depended on their financial contribution, any social stigma attached to young women working was mitigated by economic necessity and a strong belief in "hard, honest work."

Fortunately for these women, societal views of waitress work were being upgraded by the changing ethnic composition of the occupation. During World War I, the influx of young, native white women into numerous occupations previously dominated by first generation immigrants (especially women of color) greatly modified the prior Victorian conceptions of "the lady." As the career manuals of this period indicated, there was no longer so clear a distinction between women who are ladies and those who are not, but there were explicit differentiations among "classes" and "types" of working women. Now, the "character" which a woman brings to her work becomes the important focus of distinction. As one manual put it, "Girls bring their character to the work that they do, 'high class' girls elevating it, and the 'worst type' debasing it." As such, the manual describes the women who became waitresses during World War I as follows:

> The type of women who have rallied to this work is distinctly a high class and their worth is appreciated . . . Girls of this type naturally raise the standing of any occupation, and if more girls of her calibre tackle the job of waitress it will attain a real standing in the world of work (Hoerle and Saltzberg, 1919, p. 42).

In spite of the fact that less than half of these women's mothers worked outside of the home (although, some did "take in work" such as sewing or childcare to supplement the family wage), none of these women indicated that they had ever expected to not spend some time in the labor market (a few did express the expectation of not working after marriage or while raising children). In keeping with the prevalent views on proper work for women of that era, most of the women said that their initial career goals were to become school teachers. Although perhaps more indicative of things to come, was Jackie Walsh's high school desire to become a woman lecturer. After an inspirational guest lecturer visited her high school, Walsh (1980, p. 15) explained his effect on her life in this way:

> WALSH: And the only thing that I can recall, when I grew up and was in high school, that I used to think, I thought I would like to become a woman lecturer . . . For some reason, in my naive mind as a kid, I thought I just have something that I can bring about to help people if I could just talk to them and help explain things to them.

For Gertrude Sweet (1978, p. 4) growing up near Napa, California, her childhood dreams of a profession were a reflection of her very religious father's intention to send her to a "church college,"

SWEET: My obsession seemed to be in the medical field as a medical missionary as I recall it now, . . .

The significance of these childhood visions of desirable careers for women reflect not only the prevailing societal attitudes regarding those few acceptable occupations for young, unmarried women; but, also, the coherent expression of the dominant values of their parents. Perhaps the most consistent work attitude shared by all these women reflected their parents' beliefs in "hard, honest work" and caring for others. Whether their views were grounded in orthodox religious precepts, some form of secular humanism, or simply the Protestant work ethic, all of these women sought to exemplify these values and beliefs through their work. Hence, their original intentions to enter the fields of teaching or medicine with their parents' moral support and hope to send them to college to accomplish this. However, the economic reality of their lives forced them to compromise their goals of honest service to humankind, and take advantage of the opportunity for gainful employment as waitresses. As a consequence, the decision to engage in waitress work can be seen as a "negotiated response" to the various competing definitions of their selves and their work; as a waitress explains in the following:

> COMPTON: I was married seven years and their father left. . . . Well I went to work in Aberdeen as a waitress and I had no experience—I hadn't even eaten in restaurants. I must have been a terrible waitress. I was friendly; it was kind of interesting to me to have different people come in . . .
> INTERVIEWER: How did you happen to choose a restaurant to work in, to begin with?
> COMPTON: I had to, it was just the quickest way to make, to earn money . . .
> INTERVIEWER: 1929. And did you make tips, besides?
> COMPTON: I never made tips. You know, maybe thirty or forty cents in a day, but tips embarassed me.
>
> (Compton, 1978, p. 18)

Yet, not all women who went into waitress work in those days of economic deprivation would resist the opportunity to supplement their incomes through prostitution with their customers. Working in a logging camp near Portland, Oregon, Beulah Compton (1978, p. 19) describes her first encounter with the "moonlighting" activities of some of her sister waitresses:

> COMPTON: Yeah, I saw some of the girls; I made uniforms for them, in between jobs. And quite a few of them were working—I didn't realize it at the time—through their apartments. I wouldn't have recognized a house of prostitution if it had a sign on it. But I just thought that they had very gaudy tastes: lamps with all these beaded fringe on them and sofas all over the place with lots of pillows and incense burning. But they were very nice people, and they were very protective of me; they wouldn't have let me know what they were doing. But what I found out later was that what many of the waitresses did in Aberdeen was work as waitresses for a front and then . . .

What emerges from the course of these particular women's lives as waitresses is that none of them were completely satisfied with the working conditions in the restaurant industry at the time. Their descriptions of the hard work, long hours, lack of time-off or vacations, and minimal wages reveals the state of most all working-class occupations prior to the labor movement of the 1930s. As indicated in the previous section, during this period waitresses' unions were just beginning to establish themselves in major urban areas—some independently, others in conjunction with Waiters Unions. The next section examines these waitresses' decisions to become involved in their union and what meaning this activity had for their self-identity.

Becoming an Active Trade Unionist

In her dissertation, *Women in Trade Unions in San Francisco,* Lillian Matthews (1913, p. 78, 81) makes the following assessment of the effects of union organization among waitresses:

> In 1906 the waitresses formed an organization separate from the men, as they felt their problems were now distinct in many ways. Since the separation into a distinct union the waitresses have steadily added to their numbers and their influence. The waitresses union presents an interesting study of accomplishment. Its problems have been and are peculiarly difficult. It has moulded together a class of workers who are notably hard to weld. Unionism has taken this drifting, uncertain class and organized them into a body which has succeeded in establishing the occupation upon the basis of a standardized trade commanding a recognized wage and hours. Upon such a basis the work itself can command a respect which has been withheld, we should say unjustly, and with this much done, the union is in a fair position to solve some of the other problems which make life hard and said for the girl who must seek her living in a none too generous nor sympathetic society.

Thus, by the 1930s, the changing ethnic composition of the waitress occupation, combined with the effects of widespread union organization greatly improved the social status of the waitress—and the union waitress in particular. As such, involvement in the activities of the union came to be seen as an avenue of upward social mobility for industrious waitresses. Significantly, the upward mobility was not measured in terms of financial gains; in fact some waitresses would have suffered an economic loss trading their tips for straight union wages. Rather, the advancement was in terms of the status associated with being a business agent (B.A.) or elected union official—both of which provided leadership opportunities. Moreover, the position of B.A. or officer offered the chance to enact those childhood aspirations to teach and care for others. In their roles as union leaders, these women were best able to achieve their original career goals in a way that didn't compromise their values of hard, honest work in service to others. Furthermore, their position as trade unionist allowed them to maintain their standards of "lady-like" behavior—for themselves and their "girls."

From the following interview excerpts the initial view of waitresses upon joining the union indicates some variation in their prior knowledge of trade unions:

> COMPTON: Yes, that's where I joined the union, in Aberdeen. I'd never heard of the union. And the way I joined, one of the girls asked me why I wasn't a union member. So I wanted to know what a union was and she told me; and very sweetly said they wouldn't work with a nonunion girl. So I just marched myself down to the union and joined . . . I've always loved those spirited girls giving me the word (Compton, 1978, p. 19).

> WALSH: So somebody said to me, 'Why don't you go talk to the union?' And I went up to the union and filed my application. Then, when the strike was called [1937], I don't know why, but I wasn't working at that point. I just went down and reported in to do my picket duty. And I was initiated after the strike was called. So this was something it just seemed natural for me to do. Then after the strike . . . then they wanted me to come to work in the office (Walsh, 1980, p. 35).

> INTERVIEWER: Had you thought about another type of job besides looking for a waitress job?
> SWEET: No, because I had hoped that this would be temporary work until my husband could find employment that would support us. I had not thought of it as being a permanent job. Nevertheless, I wanted to be in the union, and very shortly after I became a member and attended a union meeting, I became the recording secretary. From then on I always held some sort of an office in the union (Sweet, 1978, p. 12)

From the various descriptions of their first involvements with waitresses unions, it appears that the slightest expression of initiative or interest was grounds for their recruitment into the service of the union. Once given the chance to do something extra for their union—whether in the midst of a strike or amid stability—these women distinguished themselves as eager learners, energetic and enthusiastic organizers, and eventually capable leaders of the women (*and men*) of the H.E.R.E. But how did these remarkable women move up the union career ladder to become among the nation's foremost women trade unionists of the first half century? And how did they conceive of their leadership role within the H.E.R.E., the labor movement and the women's movement? In the next section, the answers to these questions will be discussed in the context of their oral history testimony and their activities as trade union leaders.

FROM WESTERN WAITRESS TO NATIONAL TRADE UNION LEADER

Social Determinants of Leadership Among Union Waitresses

For most of these waitresses the path from union member to officer or organizer for the International H.E.R.E. was a gradual progression through the ranks as Business Agent (B.A.) to Local union officer to an official position on the Local Joint Board of the H.E.R.E. and from there to state, regional, or international officer of the H.E.R.E. According to these women, a combination of factors contributed to their rise through the union hierarchy to a position of national leadership in the labor movement. In the first place, they credit specific union officials with noticing their potential and giving them an opportunity for advancement. Their progress in becoming knowledgeable about the union and developing leadership abilities was often based on a kind of mentor-protege system of apprenticeship. Such a process based on personal loyalty created many lifelong relationships, especially later between the first generation of exceptional waitress union leaders and their successors. For many of the first generation waitress leaders, their mentor was quite often a male trade unionist, or occasionally a fellow union waitress with slightly more leadership experience.

Jackie Walsh credits union waitress, Margaret Werth, with recognizing her potential contribution to the union and getting her more involved in union activities. Below Walsh (1980, p. 64) describes how she and Werth finally decided to improve the image of the waitresses union by establishing their independence from the male unionists in the craft:

WALSH: As far as Local #48 was concerned, they were not a respected organization when I became part of the official family to the extent that nobody looked up to them. They were just women and didn't know anything. This was the one thing both Margaret and I wanted to change. They [the other waitress union officials] used to have a man come down and make up the International report. And every time they turned around they would be calling up a man to ask them how to do this, or how to do that. Well Margaret and I didn't do that. And we became very independent and we gained a recognition here in San Francisco that would compare, if not exceed, all the men's organizations in this field. We had to work a little harder at it than the men maybe. The men always wanted you to step down a notch or two. I always said I could do anything a man could, if not better, certainly just as well, and it was only their idea that I couldn't. The proof of that is I was President of the Joint Board from 1951 [the only woman that ever held the presidency] until we disbanded the Joint Board after Local #2 merged.

After a series of appointive jobs and some experience as a B.A., Beulah Compton relates how another B.A., who she greatly admired, influenced her decision to run for elected office:

COMPTON: . . . but there was a business agent that I loved; she was just a beautiful woman, a very strong woman, and she was the head of the union really, . . . And she kept trying to persuade me to come to work as her appointee as a business agent, and I was afraid of it. Some of the other women from the union I really didn't like at all; they were pretty rough and heavy drinkers. And one time, Pauline asked me why I didn't throw my hat in the ring to get elected as a delegate to the convention being held back east. And so I did, and while I was back there, she talked to me and asked me to really seriously consider coming to work for the union . . . And she explained to me that there was some dirty hands around the union and she didn't want the union to fall into them. And she told me she was dying of cancer . . . and she wanted me to get active before her death so that I could guide the affairs of the union.

While working as recording secretary for the union, Gertrude Sweet (1978, p.

12) describes her involvement in helping to get the waitresses union their first charter in October 1921:

> SWEET: We were up until then, a Waiter and Waitresses' Union. Many of the waitresses were not happy with the way the waiters were conducting the business of the union. There were no waitresses serving as officers, we were in the majority, and we decided it was time that we ran our own union. . . . Due to the fact that the waiters often were unruly at the union meeting and two or three times we had to call in police officers because they were not sober and they were fighting on the floor and we had enough of that; so we had our own union.

Recent interviews with several of the second generation of union leaders reveals an even greater emphasis on the mentor-protege system for recruiting, training, and promoting rank-and-file waitresses into positions of leadership. At Local #2 in San Francisco, both Jackie Walsh and Gertrude Sweet provided role models, moral support, as well as instructive guidance to the next cohort of waitress union leaders like Flo Douglass, Jeri Powell, and Lucy Kendall who in turn have offered their cumulative knowledge and experience to the current cadre of President Sherry Chiesa, Director of Research Cindy Young, and others. For Local #11 of Los Angeles, the extent of their influence and authority may have been more local than national, but the strength and ability of waitress leaders like Helen Anderson, Katie Hill, and Ruth Campagne incorporated and encouraged the talents of Dorothy Randazza, Dahlia Dinnocett, Betty Jane Levitt, and others. However, preliminary oral testimony from some of the second and third generation H.E.R.E. women leaders indicates an undercurrent of tension and conflict over organizational philosophies, particularly in light of the 1974 merger of all the craft unions within the hotel and restaurant industry. While it clear that nearly all of the second generation leaders are very critical of the merger, most of the third generation leaders are in favor or indifferent toward the merger issue. While all of the second and third generation leaders are aware that the merger was a decision by the International, no one acknowledges knowing the basis of their decision.[10] Interestingly, some of the disagreements over organizational strategies harken back to the original confrontations between the leadership of the A.F.L., C.I.O., and I.W.W.; i.e., issues of craft vs. industrial orientation. Moreover, the amalgamation of separate hotel and restaurant unions has brought together under one executive board the competing interests of men vs. women, Anglos vs. people of color, "skilled" vs. "unskilled" labor, left vs. right/moderate politics, centralized vs. democratized participation in decision-making, and domestic vs. immigrant (in particular, "undocumented workers"), and so on. A primary focus of future research on this union will concentrate on their efforts to accommodate such a diverse membership.

Once these waitresses succeeded in securing electoral offices in their local unions, maintaining a position of leadership at the local level or pursuing offices at the state, regional, or national level requires substantial support from the rank-and-file as well as other officers on the Joint or Executive Board. The highest official position held by a woman in the H.E.R.E. was International Vice-President which Gertrude Sweet held from 1938 until 1978. In discussing her ascent to this position, Sweet (1978, pp. 14–15, 30) offers the following account:

> SWEET: As President of the Union . . . I thoroughly enjoyed that work. . . . in 1923, a number of the members asked me to run at the next election for the office of Secretary-Treasurer which I did and was elected without any opposition and took office in January 1924. . . . I was then Secretary for fifteen years, during which time at two different elections there was opposition. I was fortunate that the members reelected me by a large majority each time and I continued in that work until Mr. Flore, the President of the International Union appointed me as an organizer in 1938.
> INTERVIEWER: Have you had any subsequent national or international appointments?
> SWEET: No, only as an International organizer and I have been elected as International Vice-President since 1938 without any opposition at succeeding conventions. I have been pretty lucky.

For Jackie Walsh, her rise to leadership within the H.E.R.E. culminated with her 1951 election to President of the San Francisco Local Joint Executive Board which she held until 1975. Her other union activities involved serving as chairwoman of the health and welfare trust funds and, also, as Trustee of the Pension Fund. In 1959 she served on International Union "Tip Committee" which successfully brought about coverage of tip earnings under Social Security benefits. In her interview she brings out another significant factor effecting whether or not a woman becomes highly active as a trade union leader (or in any leadership capacity outside the traditional boundaries of women's activities)—the full support of family and/or friends—

> KENDALL: Could you have married somebody not in the trade union movement at that point?
> WALSH: I don't know. . . . I think any marriage that would not have been in the labor movement would not have been a successful marriage because the labor movement sort of takes over your whole life when you are involved in it (Walsh, 1980: 132).

As a divorced mother of two children, the tremendous amount of traveling required by her job as International Union organizer for the Northwest region could have created serious conflicts in Gertrude Sweet's life. However, she explains that period of her life as follows:

> INTERVIEWER: How did being part of the union affect your private life, especially at first when your childern were so young?
> SWEET: That was a problem, of course. My mother cared for them for awhile, and I had a housekeeper. . . . I just wanted to be there more with them, that's all.
> INTERVIEWER: What are your children's attitudes today towards unions?
> SWEET: They are understanding and very sympathetic. One daughter was a member of the Waitresses' Union for a number of years. . . . They were always very cooperative. I had to be away from home a number of evenings, they understood it. They accepted their responsibility.

INTERVIEWER: How do they feel now about your returning to work?
SWEET: They like it. 'If that's what you want to do, Mom, that's what you should do.'
                                                                            (Sweet, 1978: 30–31)

Perceptions of Union Leadership: Issues of Class, Race and Gender

For these few waitresses who secured a historical role as leaders in their union, what were their perceptions of their official role as representatives of working-class women? The intersection of dominant ideologies, structural constraints, and the material conditions of their everyday lives shaped their definitions of their responsibilities and style of leadership as trade union officials. The history of their activities represents their "negotiated responses" to the competing demands impinging on their various decisions on how to fulfill their role expectations as labor leaders of working-class women.

In order to understand their leadership character demands consideration of their expressed beliefs about the meaning of "good" union leadership. Gertrude Sweet (1978, p. 25) put it this way:

SWEET: It was largely educational. My work really was not organizing as much as it was in training newly elected union members who were elected to offices . . . as to how to handle the work of the union for the benefit of all concerned. . . . and then someone may be elected who simply cannot grasp the work; who doesn't understand what it's all about, who doesn't even bother to read the by-laws of the union or the constitution of the International Union, and simply don't know what their duties are and what they are supposed to do, and the first thing they know, they are into trouble with the members, with employers, or they may be involved in a lawsuit.

Jackie Walsh's perception of being a good union leader comes through in her discussion of her proudest moment in the career as a union leader:

WALSH: I felt very proud of having been given the opportunity to work for the International. But my first love, I have to say, insofar as my work was concerned was the Waitresses' Union. And my years as an officer there . . . and the accomplishments in securing better benefits for women who worked in the industry and to help establish the Waitresses' Union as somebody who could stand on their own two feet and not have to ask anybody for help, that we could do our own work—this is an area that I've always felt very, very proud of in the years that I spent with the Waitresses' Union (Walsh, 1980, p. 172).

From a slightly different angle, Beulah Compton (1978, pp. 41–42) discusses her lack of regret at being ousted from her office with the Local Union in 1956 due to an incident which violated her principles of good leadership:

COMPTON: . . . Then here was a proposition that all the union business agents, and me too, should get a raise. Well that particular year we had failed to get a raise for the waitresses. And when this proposition came up it was just such a surprise to me. The business agents had gone into the executive board in my absence and made this proposal. And the executive board members just assumed that I approved; . . . Well I bucked it. In the union meeting, on the

> basis that we had failed to do anything for our bosses, the waitresses, and we didn't deserve a
> raise. And it was voted down. Well on the way out of the meeting, one of the business agents
> said we'll get you. . . . I had had the bylaws changed so that they could all be elected; I
> thought that was the democratic way to do instead of making appointments. . . . Then they
> decided that they would take over running the union.

From these comments and others made by waitress union leaders, they saw
their primary obligation as representatives of women working in the "pink-collar
ghetto" of low wages, little job security, few health and welfare benefits, and
safety violations was to improve their working conditions. Hence the union, as
well as state and national legislative, records throughout the twentieth century
show the substantial efforts of Waitresses' Union leaders like Gertrude Sweet,
Jackie Walsh, Myra Wolfgang, Lorena Showers, Mae Stoneman and others to
secure such worker benefits for women as: State and Federal Minimum Wage
laws, the Eight-Hour Day and the Six-Day Week, Social Security Benefits for
tips reported on income taxes, Unemployment insurance and Workman's Com-
pensation, maternity leave, the Fair Labor Standards Act, and Protective
legislation.

But what about their obligations to their constituency as women in a sexist and
racist society? Why have they, like so many other women trade unionists, not
participated more fully in the efforts of their sisters in the various women's/
feminist movements? Several contemporary feminist scholars have sought to
better understand the relationship between the labor organizations of working-
class women and the numerous manifestations of liberal/radical/conservative
women's social reform groups.[11] A full explication of this issue cannot be dealt
with fully in the context of this paper. Nevertheless, it is instructive to review a
few of the most indicative comments offered by these Waitresses' Union leaders
on some of the most controversial issues for them and middle-/upper-class feminists; namely, Protective legislation and the Equal Rights Amendment.

In testimony before the U.S. Senate Judiciary Subcommittee hearings on the
Equal Rights Amendment, Myra Wolfgang from Detroit, offered the following
explanation for her opposition despite her reputation as an advocate of women's
rights:

> Discrimination against women exists, but it will not be eradicated by the equal rights amend-
> ment to the constitution. Conversely, the amendment will create a whole new series of
> problems. It is a negative law with no positive provisions to combat discrimination. . . . I am
> concerned with the abolition of state minimum labor standards legislation that apply to
> women only, which would result if the amendment was enacted. . . . State laws that are
> outmoded or discriminatory should be amended or repealed on a case by case basis; the dual
> role of women in our society makes protection for them a necessity. Legislation governing
> hours of work should apply to both men and women but, until that is done, women should not
> be left without such legislation because of the state's failure to act for men. . . . Myra's
> opposition to the whole idea of a women's strike on behalf of ERA evoked an angry response
> from Betty Friedan who promptly began referring to Myra as "Aunt Tom." . . . Myra

attracted media people to Local 705's headquarters by tossing a champagne press party at which, she announced, her Ten Commandments of Women's Liberation would be unveiled. . . . Under the 'shalts' of her commandments, Myra listed, first: 'Thou shalt eliminate the 'double standard' of male supremacy in our society and shun the idea that sameness means equality.' One of the 'shalt nots' was: 'Thou shalt not seek the mirage of instant emancipation of business and professional women at the expense of working women.' (Pitrone, 1980, pp. 176, 170)

When queried on the subject of equal rights for women, Jackie Walsh's (Walsh, 1980, pp. 161–62) response reflected her lifelong view that "with rights comes responsibilities" and women (as well as people of color) have to be ready and willing to accept those responsibilities before they can demand certain rights:

WALSH: I think the equal rights issue has to be analyzed from various areas. I certainly feel that every worker is entitled to the same rights protection that any other worker is; and that it is totally wrong for the employers to hire one sex against the other because he can pay a lower rate of pay . . . most of the problems as far as equal rights are concerned is that women do not stand ready to assume the same kind of responsibility. I certainly feel that there are laws needed to be sure that they receive the same pay for the same job, or the same comparable job. . . . By the same token (and I think maybe this is partly women's fault in days gone by) they haven't been willing to accept responsibility and assume executive positions . . . I think now it is very important and very necessary that the ERA be approved and we can have it become legal. But I don't think getting it is going to end all the issues. I think we are going to still have discrimination . . .

When asked about her support of the Equal Rights Amendment, Gertrude Sweet (1978: 48–49) responded accordingly with:

INTERVIEWER: Do you support the ERA?
SWEET: Yes, if it doesn't mean that the men are going to shove women around, and I have found that there has been so much agitation. . . . And I blame equal rights agitation on ERA to some of that.
INTERVIEWER: Why do you blame . . .
SWEET: Because men didn't ever act like that before until all of this agitation about equal rights for women, and they were O.K. All right, equal rights, I may have a seat and they can stand.
INTERVIEWER: But you do see equal rights to go beyond that kind of an attitude.
SWEET: There used to be more deference for women by far in the public than there is now. I don't say the agitation for equal rights is to blame for it, and I believe in equal rights when it comes to equal pay for work done. . . . I always felt very strongly that women should receive the same pay as men when they work, and we know that even today they do not in many areas and that's why I favor Equal Rights Amendment because it will correct that, but on the other hand it has changed the attitude of the men. They are not as deferential to women as they used to be.

These comments about the ever-controversial issue of equal rights for men and women demonstrate the areas where the attitudes and values of working-class women can come into conflict with the concerns of middle- or upper-class women. Both groups of women's advocates no doubt believe that they are representing their constituencies' best interests. And for these trade union leaders of working-class women, those interests are defined primarily in economic terms—political or social. Their understanding of the double jeopardy characterizing working-class women's lives (or triple jeopardy in the case of working-class women of color) underlies their "negotiated responses" to the competing demands on their role as leaders of working-class women.

CONCLUSIONS

This paper's nascent analysis of working class women's meanings of self and work reveal the myriad ways these women developed their consciousness or understanding of their life situations. Confronted with competing ideologies and economic necessity, these women faced difficult (and highly circumscribed) choices regarding their means of survival during this period of industrialization and expansion of women's employment opportunities in the wage-earning labor force. Sarah Eisenstein's (1983, pp. 5–6) approach to the understanding and analysis of the social order impinging on working women involved the exploration of their "negotiated responses" to the conflictual cultural milieux of their time. Such an approach to understanding women's distinctive experiences contributes to a reconstitution of knowledge necessitated by a basic discontinuity in the dominant academic traditions which have tended to ignore, trivialize and suppress women's perspectives on various subject matter.[12] Reconstructing knowledge to take account of women's perspectives and experiences, therefore, involves discovering the submerged, or taken for granted, consciousness of the practical knowledge of everyday life and connecting it to the dominant social order. But how should this theory-building task be accomplished?

According to Marcia Wescott (1979, pp. 422–30):

> . . . traditional social science assumes a fit between an individual's thought and action, based on the condition of freedom to implement consciousness through direct activity. But within a patriarchal society, only males of a certain race or class have such freedom of choice or activity. Social and political constraints customarily have limited women's freedom; thus in order to adapt to society while retaining their psychological integrity women must simultaneously conform to and oppose the conditions that limit their freedom.

Therefore, in order to understand women in a society that limits their choices, one must recognize the connection between an actor's thoughts (e.g., intentions, motivations, attitudes, etc.) and actions or behavior is, at best, tenuous. As a consequence, studying women's behavior alone provides a partial depiction of their lives, and the missing aspect may be the most interesting and informative. Thus women's sphere of potentially greatest freedom—their consciousness—must be addressed in studies of their lives. Thus, from the point of view of feminist scholarship on women, oral history should involve more than simply gathering accounts from informants:

> Oral history is a basic tool in our efforts to incorporate the previously overlooked lives, activities, and feelings of women into our understanding of the past and of the present. When women speak for themselves, they reveal hidden realities: new experiences and new perspectives emerge that challenge the 'truths' of official accounts and cast doubt upon established theories. Interviews with women can explore private realms . . . to tell us what women actually did instead of what experts thought they did or should have done. Interviews can also tell us how women felt about what they did and can interpret the personal meaning and value of particular activities. They can, but they usually do not (Anderson, Armitage, Jack and Wittner, 1987, p. 104).

These feminist scholars recommend that, first, oral historians should explore emotional and subjective experience as well as facts and activities; and secondly, oral historians should take advantage of the fact that the interview is the one historical document that can ask people what they mean (Ibid, p. 112).

But on what grounds—methodologically—may feminist oral historians proceed to reconstruct theoretical accounts of society by including women's subjective experience as documented in their oral histories? The theoretical tradition of treating human actors as active subjects whose interpretations and perspectives inform and organize their courses of action began with the "verstehen" approach of social theorist Max Weber. This perspective was to culminate in the "Chicago school of sociology" characterized by its incorporation of anthropological field techniques and its use of life histories. These sociologists have been called, or call themselves, symbolic interactionists, naturalists, or humanists. No doubt, it is within the context of their interactionist perspective that Eisenstein conceives of the "negotiated response" to the conflictual or contradictory demands of the social order. Interactionists treat the social organization as a negotiated order emerging out of social interaction. Hence, social organization is seen as the culmination of these processes of intentional action, as actors collectively and individually attempt to accomplish daily activities. As a consequence, sociologists such as Erving Goffman, Howard Becker, and others have focused upon how people actively dealt with situational contingencies and resolved the problems they encountered as they pursued their goals, using and changing these situations for their own purposes as well as adapting to them.[13]

Like the anthropologist, the method of choice for the interactionist is the ethnography, often referred to as participant-observation or fieldwork. In Michael Agar's (1987, p. 210) terms,

> Ethnography is the process of showing social action in the context of one tradition to be coherent from the point of view of another . . . in ethnography, the raw material for the exercise are 'strips,' a concept introduced by Erving Goffman, used by Charles Frake, and one the usefully characterizes ethnographic data.

Following up on Goffman's (1974) use of the film strip analogy in his work on "frame analysis," Agar (1987, p. 36) defines strips as any "bonded phenomenon against which an ethnographer tests his/her understanding." As such, strips might include fieldnotes based on participant-observation in a field setting, historical documents, or interview transcripts collaborated on by the fieldworker and an informant. As a research process, Agar (1980) recommends an ethnographic approach modeled as a "funnel," where one begins with a more or less inductive perspective based on anticoherence—i.e., things are not what they seem. In other words, it is not yet clear what will count as "data" or where those data will lead you. Such openness allows for the collection of a variety of unexpected strips to be combined with a skeptical, anticoherent view of them.

In the context of "the emergence of a more fully articulated and less defensive conception of the nature of ethnographic fieldwork itself," Robert Emerson (1983, p. vii) notes that "two specific themes stand out: first, fieldworkers increasingly understand and carry out their craft as an *interpretive* enterprise; second, fieldworkers have come to devote close attention to the actual *practice* of doing fieldwork." As already demonstrated in Michael Agar's work, Emerson (Ibid) argues that current writings on fieldwork methodology emphasize an increasing sensitivity to the actual practice of fieldwork; i.e., "a concern with the process of 'doing' fieldwork." Throughout the past couple of decades several instructive monographs in fieldwork methods have appeared and can be referred to for the numerous issues and recommendations for how to achieve a more reflective, self-conscious stance toward the fieldwork enterprise.[14]

In addition, Emerson (Ibid) notes that "fieldworkers have increasingly articulated an 'interpretive paradigm' (Wilson, 1970) in which 'facts' and 'data' are understood not as 'objective entities,' but rather as social meanings attributed by social actors—including the fieldworker—in interaction with others." Adherence to an interpretive approach leads to a preoccupation with the various renderings or accounts of social action and interaction. As ethnographers and oral historians, such a focus requires systematic analysis of the available strips to gain an understanding of the past and present lives of our informants.

If the strips are primarily composed of interview transcripts, then recourse to one of several forms of discourse analysis can provide the necessary interpretive understanding of the accounts. For example, some forms of discourse analysis are designed to capture many of the devices that English speakers use to foreground and background information—phonological prominence, pause, interruptions and overlaps, and basic sentence intonation.[15] Michael Agar (1987, p. 212) suggests that discourse analysis applied to transcripts can be used to support "methodological inferences concerning the nature of the interview, symbolic inferences concerning critical biographical areas for the interviewee, and cultural inferences concerning the world in which the interviewee lives." The rudimentary inquiry of the oral history interview transcripts of the trade union waitresses discussed in this paper attempts to develop the grounds for making some symbolic and cultural inferences based on the lived experiences and meanings revealed in these few strips. Namely, symbolic inferences regarding values and meanings of self and work, as well as cultural inferences regarding the "negotiated responses" of working class women in this historically-specific context.

Nevertheless, this emerging study of twentieth century trade union waitresses' lives and perspectives requires further exploration before accomplishing a more complete understanding of their experiences and shared meanings. According to Agar (1987, p. 218), such research demands additional strips, for in ethnography, as well as history, the emphasis is on multiple strips: strips from other moments with an interviewee, from other interviewees, from other situations or settings, and from documents or other artifacts. For the goal of feminist ethnography and life history analysis is the social construction of understandings that account for the patterns across various strips. As a consequence, inferences may be constructed from, and evaluated against varying strips with an emphasis on their convergence into coherent themes of group life.

RECOMMENDATIONS

Yet, beyond a general admonition to collect, compare and contrast multiple strips for their inferential value to the historical or fieldwork venture, there needs to be an expanded notion of Emerson's concern for the "doing" of interpretive studies. In addition to the self-conscious, reflective way ethnographic fieldwork should be performed, a particular sensitivity toward the oral history interview as a communicative experience is required if the greatest potential use and value of such strips is to be accomplished.

Such a methodological concern is exemplified in the work of Eva M. McMahan (1987, p. 188) who admonishes oral historians to take Gadamer's concept of language into account when interpreting these transcripts:

> For Gadamer, language or 'languaging' is an mode of be-ing, and languaging is fundamentally conversational or dialectical in nature . . . Language is not a tool for re-presenting the world or for re-presenting some historical event. Rather, language is the means by which the account of the historical event is created. Speech and counterspeech, or the working out of understanding between historian and informant, are paradigmatic of this creative capacity of the process of conversing.

With reference to the work of Gadamer and others in the hermeneutical tradition, McMahan (1987, pp. 188–89) argues that:

> . . . insights into the creative process of oral history can be found by turning to Hans-Georg Gadamer's explication of the 'question-answer structure of the living act of dialogue as the paradigmatic site for the occurence of the constitutive or relevatory dimension of language. . . . In oral history the ideal type of that constitutive or relevatory potential is dialectical questioning or hermeneutical conversation in which 'creative negativity' is exploited by the interlocutors. Gadamer goes on to argue that the dialectical approach of Socrates may serve as a model for this form of communications. Such an approach involves a willingness to lead and to be led through the conversation; it requires that the interactants let each other speak while being guided by the subject matter.[17]

Implicit in the social construction process induced by the oral history interview are the situational identities of the participants, the interpersonal relationship between them, as well as the situational and cultural constraints applicable to the interview process, such as rules of turn-taking and topic selection. Furthermore, all of these elements will remain implicit or taken for granted until either party makes them the focus of the interaction—i.e., makes them problematic. Barbara O'Keefe and Jesse Delia (Giles and St. Clair, 1985, p. 13) observe that:

> The creative, emergent process of interpretive interaction is worked out not only within a set of socio-historically inherited constraints (socio-linguistic rules, definitions of prototypical situations and contexts, etc.) but also within a set of situationally emergent constraints. The individual must create strategies which actualize his intentions, but which do so within the constraints imposed by the contextually constituted definitions given to the situation, self, other relationships, and the focus in interactions.

Such is the complexity of the hermeneutical situation as it reveals itself in the communicative experience of the oral history interview.

As a sociolinguist, McMahan's (1987, pp. 192–95) analysis of oral history interview conversation takes "contextual" constraints—e.g., topical, goalrelated, procedural, structural—into account as well. Thus, in analyzing oral history discourse, McMahan (Ibid) focuses on "propositional and pragmatic (functional) coherence in oral interviews, [where]:

> Propositional coherence refers to the ways in which utterances can be about the same topic or theme. Pragmatic (functional) coherence refers to the relatedness of performatives—utterances that perform actions—and of their effects (perlocutions) . . .

The ways in which the interactants display appreciation for and adherence to the need for coherent, cooperative talk can provide information about how meaning is jointly constructed during the course of the interaction. The method used by McMahan and others to formulate this description of coherent talk is a variation of analytic induction.

Of particular importance to oral historians is the methodological (and thereby inferential) difference between the "challenged" and "unchallenged" interview record. McMahan (1987, p. 201) elucidates three features of talk production that characterize the unchallenged record:

> The first two aspects are related directly to R's (the interviewer) role as neutral elicitor of information. First is R's use of third-turn receipts to move talk forward on a particular topic. The second related aspect is that the third-turn receipts are propositional formulations that simultaneously serve as requests for information, clarification, elaboration, . . . the final feature . . . pertains to R's topic management; that is, R controls the agenda and the tone of the interview. R's control is demonstrated directly by R's granting or refusing E's (interviewee) indirect requests for topic shifts.

In contradistinction to the unchallenged record in which R phrases questions so as to maintain a neutral stance toward the account and the informant, the challenge record represents:

> . . . dialogue that emerges as the interviewer enacts dual roles as report elicitor and as report assessor. In so doing, R sustains E's competence as an informed respondent; at the same time, R not only evaluates E's interpretations but also introduces alternative interpretations for E's commentary (McMahan, 1987, p. 202).

For McMahan, such dialectical questioning, characteristic of the challenged record, is necessary in order to exploit fully the constitutive or relevatory potential of the oral interview form of communication. By focusing on the production of talk within the interview format, discourse analysis in the hermeneutical tradition can illuminate the language-in-use in oral history fieldwork.[18] Yet, many issues remained largely unresolved. For instance, issues concerning the effects of class, gender, race, age and other relevant social categories on speech and counterspeech must also be addressed through systematic, empirical investigations. In so doing, feminists, gerontologists, historians, ethnographers and many others can contribute to a more comprehensive understanding of the social order.

ACKNOWLEDGEMENT

Portions of this manuscript were presented at the Annual Meeting National Oral History Association Conference Long Beach, California, October 1986.

NOTES

1. cf. Jane Humphries, "The Working Class Family, Women's Liberation, and Class Struggle: The Case of Nineteenth Century British History," *Review of Radical Political Economics,* vol. 9, no. 3, 1977, pp. 25–41; and Louise Tilly and Joan Scott, *Women, Work and Family,* New York, Holt, Rinehart and Winston, 1978.

2. For the one (among several) definitive refutation of the contention that nineteenth-century women's lives were overdetermined by Victorian principles, see Dale Spender's *Women of Ideas,* London, Ark Paperbacks, 1983.

3. See Harry Braverman, *Labor and Monopoly Capital,* New York, Monthly Review Press, 1974.

4. See, for example, Herbert Gutman, *Work, Culture and Society in Industrializing America,* New York, Alfred A. Knorf, 1976; and Virginia Yans-McLaughlin, *Family and Community—Italian Immigrants in Buffalo, 1880–1930,* Ithaca, N.Y., Cornell University Press, 1977.

5. cf. Karl Marx and Frederick Engels, "The German Ideology," in Robert C. Tucker (ed), *The Marx-Engels Reader,* second edition, New York, W. W. Norton and Co., 1978, pp. 146–202; Karl Mannheim, *Essays on the Sociology of Knowledge,* London, Routledge and Kegan Paul, 1952; Peter Berger and Thomas Luckmann, *The Social Construction of Reality,* New York, Anchor Books, 1967;

170 GAIL S. LIVINGS

and John Horton, "Order and Conflict Theories of Social Problems as Competing Ideologies," *A.J.S.*, vol. 71, no. 5, 1966, pp. 701–13.

6. Obviously this situation is in the process of being remedied by the author, as well as women unionist currently in the H.E.R.E. Of particular importance is the June 1986 filing of a history dissertation by Sue Cobble at Stanford University. Dr. Cobble's dissertation is the first comprehensive attempt to reconstruct the history of waitresses unions. Unfortunately her dissertation is currently in the process stage at both Stanford University Library and University Microfilms, Inc. of Ann Arbor, Michigan, and will not be available for review for three or four more months.

7. The history of the hotel maids in this amalgamated union represents an interesting opportunity for contrasting their working-class conditions and consciousness to that of their sisters in the industry. However the paucity of organized secondary resources (the state of the primary materials is as yet uncertain) on the hotel union maids surpasses the situation of the union waitresses, and awaits the energy and efforts of this (and hopefully other) researchers in uncovering their socio-historical role in the H.E.R.E.

8. As evidence of this preoccupation with giving advice on the "proper" careers to women intending to enter the labor force are the vast numbers of manuals and monographs that appeared during the late eighteenth-century/early nineteenth-century. Among some of the more well-known were: Ella Rodman Church, *Money-Making for Ladies,* New York, Harper and Brothers, 1882; Helen Churchill Candee, *How Women May Earn A Living,* New York, MacMillan, 1900; Mary Laselle and Katherine E. Wiley, *Vocations for Girls,* New York, Houghton Mifflin, 1913; and Helen C. Hoerle and Florence B. Saltzberg, *The Girl and the Job,* New York, Henry Holt and Co., 1919.

9. It should be noted that availability of Dorothy S. Cobble's Ph.D. dissertation on the history of the waitresses' union in the twentieth century was not available from University Microfilms in Ann Arbor, Michigan at the time of writing this piece. Having since received a copy of it, I must recommend this very scholarly historical account to anyone with an interest in waitresses or working-class women and their unions. However, insofar as it does not devote significant attention to the use or interpretation of oral history interview data (relying instead on union documents as the primary source), I did not hold up publication of this monograph in order to incorporate data or findings into this manuscript.

10. In her book, *Pink-Collar Workers,* Louis K. Howe (1977, pp. 107, 128) interviewed a young waitress in New York City who told her about a ACLU lawsuit that was being filed against eleven of the most expensive restaurants in New York, as well as Local #1 of the Hotel and Restaurant Employees Union. According to Jeanne King, one of the waitresses in the case, she filed the case "after becoming fed up with the outrageous discrimination against women in first-class dining rooms." On July 25, 1976 the *New York Times* announced an agreement between the Labor Department and the Justice Department to begin a major investigation of the internal management and financial affairs of the H.E.R.E. According to one of the Waitresses' Union leaders that I interviewed, the unfavorable findings of this investigation (and fear of future discrimination charges of all kinds) prompted the International Union Executive Board to enforce the merger of all the respective craft unions. In spite of this move, I have been able to uncover subsequent lawsuits brought by various parties on a wide range of discrimination charges. However, in all fairness to the H.E.R.E., all of the union officials with whom I spoke were categorically committed to eliminating all and any traces of discriminatory practices in their Locals at any rate.

11. See for example, Ruth Milkman, *Gender at Work* (Chicago: University of Illinois Press, 1987), and ed. *Women, Work and Protest* (Boston: Routledge and Kegan Paul, 1985); Silvia Walby, *Patriarchy at Work* (Oxford: Polity/Basil Blackwell, 1986); Diana Balser, *Sisterhood and Solidarity* (Boston: South End Press, 1987); Meredith Tax, *The Rising of the Women* (New York: Monthly Review Press, 1980); Alice H. Cook, Val R. Lorwin and Arlene Kaplan Daniels, eds., *Women and Trade Unions in Eleven Industrialized Countries* (Philadelphia: Temple University Press, 1984); and Barbara Drake, *Women in Trade Unions* (London: Virago Press, 1984 [1920]).

12. cf. Marcia Millman and Rosabeth Moss Kanter, *Another Voice: Feminist Perspectives on Social Life and Social Science* (New York: Anchor Press/Doubleday, 1975); Dale Spender, *Men's Studies Modified: The Impact of Feminism on the Academic Disciplines* (Oxford: Pergamon Press, 1981); and *Women of Ideas (And What Men Have Done to Them)* (London: ARK Paperbacks, 1983).

13. Many such studies describe how social order is negotiated on a daily basis by subordinates and superordinates in organizational settings. See for example, Erving Goffman, *Asylums: Essays on the Social Situation of Mental Patients and Other Inmates* (New York: Anchor Press/Doubleday, 1961); Howard Becker, Blanch Geer, Everett Hughes and Anselm Strauss, *Boys in White: Student Culture in Medical School* (New Brunswick, N.J.: Transaction Books, 1983); Peter Berger, *The Human Shape of Work: Studies in the Sociology of Occupations* (New York: MacMillan, 1964); Harold Garfinkel, *Studies in Ethnomethodology* (Oxford: Polity/Basil Blackwell); and Robert Emerson and Melvin Pollner, "Dirty Work Designations: Their Features and Consequences in a Psychiatric Setting," *Social Problems* 23: 243–255 (1976). Their studies of the practical activities involved in the daily production of social life should inform the work of feminist scholars attempting to document and interpret historical evidence of women's everyday lives. In many respects this sensitivity to interpretive and symbolic nature of women's experiences is evidenced in other entries in this volume by Patricia Gumport, Judy Long, and Beverly Robinson.

14. cf. John M. Johnson, *Doing Field Research* (New York: Free Press, 1975); Rosalie H. Wax, *Doing Fieldwork: Warnings and Advice* (Chicago: University of Chicago Press, 1971); Barney G. Glaser and Anselm L. Strauss, *The Discovery of Grounded Theory: Strategies for Qualitative Research* (Chicago: Aldine, 1967); Martyn Hammersley and Paul Atkinson, *Ethnography - Principles in Practice* (New York: Tavistock, 1983); and Jack D. Douglas, *Investigative Social Research: Individual and Team Field Research* (Beverly Hills: Sage, 1976).

15. cf. Gilliam Brown and George Yule, *Discourse Analysis* (Cambridge: Cambridge University Press, 1983); Michael Stubbs, *Discourse Analysis: The Sociolinguistic Analysis of Natural Language* (Chicago: University of Chicago Press, 1983); Ragner Rommetveit, *On Message Structure: A Framework for the Study of Language and Communication* (London: John Wiley and Sons, 1974); Howard Giles and Robert S. Clair, eds., *Recent Advances in Language, Communication and Social Psychology* (London: Lawrence Erlbaum, 1985); and John Heritage and J. Maxwell Atkinson, *Structures of Social Action: Studies in Conversation Analysis* (Cambridge: Cambridge University Press, 1984).

16. cf. Hans-Georg Gadamer, *Truth and Method,* Garrett Barden and John Cumming, eds. (New York: Continuum, 1975); John Stewart, "Speech and Human Being: A Complement to Semiotics," *Quarterly Journal of Speech* 72 (February 1986); Hans Jonas, "Change and Permanence: On the Possibility of Understanding History," *Social Research* 38 (Autumn 1971); and Richard E. Palmer, *Hermeneutics: Interpretation Theory in Schleiermacher, Dilthey, Heidegger, and Gadamer* (Evanston: Northwestern University Press, 1969).

17. This dialogue of interpretations appears to be what Thompson advocates as "flexibility" in life history interviews. See Paul Thompson, "Life Histories and the Analysis of Social Change," in Daniel Bertraux, ed., *Biography and Society: The Life History Approach in the Social Sciences* (Beverly Hills: Sage, 1981), 294.

18. See for example, E. Culpepper Clark, Michael J. Hyde, and Eva M. McMahon, "Communication in the Oral History Interview: Investigating Problems of Interpreting Oral Data," *International Journal of Oral History* 1 (February 1980): 28–40; Michael J. Hyde, "Philosophical Hermeneutics and the Communicative Experience: The Paradigm of Oral History," *Man and World* 13 (1, 1980): 81–98; Eva McMahon, "Communication Dynamics of Hermeneutical Conversation in Oral History Interviews," *Communication Quarterly* 31 (Winter 1983): 3–11; and Ronald J. Grele, "Private Memories and Public Presentation: The Art of Oral History," *Envelopes of Sound: The Art of Oral History,* Ronald J. Grele, ed. (Chicago: Precedent Publications, 2d ed., 1985). For collections of research monographs grounded in this interdisciplinary perspective on oral history analysis,

172 GAIL S. LIVINGS

see *The Oral History Review— Special Issue: Fieldwork in Oral History,* 15 (Spring 1987) and David K. Dunaway and Willa K. Baum, eds., *Oral History— An Interdisciplinary Anthology* (Nashville, Tenn.: the American Association for State and Local History, 1984).

REFERENCES

Agar, Michael. Spring 1987. "Transcript Handling: An Ethnographic Strategy," Pp. 210–222 in *The Oral History Review—Special Issue: Fieldwork in Oral History,* 15. Lexington, KY: Oral History Association.

Anderson, Kathryn, Susan Armitage, Dana Jack and Judith Wittner. Spring 1987. "Beginning Where We Are: Feminist Methodology in Oral History," Pp. 104–130 in *The Oral History Review—Special Issue: Fieldwork in Oral History,* 15. Lexington, KY: Oral History Association.

Cobble, Dorothy Sue. 1986. "Sisters in the Craft: Waitresses and Their Unions in the Twentieth Century." Volumes I and II. Ph.D. dissertation, Stanford University.

Consumers' League of New York. 1916. *Behind the Scenes in the Restaurant—A Study of 1017 Women Restaurant Employees.*

Donovan, Frances. 1920. *The Woman Who Waits.* Boston: Gurham Press.

Drake, Barbara. 1984. *Women in Trade Unions.* London: Virago Press Ltd.

Dunaway, David K. and Willa K. Baum, eds., 1984. *Oral History—An Interdisciplinary Anthology* Nashville, Tenn.: the American Association for State and Local History.

Eisenstein, Sarah. 1983. *Give Us Bread, But Give Us Roses.* London: Routledge and Kegan Paul.

Filene, Catherine. 1935. *Careers for Women.* New York: Houghton Mifflin.

Giles, Howard and Robert St. Clair, eds. 1985. *Recent Advances in Language, Communications and Social Psychology.* London: Lawrence Erlbaum.

Gluck, Sherna. 1982. "Interlude or Change: Women and the World War II Work Experience," Pp. 92–113 in *International Journal of Oral History* 3, no. 2, 92–113.

Hoerle, Helen and Florence Saltzberg. 1919. *The Girl and the Job.* New York: Henry Holt and Co.

Howe, Louise Kapp. 1977. *Pink Collar Workers.* New York: Avon.

Josephson, Matthew. 1956. *Union House, Union Bar.* New York: Random House.

Kessler-Harris, Alice. 1981. *Women Have Always Worked.* Old Westbury, N.Y.: The Feminist Press.

Matthews, Lillian. 1913. *Women in Trade Unions in San Francisco.* (Ph.D. dissertation) Berkeley, CA: University of California Press.

McMahan, Eva M. Spring 1987. "Speech and Counterspeech: Language-in-Use in Oral History Fieldwork," Pp. 185–208 in *The Oral History Review—Special Issue: Fieldwork in Oral History,* 15. Lexington, KY: Oral History Association.

Milkman, Ruth. 1980. "Organizing the Sexual Division of Labor: Historical Perspectives on 'Women's Work' and the American Labor Movement," *Socialist Review,* no. 49.

Pitrone, Jean Maddern, 1980. *Myra: The Life and Times of Myra Wolfgang—Trade Union Leader.* Wyandotte, MI: Calibre Books.

Roberts, Elizabeth. 1984. *A Women's Place: An Oral History of Working-Class Women, 1890–1940.* Oxford: Basil Blackwell.

Rubin, Jay and M. J. Obermeier. 1943. *Growth of a Union: The Life and Times of Edward Flore.* New York: The Historical Union Association.

The Oral History Review—Special Issue: Fieldwork in Oral History, 15 Spring 1987. Lexington, KY: Oral History Association.

Ware, Susan. 1982. *Holding Their Own—American Women in the 1930s.* Boston: Twayne Publishers.

Wertheimer, Barbara. 1977. *We Were There—The Story of Working Women in America.* New York: Pantheon Books.

Wescott, Marcia. November 1979. "Feminist Criticism of the Social Sciences," Pp. 422–430 in *Harvard Educational Review* 49.

Wilson, Thomas P. 1970. "Conceptions of Interaction and Forms of Sociological Explanation." *American Sociological Review* 35: 697–710.

Wolfson, Theresa. 1926. *The Woman Worker and Trade Unions.* New York: International Publications.

"Women's History Issue." 1977. Special Issue of *Oral History,* 5, no. 2.

"Women's Oral History." 1977. Special Issue of *FRONTIERS,* vol. 2, no. 2, summer 1977.

———— *FRONTIERS.* summer 1983. vol. 7, no. 1.

ORAL HISTORY TRANSCRIPTS

Oral history interview with Beulah Compton conducted by Elizabeth Case. Program on Women and Work, Institute of Labor and Industrial Relations, University of Michigan—Wayne State University, Ann Arbor, MI, 1978.

Oral History interview with Bertha Metro conducted by Lucy Kendall. Women in California Collection, California History Society Library—Manuscript Department, San Francisco, CA, 1979.

Oral history interview with Gertrude Sweet conducted by Shirley Tanzer. Program on Women and Work, Institute of Labor and Industrial Relations. University of Michigan—Wayne State University, Ann Arbor, MI, 1978.

Oral history interview with Jackie Walsh conducted by Lucy Kendall. Women in California Collection, Hotel Strikes Series of 1937, 1941–42, California History Society Library—Manuscript Department, San Francisco, CA, 1980.

Selections from the index and biography folders of the Myra Wolfgang Collection, the Archives of Labor History and Urban Affairs, University Archives, Wayne State University, Detroit, MI.

The following oral history interviews were conducted by Gail Livings: (final transcripts in progress)
Flo Douglass, San Francisco, CA, August 1986;
Jeri Powell, San Francisco, CA, August 1986;
Sherry Chiesa, San Francisco, CA, August 1986;
Cindy Young, San Francisco, CA, August 1986;
Dorothy Randazza, Los Angeles, CA, August and September 1986;
Katie Hill, Los Angeles, CA, September 1986;
Maria Elena Durazo, Los Angeles, CA, April 1983;
Helen Anderson, Los Angeles, CA, April 1983;
L.Z., Santa Monica, CA, September 1985;
M.B., Santa Monica, CA, September 1985.

POSTSECONDARY ORGANIZATIONS AS SETTINGS FOR LIFE HISTORY:

A RATIONALE AND ILLUSTRATION OF RESEARCH METHODS

Patricia J. Gumport

PREFACE

"I refuse to be intimidated by reality anymore. After all, what is reality . . . ? Nothin' but a collective hunch." This assessment was offered by Trudy, one of Lily Tomlin's characters in the play, "The Search for Intelligent Life in the Universe" (Wagner, 1985). Trudy, a bag lady, is a creative consultant to aliens from outer space. Her "spaccchums" are here on Earth as a cosmic fact-finding committee, searching for signs of intelligent life.

Current Perspectives on Aging and the Life Cycle,
Volume 3, pages 175–190.
Copyright © 1989 by JAI Press Inc.
All rights of reproduction in any form reserved.
ISBN 0-89232-739-1

"It's a lot trickier than it sounds," Trudy explains. "We're collecting all kinds of data . . . to figure out, once and for all, just what the hell it all means." (She writes the data on post-its and they study them.) Trudy and her spacechums have undertaken an exhausting search for meaning, as she describes the process:

> All this searching. . . . All th(ese) data, and all we REALLY know is how LITTLE we know about what it ALL means. Plus there's the added question of what it MEANS to KNOW something . . . So no matter how much we know, there's more to knowing than we could EVER know. (For example, even) Sir Isacc Newton . . . secretly admitted to some of his friends (that) he understood how gravity behaved but not how it worked. . . ."

In an attempt to salvage some meaning from the ever-increasing uncertainty, Trudy suggested: "Maybe the secrets about life we don't understand are the 'cosmic carrots' in front of our noses that keep us going."

One 'cosmic carrot' for social scientists is to understand how organizations are crucial contexts for individuals through the life course. Life histories have been a way to explore individual agency within organizational constraints and opportunities, as well as to account for changing dynamics as they occur in different historical periods. Following C. Wright Mills' invitation to explore the intersection of biography, history and social structure, Gerson (1985, p. 38) advocates social science inquiry that uses life histories as an analytic avenue into understanding "the interplay between social constraints, psychological motivation, and the developing actor." Life histories provide a theoretical and empirical approach for examining the ways in which organizational settings not only affect the actions and experiences of individual participants but also frame the possibilities within which individuals may construct their own visions of the future.

An organizational setting that is especially well-suited for examining these interactive dynamics between social structure and human agency is higher education. In particular, by looking at faculty inside colleges and universities, we can see the ways in which postsecondary organizations provide the contexts for crucial career choices and aspirations over the life course.

The line of inquiry that I develop is to view postsecondary organizations as cultural or social constructions (Lincoln, 1986; Tierney, 1987). Such an inquiry leads to the investigation of the subjective experiences and perceptions of organizational participants. The development of this line of inquiry responds to the urgent call among social scientists to pay attention to the ways we think about life in universities (Morgan, 1986; Keller, 1986) and to develop multiple constructions for inquiry (Lincoln, 1986). Underlying such requests for self-conscious inquiry is the belief that researchers should sort through the range of assumptions underlying research conducted in postsecondary organizations, including one's beliefs about the social world and one's preferences for how we can know it (Gumport 1988, Silverman, 1987).

Life histories within organizational settings may be seen as part of a broader shift in social science research. As recent discourse reveals, the significant developments of this shift include widespread efforts to critique the conventional aims of scientific inquiry to discover universal, objective, abstract knowledge claims (Keller, 1986) and to construct alternative lines of inquiry that will particularize, historicize and contextualize what was previously taken as universal.

The dialogue often characterizes researchers as falling into two camps. Two camps, while admittedly simplifying the picture, is meant to suggest that they mirror the divided landscape among social scientists more broadly. There are generic scientists whose task it is to "find fundamental laws," and there are those who find limited generalizations which seem to be "not really science because they are too local, too subjective, and too hard to falsify" (D'Andrade, 1986, 26–27). Let me elaborate on each kind of inquiry.

On the one hand, there are those who do positivist social science. Even though we are in an era of post-positivist thinking, some researchers still advocate and contribute studies that have objectivist premises, scientific methods, and generalizable conclusions that often aspire for knowledge claims that have causal and/or predictive value. The promise of this approach is to yield generalizations that will not likely decay with time.

In contrast, other researchers downplay the science in social science research and reject the aim of discovering universal and objective truths, replacing it with an alternative—descriptive particularism. Geertz (1983, p. 6) describes the shift as trying to explain social phenomena "by placing them in local frames of awareness. . . . " Proponents of this approach study a particular setting, using methods that capture the experiences and perceptions of participants, aspiring to offer rich portraits of particular instances; and generalizations, if present at all, are within a particular case rather than across cases (Geertz, 1973). The epistemological underpinnings of this approach rest on more limited notions of what we can know. It is more narrowly construed as we can only learn, at best, about how people diversely construe their social worlds. And how we can know its points to reliance on, among other things, how competently one can be a participant observer, to immerse oneself in natives' views, and to be open to emergent data and emergent themes (Glaser and Strauss, 1967).

Influenced by post-positivist critiques of science, many proponents of the latter camp have taken on the label of "interpretive" research in extending the tradition of Weberian *verstehen* social science. . But it seems to me that this label is only partially useful in understanding the aims and means of such particularistic inquiry. It is useful in that "interpretive" accurately depicts the researcher as an inescapably subjective instrument of data collection who, through a filter, imposes a layer of interpretation on first-hand experience, which renders the research product, even at its best, at least one level removed from the natural day-to-day flow, the actual experiences and workings of our institutions and the people in them.

But the label of interpretive research does a disservice: it misleads us into considering this kind of research as distinctly interpretive whereas other approaches are not. We come away thinking that interpretive studies are of different (and, in this hegemonic system, of inferior) value because they embrace self-consciously subjective and particularistic elements. At the same time, it reinforces the notion that studies based on scientific method do not involve interpretation and do consist of first-order, not-having-been-translated, cleaner data.

I want to suggest that all research is interpretive, and it is inherently so. I want to develop the view that research is "a distinctively human process through which researchers MAKE knowledge, . . . in contrast to the more common view of research as a neutral, technical process through which researchers simply reveal or discover knowledge" (Morgan, 1983, 7). As knowledge producers we focus investigations on particular settings at particular times and offer a completed product that is, by definition, at least one step removed from what was (past tense) actually going on. This is true for those who do so-called scientific inquiry and those who do interpretive inquiry. The line that separates us actually connects us: we all work within limits of knowing, albeit with different angles of vision.

Having brought us to the same starting point, of viewing all research as interpretive, I will focus my remarks on doing research that embraces the complexity of life histories within postsecondary organizations. This kind of social inquiry is driven by research questions that aim to historicize, contextualize and particularize our understanding of postsecondary organizations. Such findings are arrived at through immersion in a naturalistic setting and through conducting life histories with the participants, in order to understand the multiple meanings among individuals rather than to make the quick ascent to generalize their experience.

The purpose of the remainder of this chapter is to justify and illustrate the relevance of this kind of social inquiry for social scientists who study the life course as well as those who study postsecondary organizations. First I will present a brief justification of this approach. Then I will draw on my own research experiences to illustrate the kinds of research methods that are implied in this approach.

PART I: The Rationale

There are several compelling reasons for seriously considering the value of this approach for studying postsecondary organizations as settings for life histories. Many of these reasons either implicitly or explicitly critique the use of scientific method which seeks to find law-like generalizations. I offer three reasons that this kind of social inquiry is a valuable alternative.

1. All research is particularistic regardless of its claims to be otherwise.

Within social science, the trend is to shift away from universalism. I like to think of this trend as "truth in advertising" about what research has offered historically and can offer in the future.

I have already suggested that all research is interpretive, whether it appears that way or not. Most researchers at some level acknowledge that theory at its best can only offer a partial, or selective (Knorr-Cetina, 1981), interpretation of reality, whether due to limited scope of inquiry, limited theoretical constructs, or simplifying assumptions about sampling and methods that were "necessary" limits for doing the study.

Beyond that, though, the critique of universalism is more extensive. Critical theorists and feminist scholars are among those who have made persuasive arguments that universalism in science has not been reflective of all that it claims. In pushing universalism off the pedastal, these critics characterize the approach of conventional academic science as the "dominant Western tradition," which has been produced and sustained by a privileged few in order to preserve the interests and values of the dominant group (which for the most part has had no gender, race, and class differentiation). In short, the historical context and content of academic science reveals that universalism, broadly considered, has neither been achieved nor sufficiently attempted; agreement and even intersubjective agreement reflects a history of faulty conclusions, of generalizing too far from too few.

It is important to note that proposing particularism in research is not an automatic leap to relativism (Harding, 1986). Rather it is an effort to historicize and contextualize research, to construct research that replaces the abstract, singular notion of humankind with the conceptual space for difference and ambiguity in meaning.

2. Subjectivity is especially appropriate when organizational actors are the subjects of study.

The second reason that this approach is valuable is that it makes sense to move from objectivity and to move toward subjectivity. By subjective I mean that the subject matter of our inquiry IS human subjectivity (Shweder and Fiske, 1986, p. 7). The meaning of subjective that I offer here is to collect, or acquire subjects' meanings, to learn about the experiences and perceptions of individuals from their points of view as they reflect on aging and the life course.

This is the genuine Weberian meaning of subjective, not to be confused with the actions of the researcher. In asserting that subjectivity is appropriate, I mean to suggest that the fullest understanding can come from learning people's meanings, that is, how THEY see things. What is the alternative? To treat people like objects would result in less accurate understanding and insights of people and their social contexts. Thus, I advocate the use of qualitative methods of in-depth interviewing and participant observation which give primacy to participants' subjectivity.

I also acknowledge the more conventional usage of subjective, which calls attention to the role of the researcher. To assume that the researcher is objective and value-neutral is to misrepresent the limits of research. However, to suggest that the researcher is subjective and should be self-conscious of her partiality makes good sense.

The researcher's own personal eyes, values, and meanings do frame what is collected, so that all data acquired are affected by values. Rather than this inescapable truth being a detriment to the research, it can be helpful in that it puts the researcher on equal footing with the people in the study. In establishing rapport and trust, in actively listening, the researcher gets to genuinely expose her limits of understanding and to clarify her hunches as they emerge. One can say, "I'm not sure what you mean by that; could you tell me more?" or "That isn't familiar to me, can you tell me about it?" There is an authenticity that surfaces in studying human subjectivity that can be enhanced by sharing our humble recognition of our own subjectivity.

3. An approach that rests on viewing reality as socially constructed captures the complexity in organizational life.

In moving away from an inquiry that is characterized by objectivity and universalism, a fruitful alternative lies in social construction perspectives. Such an approach stresses social inquiry within organizational settings as a human activity. The process reveals a range of experiences and perceptions and thus leaves open plenty of conceptual and empirical space for moving into uncharted intellectual directions.

The premise here is that individuals experience their social worlds differently; as social beings they are also knowers about their world (Gifford, 1983, p. 91). Informants became like narrators who help researchers understand how organizational life is complex and dynamic. This approach rejects the existence of an external, causally-determined world with attributes that are somehow independent of, and "more real" than, those that individuals reflect (Berger and Luckmann, 1967).

Given a premise of reality as socially constructed, there are far-reaching epistemological implications for the foundations of research: that is, what can we know and how can we know it? We can know or examine individual perceptions and experiences of their organizations. Viewing organizations as under construction, the researcher needs to be immersed in the daily activity and to interview (open-endedly) participants in order to "uncover and interpret the multiple perspectives of organizational life" (Tierney, 1987, p. 66). We can explore further a range of crucial life course processes for individuals, such as searching for an occupational niche (Sofer, 1970) or the different paths taken in the course of pursuing an academic career (Clark, 1987, Neyman, 1977).

Thus far I have suggested three reasons within social science research that provide a rationale for developing an alternative approach: research is particularistic, subjective data are valuable, and studying reality as socially constructed is an innovative undertaking that has potential to capture the ways in which postsecondary organizations serve as settings for individual actors choosing and adapting to their surroundings.

PART II: Illustration of Methods

I spent the last two years trying to learn about changes in academic knowledge, specifically how new knowledge emerges and acquires legitimacy. I did a case study of one area—feminist scholarship—to try to understand, among other things, how it was that people became proponents of this new knowledge before it was even recognized as legitimate. For many people (and I focus here on academic women), graduate school was a time when you could not do a dissertation on women (who thought it was important or even relevant?) and when being political (let alone being feminist) seemed irreconcilable with an academic career. How was it that something called feminist scholars emerged? How was it that a new category of academic knowledge emerged along with a new category of knowledge producers? Clearly these questions underscore the relevance of life histories as a way to study the paths of individual actors and the organizations they construct through their actions over time.

My intention in this part of the paper is to illustrate some of the pitfalls and pleasures of pursuing this topic. I will not offer recipe-like, how-to procedures. Rather I wish to tell you about my experience of conducting this kind of research.

Simply stated, I sought to explore the complexity involved in the emergence of feminist scholarship, primarily from the perspective of knowledge producers themselves. I interviewed 75 women faculty and administrators, with varying degrees of knowledge about, and involvement with, feminist scholarship. (The sample included some who ended up pursuing it and others who did not.) I wanted to understand how it is that some people come to work in a not-yet-legitimate area of knowledge given a powerful reward system that is firmly entrenched in the conventional organization of academic work/life. I was puzzled about the different paths and experiences among women faculty, about the different perceptions of what constitutes feminist scholarship, and about the different visions of whether individual, and ultimately institutional, commitments to knowledge can be redefined. These were some guiding concerns.

I want to focus on two aspects of the research process: data collection and data analysis. I learned something crucial during in each stage.

1) Emergent Data: Letting the People Speak

Prior to the study I had spent nine years in colleges and universities, so I felt pretty comfortable going into them as a research setting. I was armed with a high-level of enthusiasm, good interpersonal skills, and extra batteries for my tape recorder. The truth is that I began interviewing with a high dose of naivete. The first thing I learned was how enthusiasm and prior hunches could unintentionally constrain what people told me. In the process of interviewing, I learned how to let the people speak.

First, some background on the design (Gumport, 1987). My principle data source was women faculty. As a concession to the science in social science, I wanted to preserve some randomness in my sample. I drew a random sample of full-time faculty that was stratified across ten institutions and across three disciplines and women's studies. As a supplement to this core group, I also found informants who were knowledgeable about the emergence of feminist scholarship. Guided by theoretical sampling [Glaser and Strauss, 1967] and triangulation, the design enabled cross-case comparisons.

My main empirical anchor was to obtain the intellectual biographies and career histories of this core group of 47 women faculty. I conducted semi-structured interviews which ranged in length from 1½ to 6 hours. Interviews were taped and transcribed. Confidentiality and anonymity were assured. I made contact through letters and follow-up phone calls. (On occasion I felt that this part of the process was made more difficult due to the discrepancy between my then graduate student status and their faculty status.) I explained that my project was about women faculty, and that I was interested in their intellectual biographies and career histories. I wanted to keep it as general as possible without imposing pre-defined categories on the interview.

Once an interview was underway, I said that I was also interested in the emergence of feminist thinking in the academy, why some women (not them, per se) developed feminist perspectives and what the consequences were. I waited to ask this question because I thought, had I said it before the interview, they may have made assumptions about my valuing of feminist perspectives and then would tell me what they thought I wanted to hear. This dynamic is a problem in all interviewing, and I tried to minimize its effect.

The interviews varied in location from campus offices to residences and in one case a coffee shop. Sometimes the interviews were interrupted by personally revealing circumstances. In one case, there was a furniture delivery in which we spent half an hour repositioning the oriental rug under the new dining room table. In another case, we had to drive to the local peace activists' center to pick up flyers that the interviewee was to distribute the next day on her campus. In yet another case, we sat on the living room floor nursing the family pet, a 17 year old dog who was close to death and visibly uncomfortable just lying there. The variation in settings and circumstances reveal that the tenor of the interviews differed and that the people opened up to different degrees.

My questions were open-ended. Rather than saying "Do you think _____?" which invited a "yes" or "no" response, I phrased the questions to elicit their perceptions in their own categories: "How about if we start with you telling me about your educational background?" "How did you end up in graduate school?" "Who did you talk to?" "What was important to you at the time?" I did not ask precisely about a "decision" to go to graduate school or presuppose the significance of "planning" an academic career. This was fortunate because many of their narratives reveal having "backed into" graduate school with little or no foresight about becoming an academic. The distinctive ambivalence about academia and initial lack of investment in an academic career surprised me, since many of these women are highly successful tenured faculty at high prestige institutions.

We began chronologically as a starting point, although the interviewees weaved back and forth between stories about the present and different times in the past. Being flexible with this natural unfolding across time enabled both me and the interviewee to tie events together and to consider changes or similarities over time.

One technique, in particular, helped me to let their narratives emerge. It is the "anything else?" question. Every time I thought we were ready to move on to a new topic or if I had an idea to pursue, I asked some version of "is there anything else you'd like to add/tell me about that?" Even just a quiet "hmmm" reflecting on what I just heard was helpful. In doing this I was able to convey that I was interested, that I was listening, and that I respected the speaker, and wanted her to feel a sharing of control over the conversation.

The "anything else?" technique grew more natural as I did more interviews. But in the beginning, I was far from graceful with it. My first interview illustrates this.

I had spent three hours talking with a philosopher in her living room. We had covered a lot of ground: elementary school through graduate school to her present tenured position, her changing political involvements from the sixties through the eighties, her ongoing search for a way to do philosophy that "felt right,'' to her what-I-thought-to-be-concluding remarks about how "having an active mind is a real pleasure.'' I relaxed and sat back in the couch, thinking the interview was over.

Then I offered the "anything else?" question, mostly as a pro forma gesture and after-thought. To my surprise the question evoked a dramatic shift in tone. The woman sat silently in her corner of the couch. Her eyes filled with tears. Her voice was low, barely audible for the tape recorder: "What I haven't told you about is all the pain.'' A long pause, and then: "It has been hard, really hard, and I'm tired.'' She then told me about one aspect of her frustrations and misgivings with academia: how unhealthy it is for emotionality to be separated out so that people live out the "mind/body'' split; and how this recognition dominates her daily academic life—her teaching, her writing, and her talking with people; and how devastating it is, day in and day out, to participate in a system that she increasingly feels no identity with.

Clearly, these remarks were crucial to my understanding about her identity as a feminist scholar, her academic career history, her experience of her discipline and of that particular university setting. Initially, I did not learn about this because my enthusiasm for the research process seemed to ask for responses about how she became successful as a scholar and a teacher (at least she was by conventional "objective'' measures). I had unknowingly not made room for this disclosure. It was only in relaxing, in setting down my pen and papers on the coffee table, in explicitly setting my agenda aside, that I let the silence be there and let her speak.

As common sense might suggest, when I spent more time with someone, we achieved greater depth. These kinds of responses usually emerged as whispers or with tears or after the tape recorder was turned off, which makes them more difficult to reconstruct verbatim. Even so, the experience generates a memory that stayed with me throughout the data collection and writing, and even today. Across the board, this group was remarkably open (and articulate!), tending to reveal vulnerability and self-doubt in recalling either the past or describing current assessments of their academic lives.

2. Emergent Themes—Letting the Data Speak

I turn now to data analysis. Even though, to some extent, analysis occurs during the data collection (Glaser and Strauss, 1967), I want to focus on the stage of the research when all the interviews have been completed and transcribed, and the task is to make sense of the all those transcribed words in the huge piles of paper. I knew that I was looking for themes, common threads or different ones. I thought that, if I read and re-read the transcripts (and my field notes), I would see it more clearly.

I began with cross-case comparisons, by grouping the narratives and reading them in their respective bundles, whether by discipline of Ph.D., by prestige of current institution, by academic rank, by whether they were located in a department or a women's studies program, and even by the geographic location of their graduate schools.

I also grouped them by experiential categories that emerged from the interviews: first most generally as pre-Ph.D. and post-Ph.D., and then, more specifically, how they ended up in graduate school, how they got the first academic appointment, their involvement in politics, their experiences of feminism, and so on. Then I grouped them by categories of their perceptions: what career opportunities they thought were available, any foresight they had about academic careers, what sense they had of a tension between politics and scholarship (for self and for others), and so on.

So some categories were previously selected because they were inherent to the conceptualization of the study (e.g. political involvement) while other categories that were along more thematic lines emerged in the course of the analysis, (e.g. the experience of isolation or invisibility, and feeling they had "backed into" graduate school or an academic career).

The most dramatic shift in the data analysis was both substantive and procedural. This was the recognition that the data could be grouped by viewing the primary informants as mini-generations of academic women, based upon the time that they began their doctoral programs.

The sample was comprised of women who entered graduate school between 1956 and 1980. Given the dramatic social changes that occurred during those two and a half decades, these academic women may be grouped as mini-generations. Each generation of academic women received their graduate school training corresponding to a different stage in the development of feminism and feminist scholarship. These women became knowledge producers within particular generations, having different political contexts as well as different intellectual resources that were available within and outside their home disciplines, and within and across campuses.

Based on the year in which they began their Ph.D. programs, the women were grouped as follows: those who entered graduate school before 1964, who may be considered as Forerunners to feminist scholarship; a group of Pathfinders who entered between 1964 and 1972 and initially developed feminist scholarship; and a group of Pathtakers who began their graduate training after 1972.

Three aspects of this generational analysis must be noted. First, the choice of these years as dividing lines is arbitrary. Clearly, some women were at more politicized campuses before 1964, for example. Yet in this study, for this sample, the patterns consistently corresponded with these years. Second, the year in which they entered graduate school was selected instead of the year of finishing. The reason was that year of entry reflected the most potential for the salience of a

cohort, since the early years of a program point to the likelihood of interaction
(e.g. in coursework) to be followed by more independence in the later years of
dissertation research. Third, the generational analysis that I developed reflects a
departure from the more conventional quantitative approach used in cohort analysis, which focuses on an aggregate of individual behavior in the same chronological interval in order to investigate properties of populations rather than the
behavior of individuals (Ryder, 1965).

The data analysis explored a range of opportunities and constraints that these
women experienced and perceived in different levels of the academic system
simultaneously—in their immediate organizational surroundings on campus, in
the intellectual level of the discipline, and in the wider political climate external
to the academy. In spite of the diversity within each group, that is a diversity of
ambitions and outcomes, distinct patterns emerged. The patterns suggest dramatic differences in how some of these women became innovators of new knowledge, creatively working within perceived organizational and intellectual
constraints.

The three generations of women were historically situated within different
periods of an emerging feminist scholarly discourse. The Pathfinders created the
first visible fusions between feminism and scholarship. The early years may
reflect crude attempts in intellectual terms, even though they were daring ones in
personal terms. As feminist scholarship evolved in the 1970s, it resulted in an
increasingly visible and well-articulated agenda. This provided an already-exist-
ing paradigm for the next generation of women—the Pathtakers—who could
choose to pursue it or not. The Forerunners began their graduate training at an
earlier time when one's woman-ness did not appear to have relevance for scholarly concerns.

Once this three-stage (pre-paradigm, paradigm-developer, paradigm-existing)
approach was established, the analysis of transcripts and other data fell into
place. The generational analysis enabled me to further contextualize, and actually historicize, their experiences. The notion of a generational analysis was not
preconceived; rather it emerged from reading and re-reading the data.

Some of you may be wondering what makes this reliable as opposed to
serendipitous. I have explained the rationale for my sampling and interviewing
strategies. But you may still wonder whether this analysis is accurate and credible or whether it is merely "impressionistic." (See Lincoln and Guba, 1985,
Chapter 11, for elaboration of strategies to establish trustworthiness of data.)

Would similar patterns emerge if another interviewer asked questions in an
effort to uncover the "same" story? "No, that's not the point," I would have
said six months ago. Then I met a colleague who had used a similar approach for
trying to understand the emergence of women as educational leaders; much to
both of our surprises her categories paralleled mine: predecessors, instigators,

and inheritors (Astin and Leland, 1986). This discovery made me feel that my study had more credibility. Glaser and Strauss (1967, p. 231) support this with their assertion that multiple comparison groups help credibility.

Clearly, further research could follow-up on the accuracy of my analysis, either with more people or across more disciplines or beyond the single case of feminist scholarship to consider other emerging fields that struggle for legitimacy. But for the time being, I want to suggest that judging the credibility of the research rests with both the researcher and the reader of it. As Glaser and Strauss (1967, pp. 228–233) suggest, the reader can ". . . figure out the limitations of a . . . study by making comparisons with their own experience and knowledge of similar groups."

SUMMARY AND IMPLICATIONS

In summary, this paper has been an invitation to consider developing alternative approaches to studying academic organizations as settings for individuals to move through the life course. In proposing that we embrace complexity, I have offered a brief justification for moving away from predictive, universalizing, generalizing research and moving toward descriptive, subjective, particularistic research of life histories. I then described my experience of data collection and data analysis. I learned to approach the process with openness and flexibility, to approach with humility the people and the data to see what they can teach about multiple perspectives among individuals within organizations.

This kind of inquiry locates our discourse squarely within the "profound self-examination" that is characterizing post-positivist social science (Keller, 1985, 10; Lather, 1987). It underscores that research is essentially a human process in which people are studying people in their academic contexts or in which people are studying academic organizations through the eyes of the participants in them. This kind of social inquiry that is fundamentally rooted in human subjectivity entails a moral dimension, which has far-reaching implications for not only our methods but our vocation as well. (Adapted from Berger and Kellner, 1981, p. 11.) One of the greatest benefits of research is to learn something new or refute what I thought to be true. This kind of research is one avenue to the unanticipated.

REFERENCES

H. Astin and C. Leland. 1986. "On Behalf of Women: Issues and Accomplishments from a Leadership Perspective," Unpublished paper.

P. Berger and H. Kellner. 1981. *Sociology Reinterpreted: An Essay on Method and Vocation.* Garden City, NY: Anchor Books.

P. Berger and T. Luckmann. 1967. *The Social Construction of Reality: A Treatise in the Sociology of Knowledge*. Garden City, NY: Anchor Books.

L. Cain, Jr. 1964. "Life Course and Social Structure," Pp. 272–309 in *Handbook of Modern Sociology* edited by R. Faris. Chicago: Rand McNally and Company.

B. Clark. 1987. *The Academic Life: Small Worlds, Different Worlds*. Princeton, NJ: The Carnegie Foundation for the Advancement of Teaching.

R. D'Andrade. 1986. "Three Scientific World Views and the Covering Law Model," Pp. 19–41 in *Metatheory in Social Science: Pluaralisms and Subjectivities* edited by D. Fiske and R. Shweder. Chicago and London: University of Chicago Press.

C. Geertz. 1973. Interpretation of Cultures. New York: Basic Books.

C. Geertz. 1983. *Local Knowledge: Further Essays in Interpretive Anthropology*. New York: Basic Books.

K. Gerson. 1985. *Hard Choices: How Women Decide about Work, Career, and Motherhood*. Berkeley: University of California Press.

N.L. Gifford. 1983. *When in Rome: An Introduction to Relativism in Knowledge*. Albany: State University of New York Press.

B. Glaser and A. Strauss. 1967. *The Discovery of Grounded Theory: Strategies for Qualitative Research*. New York: Aldine.

P. Gumport. 1987. *The Social Construction of Knowledge: Individual and Institutional Commitments to Feminist Scholarship*. Unpublished Doctoral Dissertation, Stanford University.

P. Gumport. Autumn 1988. "Curricula as Signposts of Cultural Change," *The Review of Higher Education* 12(1): 49–61.

S. Harding. 1986. *The Science Question in Feminism*. Ithaca and London: Cornell University Press.

G. Keller. 1985. "Trees Without Fruit: The Problem with Research About Higher Education," *Change* (January/February) 7–10.

G. Keller. Winter 1986. "Free at Last? Breaking the Chains That Bind Education Research," *The Review of Higher Education* 10 (2): 129–134.

K. Knorr-Cetina, 1981. *The Manufacture of Knowledge: An Essay on the Constructivist and Contextual Nature of Science*. Oxford: Pergamon Press.

P. Lather. 1987. "Feminist Perspectives on Empowering Research Methodologies," Paper presented at American Educational Research Association annual meeting, Washington, D.C., April.

Y. Lincoln. Winter 1986. "A Future-Oriented Comment on the State of the Profession," *The Review of Higher Education* 10 (2): 135–142.

Y. Lincoln and E. Guba. 1985. *Naturalistic Inquiry*. Beverly Hills, CA: Sage.

G. Morgan (ed.). 1983. *Beyond Method: Strategies for Social Research*. Beverly Hills, CA: Sage.

E. Neyman. 1977. "Scientific Career, Scientific Generation, Scientific Labor Market," Pp. 71–94 in Perspectives in the Sociology of Science, edited by S. Blume. New York and London: John Wiley.

N. Ryder. December 1965. "The Cohort as a Concept in the Study of Social Change," *American Sociological Review*, 30 (6): 843–861.

R. Shweder and D. Fiske. 1986. "Introduction: Uneasy Social Science," Pp. 1–18 in *Metatheory in Social Science: Pluaralisms and Subjectivities*, edited by D. Fiske and R. Shweder. Chicago and London: University of Chicago Press.

R. Silverman. Autumn 1987. "How We Know What We Know: A Study of Higher Education Journal Articles," *The Review of Higher Education* 11, (1): 39–60.

L. Smircich. 1983. "Studying Organizations as Culture," Pp. 160–172 in G. Morgan (ed.) *Beyond Method: Strategies for Social Research* edited by G. Morgan. Beverly Hills, CA: Sage.

C. Sofer. 1970. *Men in Mid-Career: A Study of British Managers and Technical Specialists*. London and New York: Cambridge University Press.

W. Tierney. Autumn 1987. ''Facts and Constructs: Defining Reality in Higher Education Organizations,'' *The Review of Higher Education* 11, (1): 61–74.

H. Wagner. 1983. *Phenomenology of Consciousness and Sociology of the Life-World: An Introductory Study.* Edmonton, Alberta, Canada: The University of Alberta Press.

J. Wagner. 1985. *The Search for Intelligent Life in the Universe.* New York: Harper and Row.

TELLING WOMEN'S LIVES:
"SLANT," "STRAIGHT" and "MESSY"

Judy Long

INTRODUCTION

The study of individual lives is today becoming a focus of scholarly activity in a number of disciplines and genres. An explosion of interest in autobiography coincides with a revival of interest in biography, informed particularly by psychological theories. In literature, an intense interest in theory has resulted in new contributions on women and controversy about the idea of the self. In psychology, a long neglect of the person is giving way to personological theories of personality and the emergence of interest in adult development—i.e., on the whole span of a life. In sociology, large-scale cross-sectional studies are being supplemented by research on the life course. Meanwhile, the growth of oral history has produced first-person accounts of previously undocumented lives.

Current Perspectives on Aging and the Life Cycle,
Volume 3, pages 191–224.
Copyright © 1989 by JAI Press Inc.
All rights of reproduction in any form reserved.
ISBN 0-89232-739-1

Within each of these traditions, incorporation of new research on women has begun to change fundamental paradigms and theories. The contestation between feminist scholarship and traditional paradigms for telling lives is apparent in the various disciplines.

The focus of this paper is on that contestation: on the content the female subject seeks to communicate, and on the institutional context that provides her pattern, her audience and her judges. The focus of the paper is transdisciplinary; I hope that in emphasizing parallels and commonalities between social science and literary studies I can deepen our understanding of female subjectivity as constrained by the gender system, a focus of scholarly concern that is not limited to any one discipline.

As a social scientist, I will examine institutional mechanisms that affect intellectual production, reading as well as writing. The reader will find the emphasis on institutional factors quite different from the preoccupation with the subject that dominates western traditions of autobiography and biography. In literary treatments the subject is too often amputated from her historical and social location, even though her life is largely made intelligible by her biography. The sociological analysis of social control has a great deal to contribute to the discussion of women's writing and publishing, self-presentation and representation. I will argue that much of the difficulty that women experience in confronting the autobiographical act—at first glance, a relationship between the subject and her life—derives from the relationship of women to a social order that includes the genre of autobiography. To interpret the problematics of women's autobiography in intrapsychic terms is, I will argue, a misreading. To the extent that the tradition of the subject in literary studies obscures the effects of the institutional surround, it must fall under scrutiny as well.

The analysis that follows is intended to apply to social science accounts of women's lives as well as to those in literature. Ultimately, I hope that the scope of my analysis may be extended to those self-referential writings of women that are commonly excluded from literary canons, such as diaries, journals, letters and autobiographical fictions or "biomythography." Future analyses should include contributions from oral traditions: texts that are newly, barely or not written. It will become plain that I believe that published autobiographies are only the visible tip of a vast continent of women's self-referential accounts. [1]

For purposes of brevity in this paper, however, I will focus mainly on issues raised by published autobiography and biography and their intersections with the concerns of feminist social science. I will begin with a consideration of issues involved in autobiography: life-telling by the subject. Subsequently I will move to a discussion of biography: life-telling by the narrator. A comparison of men's and women's autobiographies serves to introduce issues of gender in subject, account, genre and reader.

TELLING MEN'S LIVES: AUTOBIOGRAPHY AND GENERIC AUTOBIOGRAPHY

Estelle Jelinek (1980) has anatomized the paradigmatic autobiography. Its generic features, which I analyze in terms of gender, help to illuminate some of the problematics of fitting women's lives to this form. In the critical tradition addressed to autobiography there has been little analysis of the gender significance of the kinds of patterns found, and of the fit or failure of these patterns to the raw material of women's lives. Classic critical contributions (e.g., Olney 1972; 1980; Sayre 1964; Mehlman 1971; Spengeman 1981) exclude women's autobiographies, and women's lives are not mentioned (problematic or otherwise). The generic model of autobiography in itself represents a major obstacle to telling women's lives, a theme to which we will return several times during this discussion.

A defining characteristic of male autobiography is the interpretation of a life in terms of a pattern. Jelinek says that the autobiography is shaped to make intelligible some single theme, outcome or destination.

Northrup Frye characterizes autobiography as "inspired by a fictitious impulse to select only those events and experiences in the writer's life that go to build an integrated pattern (1957)." Olney (1980) and Spacks (1980) echo this language, notable for its emphasis on the imaginative construction of a life, and for the sense of agency attributed to the life-writer. Mimesis gets short shrift in critical writing about autobiography. As Spacks says, "To understand one's life as a story demands that one perceive that life as making sense; autobiographies record the sense their authors hope their lives make (p. 131)." The notion of the autobiography as an imaginative construction, shaped to the subject's purpose, is well accepted. An issue for female autobiography is what stories, what paradigms are available for telling one's life.

One script for male autobiography is the solitary hero en route to this destiny. Sidonie Smith (1987) points out the link between the privileged cultural fictions of male selfhood and the judgment of "significance" in life stories (p. 93). Paradigmatic accounts minimize connectedness between the hero and others, hewing to a tradition of individualism which is itself a fiction. The hero in modern literature is little influenced by others; the unilinear narrative is achieved at the expense of substantial pruning.

The "destination" of the published life is a second element with strong gender significance. Lives that have been celebrated in published autobiographies most often have a destination in the public sphere. From many public lieux, women have long been barred by law and custom. The lack of barriers, the easy entree into the public sphere has its effects in male autobiography. In writing his life, the male autobiographer is able simultaneously to place himself in history and lay claim to a public heritage. By his narrative, the subject thus links himself to some reality "larger" than his personal life, magnifying thereby the significance of the latter.[2] This process reinforces the masculinity of history, itself a consideration for the woman seeking to add her name to its pages.

Jelinek notes that the paradigmatic autobiography is invested with universal themes, by male author and male critic alike. It becomes larger than life, and so its subject appears. This mutual magnification is a consequence of the gender system, which regards male as the norm, and places male experience at the center of knowledge, culture and history.

It should be noted that these effects of gender are by no means limited to autobiography. Mary Helen Washington, in comparing the critical reception of Gwendolyn Brooks' *Maud Martha* with that of Ralph Ellison's *Invisible Man,* noted that critics were able to see Ellison's protagonist as Everyman, and compared his author with Dostoevsky, Faulkner and Richard Wright. Brooks' protagonist could not be universalized by these critics; they were unable to connect Maud Martha's personal experience with the history of her people. Her story remained a personal and private one, hence lacking in significance.

The ability to see one's life as mirroring the themes, the heroes, the tradition of the common heritage is a significant component of the autobiographical act, and one that seems particularly difficult for women to achieve. Current writing on female autobiography, as we shall see, notes the problem women have with this form of self assertion, but offers little in the way of explanation. The limitations of an intrapsychic strategy of interpretation will be discussed in a later section.

A third feature noted by Jelinek also has gender significance. Jelinek observes that in the distinguished autobiographies of the canon, self revelation and intimacy are surprisingly absent (p. 10ff). Annette Kolodny (1980) observes that omissions and deletions are central to the genre (p. 240). In thinking of how autobiography is gendered, it is significant that the autobiographical subject excises parts of his self and life, and it is significant which parts they are. Specifically, painful and/or intimate happenings are suppressed, as well as romantic attachments, familial relations and personal idiosyncrasies. In the generic autobiography, the world of emotions, of home, of relationships, the processes of work and struggle in relationships—in short, the "female heart" of Domna Stanton (1984)—are minimized. Thus Henry Adams' canonical autobiography is told without any mention of his beloved wife Clover, after her suicide. The omission is a striking one from a commonsense standpoint, but is not anomalous within the conventions of generic autobiography.

The excision of the personal and emotional, considered good form in biography and autobiography, mirrors the gender script for men. The excision of the personal and the emotional is also considered a virtue in orthodox social science research procedure. The issue is one of the sites of contestation between orthodox and woman-centered autobiography, and between orthodox and feminist methodology in social science.

These characteristics of masculine self presentation have only recently been analyzed in terms of gender significance (see Evelyn Fox Keller 1985). Research on masculine personality underscores the premise that, far from being distinct from the "real world," our intellectual traditions are shaped by gender. Elements of masculine culture, with its emphasis on suppression of affect, its impersonality, its predilection for asymmetrical power relations, can be discerned in the strictures of literary genre or social science methodology.

Gender, unacknowledged or even disavowed, informs the obvious but underestimated divergence between the life as lived and the life we read. We turn now specifically to the consideration of generic autobiography.

Generic Autobiography

The autobiography, then, is an account of a life which is shaped in certain ways and which follows certain conventions. It is shorn of dailiness, confusion and interpersonal hubbub, with only the intentional design revealed. Its trajectory is smoothed toward a known outcome. It discovers personal destiny and links it with history. The related processes of elision, excision and editing, engaged in by the subject and accepted by the reader, raise several issues for life-telling. First is a lack of fit between the life as lived and the created account of it. Only the latter survives. Dorothy Smith (1979) might have been speaking of autobiography or biography instead of life history when she warned,

> . . . anyone's life situation is always complex, rooted in others' lives, and multifaceted. Through the work of those who reconstruct the patient's life as a case history, it is obliterated as it was experienced and lived (p. 169).

A second issue inheres in the fact that it is on this creation that we as readers (and we as critics) focus, rather than on the life which was its raw material. It is important to bear in mind that it is the life once shaped, fictionalized and universalized that we read. It is generally through this account, and only through the account, that we come to know the life, or imagine that we do. The generic autobiography does not attempt or claim to portray the life as lived. The accepted convention is to "clean up the act" in the ways specified above. The result is that the public man, the historical figure appear before us shaven and shorn. But herein resides a dilemma. Although we know that the "shaven and shorn" figure presented to us is a mask, produced through human contrivance, we generally lose sight of this fact, and act as though the life and the account were one. What is significant here is that the historical figure looks normal to us shaven and shorn; we forget, or never knew, what the life was like.

With much that is personal shorn away, the paradigmatic autobiography may give us the semblance of a generic person. In my view, the generic person is mutilated, a disembodied if not disembowelled cipher. It is a distortion that serves neither women nor men well.

If the generic person is a distortion, it underlines the contention that conventional methods are inadequate for telling lives, not only women's lives. In both social science and literature, the critique occasioned by the attempt to tell women's lives is significant for raising issues of validity. Nevertheless, generic autobiography follows the conventions of male autobiography. The life that merits a place in history retains a male outline. Certainly its lineaments do not accommodate, and for that reason negate, women's lives. In the telling, publishing, reading, discussing, teaching and citing of male lives, a template for telling lives is developed. The building up of the generic tradition of autobiography, which the would-be-subject must confront, is a process of increasingly concentrated masculinity. As Marcia Westkott (1979) reminds us, "The concept of the human being as a universal category is only the man writ large (p. 423)." When I refer to generic autobiography, then, I mean to remind the reader that we are speaking of an account of a life that is shaven and shorn, and one that is shaped along masculine lines. It is masculine in its claim on history-with-a-capital-H, on posterity, its claim of universality. It is shaven and shorn in that human characteristics have been intentionally removed.

Sidonie Smith writes, "Autobiography, an androcentric genre, demands a public story of the public life (p. 93)." The generic expectations of life stories include "significance," requiring a narrative that will resonate with the privileged cultural fictions of male selfhood (p. 93). Smith's interest in autobiography is in the problematic of how the female subject positions herself, how she establishes the authority to interpret herself publicly in patriarchal culture and androcentric genre (p. 82).

The accepted surface of generic autobiography seems to distract the reader's eye from the possibility that men's lives, like women's, are in reality "messy." The autobiography, in its accepted form, "cleans up" the life and introduces major distortions. It is from these distortions, and their hegemonic status, that many of the problems encountered by female life-writers derive. When the shaven and shorn life is placed in the center of our orthodoxy and our perception, the reality of women's lives is further distorted by being pushed to the periphery and being subjected to an exaggerated contrast with the fiction of men's lives.

The shaping and "fictionalizing" of the life is commonly acknowledged by authorities in the field of autobiography. What is often overlooked is that, although we accept the edited life as merely conventional, it acquires active normative force. In Kuhn's (1979) sense, it gives rise to, it requires judgments. Attempts at life-telling are assessed against the generic norm. Indeed, the project of telling lives may be shaped very strongly by the norm; even the intention of telling one's life. As Bakhtin (1981) observes, cultural stories of selfhood are populated—overpopulated—with the intentions of others (p. 284). Sprinker (1980) opines, ". . . every subject, every author, every self is the articulation of an intersubjectivity structured within and around the discourses available to it (at) any moment in time (p. 324)."

Challenges to the attributes of generic autobiography arise within the tradition of women's autobiography. Equally, women's autobiographies challenge the whole project of genre.

Women's Autobiography

Women's autobiographies, when contrasted with those of men, sometimes appear smaller than life-size. When seen from the perspective of the male tradition, women's lives are miniaturized before pen ever touches paper. Women's lives, and hence accounts, are presumptively limited to the domestic sphere. Women's themes are seen as restricted to interior landscapes, personal relationships and emotional dramatics. Jelinek observes that, in contrast to the large themes of men's lives, women's lives are pictured in terms of *impotent emotions* (1980, p. 5; italics mine). Whatever the validity of this characterization, its significance in gender terms is apparent. Emotions are part of what is selected out in the generic autobiography; what is selected in follows a masculine sexual script of puissance.

Such characterizations of women's autobiography have the aspect of a projection or a caricature, the "other" as seen from the subject-positioning of the male norm. Characteristics of female autobiographies are seen through the lens of this subject-positioning, are inevitably placed in the periphery and reacted to as "different." As the foregoing makes clear, the dominant perspective exercises an important influence on women's autobiography, but it inevitably distorts what women are trying to recount: the elements of women's lives and the relations among them.

In order to read the content that is intensely and singularly urgent for women's accounts of their lives, and the dilemmas that affect their life-writing, we must consult women's self-referential writing. We do well to remember that published autobiography is only the visible tip of the submerged continent of women's self-referential writing. Some themes that are muted in the public work sound more clearly in women's more private voices. Other themes mirror, in first-person accounts, the institutional factors we have discussed above.

In terms of the content in women's autobiography, I will emphasize daily life, ideas of the audience, the desire to be known, and the split subject. The first of these, women's daily life, contrasts strongly with male or generic autobiography. Women's autobiographical writing often portrays life as lived—its dailiness, the web of tasks, the web of relationships, the lack of closure, the nonlinear experience of time, interruptions, the psychological and emotional realities. Such "messy" accounts make no attempt to shear or to shave the life; indeed, their authors often attempt to retain what is conventionally eliminated in generic biography and autobiography. The importance of everyday life to women autobiographers is indicated by the volume of such accounts. The repetitive character of women's daily work is concretely represented in women's journals, letters and diaries; some attempt to claim a place for it is found in autobiographies as well. The importance of women's repetitive labor could not be conveyed by use of ditto marks or "etc.;" the repetition itself makes a statement. It seems to me that the presence of the details of daily life in women's writing is a contestation of their elimination from the approved generic account. The space devoted to this content in women's autobiographies is in proportion to the volume of content omitted in those of men. The presence of this content, and the form its expression requires, contest the insignificance attributed to the activities and sites that make up the woman's daily life.

The woman's daily life has the additional existential significance of situating her, of stating the material and bodily basis of her subjectivity. This is in contrast to the "abstracted conceptual mode" which renders her work and its significance invisible; indeed, suppressing the local and material reality is a prerequisite for operating in the abstracted conceptual mode (Dorothy Smith 1979). Smith notes that viewed from the position of the ruling apparatus,

> the local and particular, the actualities of the world which is live, are necessarily untamed, disordered, and incoherent (1979, p. 174).

The housewife's life provides ideal subject-matter for a divergent form of account: "Over a lifetime and in the daily routines, women's lives tend to show a loose, episodic structure that reflects the ways in which their lives are organized and determined externally to them and the situations they order and control (Smith 1979, p. 152). Dorothy Smith comments on the invisible organization of women's work which may be chronicled in a woman's life story:

> A housewife, holding in place the simultaneous and divergent schedules and activities of a family, depends upon a diffuse and open organization of consciousness available to the various strands, which are coordinated only in her head and by her work and do not coordinate otherwise in the world (152).

Women's lives are particularly apt subject-matter for the creation of new forms, precisely because they do not resemble the goal-directed, unidirectional trajectory of the smoothed and simplified male model. The resulting challenge to form is found not only in autobiography but wherever women's lives became the focus of study.

It is interesting to note that written accounts of women's lives in forms other than autobiography exhibit similar structural elements and are also seen as deviant from the norm. Critics of women's plays complain about the lack of plot and loose, episodic structure. Lucy Lippard characterizes women's poetry as open, intimate, particular, involved, engaged, committed. The form of the poetry is an extension of the self. Thus Alta, a poet who is also a mother, produces a text that is full of interruptions and alternations of voice.

The attempt to convey women's lives as lived seems inescapably to lead to innovation in form, whatever the genre. The tradition of the heroically isolated subject is a poor fit for telling women's lives for a number of reasons. Friedman (n.d.) challenges Gusdorf's stress on individuality in the autobiography, pointing out that where a group identity is imposed, the tradition of the self is misapplied. The female subject must contend with her collectively held alterity, and with her collective identity with other women as well as with her own subjectivity. A number of authorities have remarked on the difficulties inherent in inscribing the split subject ascribed to women. One possible form for women's autobiography is collective; here a potential parallel with the revival of prosopography has possibilities. In its narrative strategy the woman's autobiography may retain the lineaments of conversation and its intimacy. In so doing it reasserts a fundamental fact of women's lives: the self implicated in relationships with others. The inscription of the split self may require autobiographical forms that are more collective. (Friedman n.d.). Kate Millett, in *Flying,* violates many conventions of generic autobiography. Maxine Hong Kingston and Audre Lord pursue autobiographical projects in the form of fiction.

In women's self-referential writings the failure to approximate a smoothed trajectory, a shaven and shorn surface signals openness, engagement, reciprocity, and an acceptance of the impossibility of unilateral control in relations with others. These characteristics are shared by the feminist critique of social science. In both instances, we see once again that the personal is political: insisting on the representation of women's lives in the genres of scholarship is a political challenge of a fundamental kind.

A second content focus in women's autobiography is prefigured by the foregoing discussion. Authorities on women's autobiography have emphasized the intrapsychic splits in the writer (Stanton 1985; Spacks 1980). Stanton sees female autobiography as problematic precisely because of the difficulty of inscribing the split subject. One dimension of fracture is the sense of trespass involved in positioning the female self in the public sphere, in history. The intrapsychic conflict thus reflects the desire for self assertion confronting the forbidding posture of the gatekeepers of fame and history. Stanton observes that the ''autogynographical narrative'' is marked by conflicts between the private and the public, the personal and the professional (1984, p. 14). This split can be seen in the lives studied by Patricia Spacks. Spacks, analyzing the autobiographies of public women (Eleanor Roosevelt, Golda Meir, Emma Goldman, Dorothy Day, Emmeline Pankhurst), notes several strategies they employed in telling their lives. There was a negation of the individual self, and an identification of the self with a cause. Their accounts offered a self definition in terms of a public cause; the personal self was effaced. Subjects appropriated a sense of virtue, of being good or doing good, from a cause, not claiming its origin in the self. Themes of accomplishment and personal destiny (both of which would link public achievement to the self) are lacking from these narratives. In their accounts Spacks reads a hidden doubt about the subjects' own goodness, value and integrity. The doubt may be camouflaged by the recital of public achievements, accomplishments that are approved (but not for women); or by the recital of domestic decorum (which excites no public acclaim). Spacks concludes that her subjects' texts are profoundly contradictory.

In treatments such as Spacks', doubts and conflicts are always interpreted from within. It is my contention that Spacks' "rhetoric of uncertainty" reflects, not an unfledged self, but the approach/avoidance conflict so familiar to students of Matina Horner's research on the Motive to Avoid Success in women. Spacks' "selves in hiding," reinterpreted, are not so elusive: merely the selves which society forbids women to be. They need to be seen in the discursive context of images of heroism, virtue, authority, wisdom, all male. Doubts and division might well assail a woman confronting the challenge of telling her own life, having as models only male autobiographies, seeking to resonate with male history, theology etc.

An analysis limited to the individual level has the well-known (if unintended) consequence of "blaming the victim (Ryan 1971)." A different style of interpretation is adopted by Marcia Westkott. Westkott, in a feminist rereading of Karen Horney's psychology of women, is concerned explicitly with the mystification of "feminine traits" which, stripped of their context, lend themselves to essentialist fallacies. The psychology of women should be a social psychology (1986, p. 12), based on experience not "essence." The relevant context for the familiar feminine virtues was a 19th century culture in which women were systematically devalued, relative to men. Both societal devaluation and resistance are incorporated in women's personality. The crux of Westkott's conception is the female alienation that results from systematic devaluation. Westkott identifies female development as a dialectical relationship between the self and the social. Feminine nurturance is related to female subordination, affiliation to powerlessness, "sweet compliance to the rage to triumph (1986, p. 4)." In this way Westkott explores the underside of culturally valued "feminine" traits. The virtue/underside unity corresponds to the psychological concept of the motivational approach/avoidance conflict. The "dual and contradictory impulse," apparent even in the earliest female autobiographies, can be best understood as an instance of approach/avoidance conflict: the subject anticipates both gratification and punishment from the very same state or behavior. Avoiding the punishment means giving up the gratification, and vice versa. Such a conflict cannot be "reconciled;" approach and avoidance cannot be averaged.

Critical analysis of the split subject exhibits the limitations of the intrapsychic fallacy. The critical literature on women's autobiography is plagued by a failure to distinguish between the "split subject," the "divided self" and "divided consciousness," but this may be a minor failing in the light of overall weakness of conceptualization. It could be argued that the split subject is the appropriate way to conceptualize women's lives, unarguably composed of multiple roles which frequently conflict. Conceding that, one might want to avoid an essentialist notion of "a" or "the" self, which casts women's telling of their lives as problematic for the tradition of autobiography. The difficulty inherent in casting the split subject as problematic is illustrated in Spacks' linking the auto-biographical subject's "divided consciousness" with a "rhetoric of uncertainty" in her text. The example of Elizabeth Cady Stanton better illustrates approach/avoidance behavior than it does an impacted intrapsychic state without external referents. Spacks tells us that Stanton intended to record her private life, emphasizing her skill and pride in domestic achievements. But her account kept veering toward the public career. Keeping in mind the multiple audiences that female autobiographers keep in mind, we might interpret this inconsistency not in terms of uncertainty (an intrapsychic construct) but in terms of a behavioral alternation between contradictory communicative efforts. The display of domesticity might divert male criticism, but the account of her career might be of greater importance to women readers. In the light of the importance of female readers to the female subject (see below), this interpretation cannot be dismissed lightly.

A third hallmark of women's autobiographical writing is a profound ambivalence about disclosure and concealment. A major theme in the content of women's autobiographies is the desire to be known. The desire to be recognized as an individual burns strongly for the woman whose consciousness is effaced or distorted in the patriarchal arrangements under which she lives. We have noted earlier the difficulty of finding a voice where one does not see herself reflected in the literature that went before. The unpublished writings of women are sometimes the only guide to what women wish to say. Westkott (1979) emphasizes the importance of women's records of their consciousness, which may be their only sphere of freedom in a patriarchal social order where their behavior is tightly controlled. Westkott emphasizes the importance of personal documents of women—journals, diaries, memoirs and letters—in a society where behavior is substantially constrained and consciousness is women's sphere of freedom. Interpretation of women's behavior, she insists, must have as its referent women's consciousness.

> To ignore women's consciousness is to miss the most important area of women's creative
> expressions of self in a society which denies that freedom in behavior (1979, p. 429).

Westkott, as a sociologist, is unlikely to commit the fallacy of individualism and overlook the influence of constraint on behavior. "Considering behavior alone is insufficient to understanding women in patriarchy, nor is it adequate to link women's behavior only to the dominant, male-created ideology (Ibid.)." Westkott warns us not to neglect women's interpretation of their own conforming behavior and the effect of that interpretation on their behavior. This understanding requires privileged access to women's consciousness.

To be sure, the desire to be known wars with the sense of trespass and fear of retaliation or ridicule in women's autobiographical writings. Women's writings reveal their anticipated rejection or ridicule at the hands of the male reader and male critic, not surprising in the context of the institutional masculinity previously discussed. Yet in the writings there is also an express yearning for the female reader. The imaginative reaching toward the anticipated audience colors the third content area.

The Subject-Reader-Text Triad

In speaking of women's autobiography it is impossible to ignore the reader. In women's autobiographies it becomes clear that the reader is an essential partner in life-telling. The idea of an active collaboration between the subject and the reader is also found in current work on reader-response. Reader-response theory claims for the reader an active role in constructing the test she reads. The meaning of the text is not, in this way of thinking, controlled and fixed by the author. A text, it is said, may have as many readings as it has readers. The reader completes the text as she claims it. If this is so, texts do not have the fixity we were once taught, nor, perhaps are "the classics" eternal in their verity. The "universals" that privilege some texts at the expense of others begin to be "situated," in the sociological sense, and lose their majesty. Not only is authorial authority challenged by reader-response theory, but also the power of the critic, the legitimacy of the canon and "official" interpretation of a text. Reader-response opens the door to the claims of readers who have traditionally "not counted," the "invincible mediocrities" who may seek to elevate "minor" figures to the canon.

Schibanoff (1983) suggests that, at least since medieval times, there have been two classes of readers: the common reader, or consumer, and the primary reader, a gatekeeper or critic. These two classes of readers have quite different significance for women writers. Women writers, Schibanoff maintains, have been pessimistic about men's ability to hear women's voices, specifically about the primary audience's ability or motivation. Schibanoff detects a desire in women writers to circumvent the male gatekeepers and make direct contact with their female readers. There is a sense in which the idea of a female readership is a lifeline to the woman writing her life. Virginia Woolf speaks of readers as hearers of the long forgotten appeal, rescuers, life-givers (1925). Later we will see this theme of active succorance echoed in the thinking of biographers writing about life-telling.

Reader-response theory also directs attention to differences among readers as a major factor in the meaning of texts. In reading autobiography, it seems clear, gender affects the relationship of the reader with the subject. Brownstein (1982) proposes that women read novels in search of heroines, that is, someone whose life is worth writing about, a woman who is attractive and powerful and significant. The same search leads women to read autobiographies and biographies.

Brownstein contends that female readers search literature for the inscription of a female destiny that is heroic, that inspires them, that certifies a life of worth. The woman reader seeks a world that revolves around her, that makes being the way she is make sense (1982). I am reminded of Kazin's observation, "To have a sense of history one must consider *oneself* a piece of history (1979, p. 85)." The portrayal of women makes the world easier to claim, as their omission makes it more difficult.

The theme of a female reader/savior can be found in women's more "private" writings as well as in autobiographies intended for a broader public. The message in women's private writings may, like Elizabeth Foote Washington's diary, be intended explicitly for a daughter. The purpose of Elizabeth Foote Washington's manuscript changed over her lifespan, as did its intended recipient. In 1784 she wrote,

> I once had a thought of being more particular, and to have kept a journal of my life,—but that I could not have done faithfully, without speaking of all the ill treatment I ever met with, and that I did not wish to hand down. Therefore, whatever memorandums I have made in times past, I now shall destroy them all, and let only this manuscript book remain,—because should I have children, and especially daughters, it can be no disadvantage to them for to know something of my general conduct in my family (Evans, p. 346).

> What I have wrote her in this book, was done to let my child see what was my thoughts at the time I was going to change my state, and what was my thoughts at the times that her sister and self was born—also my method of behavior to my domestics—should I not live to give her direction and instruction thereon (Evans, p. 348).

Elizabeth Foote Washington's often-revised manuscript had as a practical goal the instruction of her daughters (then unborn) in household management and philosophy. She also felt that a record of feelings was an important and worthy content for communication with a surviving daughter. Elizabeth Foote Washington was much concerned about living a godly life, and imparting its precepts. As time went on, she survived her daughters, and it became clear that any future reader would be a relative, or even someone unknown to her. Ultimately, in 1792, she addresses this stranger directly:

I have thought should I meet with a female relation who appeared to think favourable of religion I might give away my manuscripts before my death, as I do not expect my dear partner will ever see again. But should I not do it—who ever you are that they fall into the hands of, I could wish for your own sakes that you would give them more than one reading—who knows but what it may please God to cause a thought to arise in your hearts that you may be better for it ever after (Evans, p. 355)."

Often female diary-writers or journal keepers mention—as Elizabeth Foote Washington did—that their husband has never read their private writings, or is unaware of their existence.

"I can declare with truth that my dear Mr. Washington has never seen any of them, so that was it to please God to restore him to his eye sight and take me from him, he would be surprised when he looked over my papers, though I trust he would set great value on all my scribble, for altho my heart has been much taken up with religion, yet very little have I talked about (1792) (Evans, p. 354–5).''

The prominence of the female reader is key to deeper significance for the autobiographical subject. Many anthologists have recognized that women's writing has survival value (cf. Hoffman and Culley 1985). A woman who dares not speak out can still speak up in her secret writings. In private she can oppose, accuse, spurn, revile, curse and threaten, doubt and challenge. She can lay down upon the page hurts she must conceal each day, despair she cannot declare. Fanny Fern urged women to write, for the sake of sanity. She felt that the domestic round into which women were locked, the overlooking of their finer sensibilities in marriage, could lead to madness without this outlet. She thought writing would be a safe outlet for thoughts and feelings that could not be heard even by those nearest and dearest to the writer. Fern felt it was not safe to repress so much, but that men were poor confidants for what women saw in their world (Wood 1971).

In addition to its function for catharsis, women's secret writings are recognized to have potential for future revenge.[3] Equally poignant, the writer of the secret diary could anticipate the amazement of her reader, and the recognition that in life (he) did not know her.

While secrecy and privacy have been the requirements for much autobiographical writing, the express desire to share thoughts and feelings is prevalent. Even secret writing entertained a future audience, and more than one diarist flirted with the idea of publication.

Autogynography and Posterity

One of the dilemmas with which female autobiographers grapple is that of posterity/obscurity. Unsure of their welcome in the male world of publishing, they are uncertain about their posterity. The celebrated women about whom Jane Marcus writes were famous in their lifetimes, but felt they had to publish their memoirs as a ''hedge against certain deflation of their reputations (1986, p. 7).'' Elizabeth Foote Washington, abandoning her plan to entrust her diary to a daughter, wrote in 1792,

> Now that it is probable that I may not have any children, so of course this book and all my other manuscripts must fall into the hands of some relative, who may laugh at them and think as my servants, and may not give them a reading—but throw them into some old drawer as waste paper, or give them to their children to tear up, as is too often the case with many people, they give their children books to play with and destroy. Tis certain I have wrote and copied together a great deal, an abundance I have destroyed. But what I have now I shall keep for my own satisfaction and comfort. Let them that comes after me think what they will, I do not keep them with an ostentatious view . . . (1792) (Evans p. 354).

Posterity is much on her mind when she writes, in 1792,

> Tis certain it would be a pleasing satisfaction had I a relation that I could think would set a value on my religious books of every description, and would read them all through at least once, if not oftener. I feel sorry when I reflect the time will arrive when they will be thrown by (p. 355).

Again, the survival of her subjectivity is Elizabeth Foote Washington's concern when she writes,

> Somehow I feel a greater desire to have had someone to have given this book to, than I do of any of the others. I suppose the reason is because this book is all my own thoughts and reflections. Though I am sensible there is great imperfections in the book. Yet I am desirous some one should have it—oh the weakness of human nature is great. But as I do not expect to have children now, if I leave this book behind me after my death, think I ought before that happens to write it over again, to correct what errors there may be in the diction of it—but whether I shall ever have the time, I cannot say. What pleasure a child would have taken in reading this after their mother was gone—but let me be dumb, and not say another word (1792, p. 357).

The other side of the disclosure/concealment dilemma is fear of penalty. The anticipated consequences of "going public" in an autobiography are intimidating for the female subject. The horns of her dilemma are trespass and silence.

Institutional Factors in Telling Women's Lives

In women's autobiography relationships become problematic that remain implicit in male autobiography and its criticism. The first of these is the reception anticipated by the woman who seeks to "go public" with her life. Generally, writers dealing with autobiography have not problematized the presumed easy relationship between the male citizen and the public world on whose attention he makes a claim by writing his autobiography. This issue cannot be ignored when female autobiography comes under consideration. In a cross-disciplinary parallel, Marcia Westkott pointed out the affinity between a patriarchal social science and the assumption of congruence and harmony between the individual and the social system (or, as it is frequently rendered in beginning sociology texts, "Man and Society"). In this "interpretive fallacy" the person and society mutually support each other. This, however, is not the world experienced by women. The alterity of women, as subject or as subject-matter, is reinforced by the masculine quality prevailing in discourses of life-telling. It is the male account, shaven and shorn, that lays claim to the territory of generic autobiography.

The anticipated consequences of "going public" in an autobiography are intimidating for the female subject. Sidonie Smith observes that a woman's reputation is "founded upon public silence (p. 96)" but "autobiography, an androcentric genre, demands the public story of the public life (p. 93)." A woman in "Choosing to write autobiography, therefore (she) unmasks her transgressive desire for cultural and literary authority (p. 90).'' Again, "The woman who does not challenge her culture's ideology of gender and its ideals about women's life script, textual inscription, and speaking voice does not tell her story in public (p. 91)."

"Going public" in this way claims a certain authority for the subject. The formal autobiography, a story of self written with the aim of dissemination, also implies a claim of significance. When women write autobiography, they know they are writing for public consumption, and they anticipate the hostility and ridicule of male critics and mandarins.

Our focus here is not the conflict experienced subjectively by the woman autobiographer, but the environing social facts to which she reacts. The concern of the woman autobiographer reflects the social structural position of the woman who wants to write, the interdictions and intimidation she faces. Gilbert and Gubar (1979) have indicated the ways in which male preemption of authorship and authority, interdict women's writing. Russ (1983) specifies initiatives men undertake to undo women's writing once written. In a parallel manner, social scientists anatomize institutional discrimination (Laws 1979), "microinequities" (Rowe 1975), and distortion of the evaluation of women (Nieva and Gutek 1984), all factors that penalize women's participation in male preserves. Foucault on censorship (1978) is discussing institutional, not individual force when he says,

> This interdiction is thought to take three forms: affirming that such a thing is not permitted, preventing it from being said, denying that it exists (1980, p. 84).

The traditions of the genre are intimidating to the female autobiographical subject. The issue of censorship is closely related to the relationship between the female autobiographer and the male critic. The institutional approach focuses attention on the canon, as a standard and as a source of models for autobiography, and on the role of critics as gatekeepers. Jelinek documents (1980; 1986) that the literary tradition of autobiography neglects the autobiographies of women. The paucity of women is even more true of the body of critical writing than it is of the population of work upon which it feeds. To give only one example, Sayre, in a book devoted to the autobiographies of Benjamin Franklin, Henry Adams and Henry James, avers that they are compelling for him because they "comment profoundly on the American experience." The female reader wonders how profound an understanding is gained that leaves out half the American experience; and how valuable a critical assessment for which the universalizing "*the* American experience" remains unproblematic. One of the institutional elements in the "invisibility" of women's lives is the lack of mention of women's autobiographies in such compilations as Olney (1980), Sayre (1964) and Krupat (1986). The neglect of women's autobiographies even in such recent efforts as Gunn's (1982), Egan's (1984) and Krupat's (1986) speaks for itself. It is particularly regrettable that Krupat, whose book is designed to counteract the neglect of Native American voices in the canon of autobiography, feels justified in neglecting those of women.

The critical literature on autobiography also evinces an anomalous asymmetry in scholarship. In the literature of any particular period, the reader will find that work on women's autobiography (the "special case") scrupulously reviews the corpus of masculine and 'generic' autobiography. However, authors writing about men (the general or inclusive treatment) fail to cite even the contemporaneous scholarship that deals with women.

The critical tradition thus contributes to the masculinity of the genre through its selective emphasis on male autobiography.

The masculinization of autobiography is not merely faulty scholarship, affecting the adequacy of understanding available to scholars in the field. Nor are its effects limited to the penalties female autobiographers suffer at the hands of male critics. Its effects are proactive as well: it casts an inescapable shadow over women considering, attempting or struggling to write their own lives. Ironically, it may be that women who have been privileged to study in universities are most hampered in life-telling by exposure to the male canon. As Dorothy Smith says, "The penetration of the society by the ideological process includes, particularly for the highly educated, an "in-depth" organization of consciousness (1979, p. 144)."

Another impediment to telling women's lives is the relation of women writers of women's lives to the canon. Being "invisible" to the critical establishment, the women's writing tradition does not appear in the canon, is not presented to aspiring writers as exemplary literature, is not manifest as the standard to be met or extended. Parallels exist here between in literature, social science and behavioral science. In each, women's writing is pushed to the periphery, and its content devaluated. The "otherness" by which women are defined in scholarly work is a deviant status, created by the dominant group for its own advantage (see Laws, 1975, 1979; Schurr, 1984). One consequence of the peripheralization of women's lives in literary studies is the effect on the woman striving to tell women's lives, whether her own or those of others. As Russ observes, women denied access to the feminine literary tradition carry the heavy burden of having to reinvent it in every generation. Every woman writer must then be a pioneer (p. 63ff.). A painful consequence of the maleness of the literary cannon is the perception that there have been no "great" women writers.

Dorothy Smith might have been describing this critical tradition when she wrote

Men attend to and treat as significant what men say and have said.
The circle of men whose writing and talk have been significant to one another.
extends back in time as far as our records reach.
What men were doing has been relevant to men,

was written by men
about men
for men.
Men listened and listen to what one another say.
A tradition is formed,
traditions form,
in a discourse of the past with the present.
The themes, problematics, assumptions, metaphors and images
form
as the circle of those present
draws upon the work
of those speaking from the past
and builds it up
to project it
into the future (1979: 137)

The interplay of gender in reader and writer, reader and critic has important consequences for reading and writing women's lives. First, if women's lives to not fit the pattern (and all evidence suggests they do not), then writing a woman's life, whether by the subject or the narrator, is problematic. Second, if women's autobiographies (and biographies) convey the pattern of women's lives, they will not fit the template, and this lack of fit will affect their critical reception and their relationship to the canon. The woman autobiographer, in seeking to make her life public, confronts a dual estrangement: from the male lifespan, different from hers in fundamental ways, and from the generic standard, which is masculine in outline though eviscerated.

These factors contribute to the problematic of telling women's lives, and contribute also to the urgency with which women seek a voice. In the context of institutional autobiography, telling women's lives has the significance of contestation.

Autobiography as Battleground

As the foregoing has made clear, a revolt of the readers is currently challenging the hegemonic structure of autobiography. Many a female reader might dissent from Leslie Stephen's comment, quoted by Olney, that no man has ever written a dull autobiography. Sayre, in a book devoted to the autobiographies of Benjamin Franklin, Henry Adams and Henry James, avers that he chose *the best and most revealing* (italics mine). Franklin, Adams and James are compelling for Sayre because they "comment profoundly on *the* American experience." The female reader wonders how profound an understanding is gained that leaves out half the American experience, and how valuable a critical assessment for which the universalizing "*the* American experience" remains unproblematic.

At present there is a challenge from female readers that parallels that from female subjects, and has the same root: the shape and content of women's lives and the insistence on making them known. It is evident that contemporary

women's autobiography embodies fundamental challenges to genre and its en-
forcers, and to the tradition of a generic subject in literature. Women writers and
readers appear to have the same agenda: contesting "ventriloquism," "female
impersonation" and stereotyping of women in literature, and "telling it
straight" instead.

As we have seen, there is more than a suggestion of collaboration between
autobiographical subject and reader. The reader is an acknowledged and a welcome presence in women's autobiography but often denied, effaced in men's
autobiography.

Among critics, there is a clear divergence that suggests gender interests.
Positions on the battleground seem straightforwardly related to subject positioning. This comes as no surprise to social scientists, for whom the sociology of
knowledge tradition teaches that the production of knowledge is socially situated
(cf. Mills 1940; Merton 1973). The gendering of criticism and genre is but a
further reminder of the importance of institutional factors in the telling and
reading of lives.

The writings of many female critics emphasize the conventional nature of
autobiography as a genre, and the constraints it imposes on female life-telling.
Bruss (1976) sounds the note of convention in observing that genres represent
stable conventions in a community that help define *what is permitted a writer
and expected of a reader* (italics mine). This pact is one of the sites of contestation, among autobiographical subjects and among critics. Stanton affirms the
force of convention, noting that every autobiography assumes and reworks literary conventions for writing and reading (1984, p. 10). Its discursive context
confines and perhaps defines the 'speaking subject.' However, Stanton questions
the whole system of genres, stratified as they are, and preoccupied with the
politics of inclusion/exclusion, mirroring the patriarchal gender system. In Stanton's critique the social control function of genre is emphasized.

If women view autobiography as too restrictive and too limiting, at least one
leading male critic deplores a lack of limits. James Olney says of current autobiography, "Here all sorts of generic boundaries *and even lines dividing discipline from discipline* are simply wiped away, and we often cannot tell whether
we should call something a novel, a poem, a critical disseration, or an autobiography (1980, p. 4, italics mine)."

Gunn (1982) notes that only the decline of formalism permitted autobiography
entree as a literary genre; previously it had been "too unruly" to fit in. One line
of division on the contemporary battleground is precisely the issue of how
"messy" autobiography is or should be. There is a suspicion that women and
other outsiders may be drawn to this genre precisely because of its messiness.
Olney opines, with apparent regret, ". . . there is no way to bring autobiography to heel as a literary genre with its own proper form, terminology and
observances." Elsewhere he states, "In the hands of other critics, autobiography
has become the focalizing literature for various "studies" that otherwise have

little by way of a defining, organizing center to them. I have in mind such "studies" as American Studies, Black Studies, Women's Studies and African Studies. According to the argument of these critics (who are becoming more numerous every day) autobiography—the study of a distinctive culture written in individual characters and from within—offers a privileged access to an experience (the American experience, the black experience, the female experience, the African experience) that no other variety of writing can offer (1980, p. 13)."

Elsewhere Olney seems to be aware that the genre of autobiography has privileged some self-referential accounts and excluded others. Olney observes that black history tends to be preserved in autobiography rather than standard histories—but does not analyze why this might be the case, and whether the boundaries of literary genres might constitute similar filters in regard to other groups. For Olney, neither the identity of the subject nor its referential status is problematic.

The issue of whether the Procrustean bed of genre is sacred is another issue along which critics diverge: should the narrative be lopped and chopped, shaven and shorn to fit the form, or should the form be expanded to encompass the accounts? Gunn comes out forthrightly for the position that genre is an instrument for reading, not for writing (1982, p. 21). It links the text with the world readers inhabit (and, I might add, the subject as well).

The parallelism between the gendered discourses of the culture and those of literature has been remarked by many feminist critics. Nancy Miller, in speaking of the novel, notes that the culture underwrites (only) certain narrative strategies (1980). Stanton points out that the discursive context confines and perhaps defines the 'speaking subject.' Sidonie Smith more bluntly views autobiography as a generic contract which reproduces the patrilineage (82). Jane Marcus defines autobiography as the male genius' obligation to the patriarchy, wondering, as we do here, what is the obligation of the female genius, and in what form is her life to be told?

Controversy about form is a central feature of current autobiography. Nonstandard content finds expression in innovation of form. Friedman (n.d.) challenges Gusdorf's stress on individuality in the autobiography, pointing out that where a group identity is imposed, the tradition of the self is misapplied. The split self may be inscribed in autobiographical forms that are more collective. (Friedman n.d.). The form for which women's self-referential writing is most often criticized—discontinuous, fragmented, digressive—is, in Stanton's thinking, the appropriate one for inscribing the split subject of women's autobiography. The texts of women's lives may retain the lineaments of conversation, may speak in intimate tones, may portray the self implicated in relationships. However, if Stanton is correct, the "personal" quality critics find in women's narratives is ambiguous: in revealing what is permitted and expected of her (e.g., a "domestic dailiness," a prosaic and limited round, a preoccupation with close relations), the female autobiographer is simultaneously concealing what is not permitted.

The "typical" female narrative is thus a partial inscription, perhaps an intentional disguise.

The split self necessitates the autobiographical act: Stanton asserts that (only by) "autogynography" is the fracture healed. That act of assertion, once achieved, may in fact breach the 'line of fault' between the self and the world of consequence. Stanton provides the basis for an argument that writing heals the split.

Current thinking about women's autobiography has yet to forge the link between the "split subject" and the split reader. That link might well be found in the communicative strategies adopted by writers who are aware of two classes of readers of their lives.

STRATEGIES FOR TELLING WOMEN'S LIVES

Three strategies used in first-person accounts of women's lives can be viewed as responses to the impulse to tell one's life in a context of threat: "telling it slant," "telling it straight" and telling it "messy." Differing dynamics of form, content and their relationship can be seen in the three approaches. Specifically, they represent three strategies for telling women's lives, given a critical establishment and a literary tradition that are not receptive to women's voices. Not surprisingly, then, women's autobiography exhibits an ambivalence about disclosure and concealment. The desire to be known wars with the sense of trespass and the fear of retaliation or ridicule. Different strategies for telling women's lives can be seen as attempts to compass this dilemma. All contain some degree of challenge to the all-male tradition of autobiography, and all imply some difficulty in being "read" or "heard" by a male audience. My reading of women's self-referential writings suggest that there may be at least three narrative strategies. Any one writer may employ any or all of these.

"Telling it slant"

Since Fanny Fern, American women writers have prefaced their work with self-deprecating disclaimers, have laid claim only to a readership that they define as peripheral and negligible (Wood 1984). The sense of trespass is palpable, in their writings and in the critical response. Hawthorne denounced the "scribbling women" who were invading the profitable field of writing for the popular press. Fanny Fern's works were controversial, and she had received her share of critical attacks on the grounds of insufficient femininity. While women writers tried to placate the critics by denying that they were men's competitors, Hawthorne was asserting himself in battleground metaphors (Wood 1984, p. 9). Early American women in print disclaimed any intention to compete with men, to pretend to literature. Caroline Lee Hentz, a popular southern novelist of the nineteenth

century, disavows all literary pretension, all skill, intention or intellectual effort
(asserting at the same time the theme of intimate revelation):

> Book! Am I writing a book? No, indeed! This is only a record of my heart's life, written at
> random and carelessly thrown aside, sheet after sheet, sibylline leaves from the great book of
> fate. The wind may blow them away, a spark consume them. I may myself commit them to
> the flames. I am tempted to do so at just this moment. (*Ernest Linwood,* Boston: JP Hewett &
> Co., 1856, p. 69)

Women's autobiographical writings thus contain defenses against anticipated
attacks: disguises, disavowals, self-deprecation. Nord (1985), in discussing
female traditions of Victorian autobiography, notes the predilection of women
for the humbler memoir, which lacks the personal and historical assertion of the
paradigmatic autobiography. Nord points out other means of "telling it slant"
intended to minimize threat and consequently retaliation. In the seventeenth
century, female memoirs were often disguised as biographies of a husband.
Religiously motivated memoirs written by women were also comparatively safe
from accusations of presumption, since their purpose was to draw attention to the
Creator and not their humble creatrix. Another conventional shaping of the
account of a woman's life placed causes or travels at the center, weaving an
incidental female life into the border. In anatomizing the autobiography of Beatrice Webb, Nord points out the boldness the autobiographical act requires in a
woman.

Negation of self, denial of affect, loyalty oaths and disclaimers of all descriptions—all are ways of "telling it slant," of avoiding a head-on confrontation
with male prerogatives. Social scientists have found parallel behavior in the
subjects of oral histories of women, and social researchers in general, encounter
similar disclaimers in attempting to induce women to tell their lives. Some critics
have suggested that a woman can tell her life more truthfully in fiction than in
autobiography (Huf 1983, p. 13). Thus the need for indirection counselled by
women's fears of male scorn is met by "telling it slant."

Women students respond with instant comprehension to a poem by Emily
Dickinson. Tillie Olsen and Adrienne Rich invoke the same poem, in their own
work on self-expression and secrecy. Clearly, it capsulizes a tension in women's
self expression. This is Emily Dickinson's text:

> Tell all the truth but tell it slant
> Success in circuit lies
> Too bright for our infirm Delight
> The truth's superb surprise
> As lightening to the Children eased
> With explanation kind
> The truth must dazzle gradually
> Or every man be blind—

(p. 1129)

Relatively little systematic analysis of "telling it slant" as a communicative strategy has appeared in studies of literature. From a sociological point of view, "telling it slant" comprises communicative behavior of a group in a less powerful position vis à vis a dominant group (Mitchell 1984). Characteristics of the communication reflect aspects of the climate of interdictions and intimidation women speakers and writers confront.

What are the characteristics of the text produced 'on the slant?' Feminist literary critics have recently suggested that women writers and readers share a common code, based on a separate literature (Kolodny 1985). Women's writing is thus "in code," which permits covert communication via a subtext, often rebellious or subversive, lurking beneath the manifest text which is written in the King's English (Miller 1980). Women's writing, by this argument, constitutes a self-referential canon which remains outside the awareness of the critical establishment. Male readers will ordinarily be unable to read a woman's text in the way women do.

By emphasizing the constraints that exist outside the writer herself, I do not mean to suggest an iron necessity to the communicative strategy of indirection. To view "telling it slant" as an imperative under patriarchy might lead to undesirable outcomes. One effect might be acceptance of the ghettoization of women's work and its alienation from the canon. Other possibilities exist, including other narrative strategies employed by women in their autobiographical writings.

"Telling it Messy"

I have suggested above that from the point of view of orthodox literary criticism, accounts of women's lives must be viewed as different, hence deviant, hence positioned at the periphery. There is pressure toward portraying women's lives as similar to men's as may be; and precious few models for telling lives any other way.

A different strategy for telling women's lives, which I call "telling it messy," points emphatically to a different direction. I have suggested that there exists a feminist challenge to the generic paradigm for autobiography and that the pressure to communicate women's lives as lived necessarily creates innovations in form. "Telling it messy" embraces complexity and open-endedness, in content and in form: it portrays lives as lived—the dailiness, the web of tasks, the web of relationships, the lack of closure, the nonlinear experience of time, the psychological and emotional realities. "Messy" accounts make no attempt to shear or to shave; indeed, their authors insist on retaining what is conventionally eliminated in generic biography and autobiography. Women's lives are particularly apt subject-matter for this project; the rhythms and routines of women's work contradict the smoothed and simplified male model. As Dorothy Smith has said, women's work depends on a diffuse and open organization of consciousness.

This second strategy for telling women's lives poses a more direct challenge to the canon. Accounts of women's lives concede baldly that women's lives are messy: they involve dirt, diapers, infections, repetitive labor, interruptions, lack of closure, obligations, intensity, vigilance, minutiae, endless process with no product, and lack of unilateral control.[4]

A number of writers and critics are advocating new forms capable of retaining the complexity of women's lives (see, e.g., Kolodny on Millett; Juhasz on Kingston). The dominant conventions of autobiography are rejected (Jelinek 1980), and some narrative conventions are rejected. If we include narratives that remain within the oral tradition, the narrow confines of the genre are further strained (Sterling 1984).

Kolodny's evaluation of Millett's *Flying* and of the critical response to it demonstrates the significance of the reader's subject positioning. To some critics Millett's autobiographical effort appeared "an endless outpouring of shallow, witless comment," interminable, tedious. The subject was unworthy, the content without interest, the form "confessional," not truly autobiographical. The content *might* have been transformed into a work of literature by one with the requisite (masculine) slant, as indeed many of the reviews cited by Kolodny confirm. It remained for Kolodny's reading of *Flying* to demonstrate that Millett's "formlessness" was a new form, consciously designed to convey content that had previously been denied entry. Millett explicitly challenges the critics and reinforced Kolodny's reading.

We have already seen that "messiness" in autobiography excites a negative critical response. This is unsurprising, in the light of my argument that the generic autobiography is shaven and shorn. Where absence is the norm, presence ("messiness") grates. Nor is the rejection of the "messy" content of women's lives limited to critical commentary on autobiography. The early critique of Sylvia Plath's poetry contained a similar rejection of the personal. Here, too, the style of criticism reflects a commitment to the masculine/generic outline, and a distrust of feminine concerns with domestic settings, with emotions and with relationships—all of which are "messy." Plath's poetry, termed "confessional," was judged lacking in artistry; its domestic content of dubious propriety; a fad with no future, the ravings of a madwoman.

There are two issues associated with the exclusion of women's subjectivity from literature (and social science, for that matter). Content that was previously suppressed, omitted, denied, sheared requires for its expression language and concepts that do not seem to be available. Writers in many fields are now grappling with the problems presented by the lack of a language to express women's subjectivity. The other issue, illustrated by the rejection of "confessional" poetry, is the problem of "intelligibility."[5]

Dorothy Smith raises both issues in observing,

... in opposing women's oppression we have had to resort to women's experience as yet unformulated and unformed; lacking means of expression; lacking symbolic forms, images, concepts, conceptual frameworks, methods of analysis; more straightforwardly, lacking self-information and self-knowledge (p. 144).

This suggests that inscribing women's lives is not a simple matter of using existing language, or even of "translating" unfamiliar experiences into familiar terms. "Telling it messy," as a means of communicating previously excluded content, will at least in the short run continue to be unintelligible from the subject positioning of the dominant group.

"Telling It Straight"

Feminist critique of renderings of women's lives, whether in social science or literature, has given rise to the desire to set the record straight, to correct the distortions, and to create a new, valid body of work. "Telling it straight" seems, paradoxically, the most difficult approach to telling women's lives. Writer, reader and subject are all implicated, to varying degrees, in the web of sexism that makes telling women's lives so problematic.

"Telling it straight" is the only strategy with the potential for healing the "line of fault" Dorothy Smith (1979) discovers between one's experience as a woman and the scholarly tradition to which one is devoting one's life. But "telling it straight" is not simple, not a direct connection between subject and reader; not a "natural," untutored communication. "Telling it straight" requires "straightening," the corrective of feminist analysis.

One method for "straightening" is simply to move women's experience from the periphery to the center. For this purpose, no enterprise is more strategic than telling women's lives. For if Stanton notes that only "autogynography" can inscribe the female, it is also true that it is only in women's writing that the female attains the status of subject. In men's writing, Stanton notes, women remain the inessential 'other,' an object.

Although Stanton sees female autobiography as problematic precisely because of the difficulty of inscribing the split subject, she also provides the basis for an argument that writing heals the split. One dimension of fracture is the sense of trespass involved in positioning the female self in the public sphere, in history. Writing the autobiography, says Stanton, is a conquest of identity through writing. That act of assertion, once achieved, in fact breaches the 'line of fault' between the self and the world of consequence. It does more than merely connect the previously disjunct, however; the fact of self assertion changes the position of figure and ground, of female and male. It is the revenge of the dancing dog: if dogs can indeed dance, then thinking about dancing is changed, as well as thinking about dogs. The act of assertion, once achieved, is in fact not only substantive but methodological; methodological upheavals are an inevitable accompaniment of feminist scholarship.

Stanton holds that writing the *autos* is an act of self assertion that is essential to the denial and reversal of the status assigned to women under patriarchy. The same kind of assertion of behalf of another woman is an underlying motivation in the telling of women's lives by women—e.g., in biography or social science.

"Telling it straight" requires a feminist writer and a feminist reader (if not a feminist subject). There is, in addition, a role for feminist literary criticism, mediating the relationship between the author and the reader, and that between the work and the canon. Increasingly, feminist scholars have come to see that an unavoidable part of setting the record straight is exposing the role of their own disciplines in discrediting woman-centered work (e.g., Showalter 1985). Feminist criticism is required to set the record straight, as in the case of Millett's or Plath's work. To discern and to communicate the artistry and the structure in work about women's lives is a task to which the literary establishment has proven itself unequal. Not surprisingly: understanding women's experiences, or accounts of them, requires one to understand women's place in society and history (Hilda Smith 1976).

New Directions in Life-Telling

It is fundamental to my critique that "generic" autobiography and biography do not offer adequate models for telling women's lives. To begin, we must burst the smoothed and mythic trajectory of the hero's life. We must insist on full scope for the shape and pace of women's lives. New forms required by the new content must emerge from an understanding of women's first-person accounts of their lives.

The importance of autobiography cannot be overestimated. Nevertheless, others have an important role in telling women's lives. Much actual and potential innovation in telling women's lives stems from disciplines with a third-person tradition: anthropological life history, biography, psychoanalysis, history, sociology. Current work in a number of these fields abandons the researcher's traditional posture of distance and control and the convention that he is not part of the situation. The acknowledgement of the narrator's presence seems to open up the possibility of acknowledging other, non-cognitive attributes that can enhance the research process (see below).

The view of the text as a natural object (or "data") is giving way to an appreciation of the account as a social emergent. Rather than thinking of the life history (for example) as something that is "in" the subject that the researcher tried to "get out," it may be thought of as a record of meaning constructed in the dialogue between subject and narrator (Reinharz 1984).

Approaches newly developed in a number of disciplines offer some promise of restoring life-telling to life. Above all, they place the text of a woman's life within the interpersonal relationship which produced the account. Most radical of all, they reintroduce and situate the narrator in all her historical, social and biographical particularity. The new approaches contradict the separation of subject and object, of 'subject' and narrator. They assert the essential role of emotion, of the subjective, in understanding a life, and recognize the emotional resonance between subject and teller. A few examples will serve to indicate some of the innovative approaches.

John Kotre, a psychologist, works within the tradition of life history. Kotre's work focuses substantively on the phenomenon of generativity—the psychic and emotional involvement of individuals with the future. Kotre's book is entitled *Outliving the Self,* in reference to his definition of generativity: "The desire to invest one's substance in forms of life and work that will outlive the self (p. 10)."

Because his research focuses on generativity, Kotre sees himself, the researcher, as a "generative target" in the life history interview. He is a symbol of posterity, with whom the life story is shared, and a catalyst for the life review (p. 30). Kotre's presence becomes for his subjects a mirror before which to puzzle over their lives, a record to be set straight, a means for making their life permanent. He is a witness to the hope that life has had value; the occasion for creating their own image and likeness (p. 30).

Kotre speaks of "generative transference" in the subject, because the life-storytelling interaction and the account produced become generative, in the light of their purpose. Although its form is retrospective, the account has a future referent.

Kotre neglects to speak of "generative countertransference," although he might well do so within his conceptual framework. Biographers not uncommonly refer to the life-teller's generative aims. Pachter, for example, speaks of the biographer's "near-missionary drive to save, if not a soul then a personality for the company of future generations (1979, p. 4)." The biographer explicitly acknowledges her/his emotional investment in making the subject outlive her or his mortal span. The life-teller sees a past life being connected with the present and with the future through her/his agency.

If Kotre focuses on the subject's psychological processes as they involve the researcher, Loewenberg focuses on those of the researcher, as they inevitably affect the portrait of the subject that is produced.

Peter Loewenberg (1983), both a historian and a psychoanalyst, calls his method "psychohistory." His approach gives prominence to an element of the research exchange—here, the taking and giving of a life history—that is often ignored or hushed up. Loewenberg calls this factor "countertransference: the emotional and subjective sensibility of the observer." Countertransference is the process by which the observer perceives the unconscious of the subject via her own unconscious (1983, p. 4). As might be expected from the psychoanalytic framework, Loewenberg urges attention to conflict, in addition to content; he urges the analysis of personal style, and latent or unconscious themes in the life. Loewenberg asserts that the researcher's emotional and subjective resonance with her subject's life, far from confounding the work, contributes depth and conviction.

Both the historian and the psychoanalyst, according to Loewenberg, rely on immersion in their subject-matter. Both history and psychoanalysis place the observer in the midst of the field she analyzes, requiring a special mixture of identification and detachment. The analyst must become an integral part of the subject's psychological field, and experience it fully, without either resisting it or being damaged by its tensions. The analysts's scrutiny of her own countertransference presumably provides the balance she needs in order to avoid being taken captive in the psychic terrain in which she is sojourning.

Countertransference, like transference, involves bringing patterns of feelings, loves, hates and fantasies from other periods and persons of one's life into the working present.

Freud saw the analyst's unconscious as a sensitive scanner used in the process of understanding. He rejected any a priori basis, such as that furnished by theory, for selecting any realm or item as significant. The analyst, he felt, must be able to tolerate this lack of control.

Loewenberg sees the thoughtful incorporation of emotion into the researcher's work not as a desirable option, but as a necessity. No phenomenon has any inherent meaning; it becomes a datum only via the intervention of the researcher, whose personal psychodynamics affects its every transformation (1983, p. iii). "Distortion arises from the failure to account for the observer *in each act of knowledge* (1983, p. iii, italics mine)."

It is Bell Gale Chevigny who has joined awareness of the narrator with awareness of the subject and developed a theory of the reciprocity in their motivations. Chevigny's starting point was a sense of unfinished business that persisted once her biography of Margaret Fuller was completed.

Chevigny, in common with other biographers, had experienced a profound identification with her subject. Chevigny experienced the intensity of her engagement with Fuller as alien and troubling; she feared introducing distortion through her own feminist commitments. She also experienced her involvement with her subject as "promising;" and, indeed, the psychic resonance which she permitted to capture her attention proved the basis for a profoundly important feminist theory of biography. The dangers were balanced by the promising possibility that by engaging the identification with the subject (rather than fleeing it) one can achieve a deeper and clearer understanding of her than that vouchsafed by "objectivity (p. 81)."

Chevigny came to identify the psychic energy source in her professional involvement with Fuller as a symbolic mother-daughter bond. The clue which eventuated in Chevigny's insight emerged from her sense of a link between her own biographical act and Fuller's autobiographical act.[6] Fuller's autobiographical act was a search for a precursor, as indeed was Chevigny's biographical act.

Chevigny found herself haunted by an unfinished manuscript fragment that oddly distorted the facts of Fuller's childhood and her mother. This voice from the past spoke to her, compellingly but without clarity. Chevigny placed this effort beside two other attempts that Fuller made to tell her life. Chevigny offers several possible readings of Fuller's "manuscript fiction." More important than the specific interpretation is, however, Chevigny's recognition that what Fuller was doing (and continued to do) was to grapple with her relationship with her mother, seeking to wrest from it what she needed. In becoming aware of a parallel process in her engagement with Fuller, Chevigny came to the realization that she was creating in Fuller a mirroring self, and simultaneously providing such a mirror for her subject. Chevigny's position is that no woman has adequate models (particularly for autonomy); hence generational separations break down and we must all "mother" each other. In such a symbolic relationship that is both intellective and empathic, Chevigny says, an author can be in possession of, and possessed by, her subject in such a way that the embrace enhances the autonomy of both (p. 81).

CONCLUSION

In this chapter women's autobiography has been the focus of an analysis of how gender influences women's self-referential writings. Women's autobiography is a primary means of access to women's subjectivity. That subjectivity, however, is continuously influenced by the gender system; it is not "purely personal" but resonates with the expressed experience of other women. One focus of the chapter, therefore, was emerging themes of women's autobiographies. One cannot discuss the concerns of women autobiographers—their daily life, relationships with others, a yearning to be understood, the fundamental fact of interpersonal connectedness, the fear of masculine ridicule, the fear of oblivion—without understanding them as a reflection of the subject's social world. A thoroughgoing analysis of the gender system is fundamental to understanding the social worlds of women (and men). In the emphasis I have given to social and institutional forces affecting women's autobiography, it is understood that the analysis of gender is a first—and overdue—step toward adequately specifying the social location of subjects.

When we examine the history of published autobiographies of women we find ourselves face to face with the politics of publication, evaluation, preservation. Critical evaluation is solidly rooted in the conventions of genre. In this chapter these too are subjected to a gender analysis.

It is clear that women's autobiographies represent a challenge to the established canon, even if only at the level of content. That challenge is sharpened by the feminist critique that has arisen within the disciplines. The potential impact of a feminist critique of the disciplines is magnified by communication and collaboration across disciplinary lines. Telling women's lives, according to the many existing traditions, is an extremely strategic focus for trans-disciplinary work whose impact on the disciplines will be considerable.

The feminist critique of "generic" and masculine autobiography underscores the need for women's accounts, with their counterpoint and inevitable corrective effect. Telling women's lives is vital to the future project of a more valid body of knowledge and theory.

An emerging feminist methodology has made signal contributions to the re-conceptualization of the role of the researcher/narrator and of the relationship that connects subject and narrator. Future theoretical development can undoubt-edly extend that same thinking to the relationship between subject and reader, and that between narrator and reader. Limitations of space in the present essay prevent a full discussion of the contributions of feminist methodology to telling women's lives (but see Barry 1987; Hall, 1987; Long 1987; Reinharz 1984).

NOTES

1. Sidonie Smith notes that the woman who does not want to challenge overtly her culture's ideology of gender may choose as the language of self-expression diaries, letters, journals (1987).

2. We may speculate that the male writer of autobiography feels more at home, and more welcome, in politics, religion, economics, the academy, the military and indeed in history, than does his female peer. He has only to make the claim, to assert the connection between his life and the congenial tradition of his chosen field of endeavor.

3. Fanny Fern reveals the fantasy of revenge in predicting the reactions when the secret diary is posthumously read. Women's writing would someday accuse the men who ignored or brutalized them. Phyllis Rose (1983) recounts the successful execution of this project by Jane Carlyle, whose secret diaries reproached her husband as she had never done.

4. Male literary critics and social scientists often have visceral reactions to the material of women's lives. In social science, for example, men remain unattracted by the topics of domestic labor and of everyday life, which are compelling for feminist scholars.

5. "Intelligibility" is an attribute of the reader, not of the text. A failure or refusal to com-prehend a text is an element in the process of the social production of obscurity (Long 1987). By means of specific mechanisms of perception, evaluation and attribution a woman's work and life can be relegated to obscurity.

6. Chevigny views autobiography as an imaginative construction whose purpose is to justify the writer's current sense of self. This effort is particularly crucial for women who are living uncharted lifestyles, whose lives have changed direction or whose sense of self has changed substantially (p. 83).

REFERENCES

Alta. 1974. *Momma: A Start on All the Untold Stories.* New York: Times Change Press.
Bakhtin, M.M. 1969. *The Dialogic Imagination: Four Essays.* Holquist, Michael, Ed. Emerson, Carlyle, Tr. New York: Russell Sage Foundation.
Barry, Kathleen. 1987. "Theory of Women's Biography." Paper presented at the annual meetings of the American Sociological Association, Chicago.
Brownstein, Rachel M. 1982. *Becoming a Heroine: Reading About Women in Novels,* New York: The Viking Press.
Bryant, R.K. Aug. 1974. "Drowning in Claustrophobia." *National Review* 30: 990.

Chevigny, Bell Gale. 1983. "Daughters Writing: Towards a Theory of Women's Biography." *Feminist Studies* 9, (1): 79–102.

Chicago, Judy. 1975. *Through the Flower: My Struggle as a Woman Artist*. Garden City: Doubleday.

Cohler, Bertram J. 1982. Personal Narrative and Life course. In *Life Span Development and Behavior V.IV*. NY: Academic Press.

Egan, Susannah. 1984. *Patterns of Experience in Autobiography*. Chapel Hill: University of North Carolina Press.

Evans, Elizabeth. 1975. "Elizabeth Foote Washington." In *Weathering the Storm: Women of the American Revolution*, edited by Elizabeth Evans. New York: Charles Scribner's Sons.

Foucault, Michel. 1980. *The History of Sexuality, V.I*. New York: Random House.

Friedman, Susan. forthcoming. "Women's Autobiographical Selves." In *The Private Self: Woman's Autobiographical Writings*, edited by Shari Benstock.

Frye, Northrup. 1957. *Anatomy of Criticism*. Princeton: Princeton University Press.

Gilbert, Sandra and Susan Gubar. 1979. *The Madwoman in the Attic: The Woman Writer and the Nineteenth Century Literary Imagination*. New Haven: Yale University Press.

Gunn, Janet V. 1982. *Autobiography: Toward a Poetics of Experience*. Philadelphia: University of Pennsylvania Press.

Gusdorf, Georges. 1956. "Conditions et limites de l'autographie." *Formen der Selbstdarstellung* edited by Reichenkron, G. and E. Haase, Berlin: Duncker & Humbolt.

Hall, Jacquelyn Dowd. Spring 1987. "Second Thoughts: On Writing a Feminist Biography." *Feminist Studies 13*, 1: 19–37.

Hoffman, Leonore and Culley, Margo. 1985. *Women's Personal Narratives: Essays in Criticism and Pedagogy*. New York: MLA.

Huf, Linda. 1983. *A Portrait of the Artist as a Young Woman*. New York: F. Ungar Publishing Co.

Jelinek, Estelle. 1980. *Women's Autobiography: Essays in Criticism*. Bloomington: Indiana University Press.

————. 1986. *The Tradition of Women's Autobiography: From Antiquity to the Present*. Boston: Twayne Publishers.

Johnson, Barbara. 1984. Metaphor, Metonymy and Voice in *Their Eyes Were Watching God*. Pp. 205–219. *Black Literature and Literary Theory* edited by Gates, H. L., Jr. New York: Methuen.

Juhasz, Suzanne. 1980. "Towards a Theory of Form in Feminist Autobiography: Kate Millett's *Flying* and *Sita;* Maxine Hong Kingston's *The woman Warrior*". Pp. 221–237. in *Women's Autobiography* edited by Estelle Jelinek Bloomington, Indiana: Indiana University Press.

Kolodny, Annette. 1980. "A Map for Rereading: Or, Gender and the Interpretation of Literary Texts." *New Lit. His. xx.*

————. 1980. "The Lady's Not for Spurning: Kate Millett and the Critics." Pp. 238–259 in Jelinek, Ed.

Kazin, Alfred. 1979. "The Self as History: Reflections on Autobiography." Pp. 74–89 in *Telling Lives: The Biographer's Art* edited by Marc Pachter. Washington: New Republic Books.

Keller, Evelyn Fox. 1985. *Reflections on Gender and Science*. New Haven: Yale University Press.

Kingston, Maxine Hong. 1977. *The Woman Warrior: Memoirs of a Girlhood Among Ghosts*. New York: Vintage Books.

Kotre, John. 1984. *Outliving the Self: Generativity and the Interpretation of Lives*. Baltimore: The Johns Hopkins University Press.

Krupat, Arnold. *For Those Who Come After: A Study of Native American Autobiography*. Berkeley: University of California Press.

Kuhn, T.S. 1979. *The Essential Tension: Selected Studies in Scientific Tradition and Change*. Chicago: University of Chicago Press.

Laws, Judith Long and Schwartz, Pepper. 1977. *Sexual Scripts: The Social Construction of Female Sexuality*. Hinsdale, IL: Dryden Press.

Laws, Judith Long. 1975. "The Psychology of Tokenism: an Analysis." *Sex Roles 1*, (1): 151–174.

Laws, Judith Long. 1979. *The Second X: Sex Role and Social Role.* New York: Elsevier.

Loewenberg, Peter. 1983. *Decoding the Past: The Psycho-historical Approach.* New York: Alfred A. Knopf.

Long, Judy. Aug. 1987. "Telling Women's Lives: The New Sociobiography." Paper presented at the annual meetings of the American Sociological Association.

Lord, Audre. 1982. *Zami: A new Spelling of My Name.* Trumansburg, N.Y.: The Crossing Press.

Marcus, Jane. April 1986. "Invincible Mediocrity: the Private Selves of Public Women." Paper presented at the Conference on Women in Biography and Autobiography, Stanford.

Merton, Robert. 1973. "Perspectives of Insiders and Outsiders." In *The Sociology of Science: Theoretical and Empirical Investigations* edited by Robert Merton. Chicago: University of Chicago Press.

Miller, Nancy K. 1980. *The Heroine's Text: Readings in the French and English Novel, 1722–1782.* New York: Columbia University Press.

Mills, C. Wright. Situated actions and vocabularies of motive. *Am. Soc. Rev.* 5: 904–913.

Mitchell, Gillian. 1984. Women and Lying: A Pragmatic and Semantic Analysis of 'Telling it Slant.'' *Wom. Stu. Int. Forum 7,* 5: 375–83.

Nieva, Veronica and Gutek, Barbara. 1984. Sex Effects in Evaluation. *Academy of Management Review* 5, 2: 267–276.

Nord, Deborah Epstein. 1985. *The Apprenticeship of Beatrice Webb.* Amherst: U. Mass. Press.

Oakley, Ann. 1981. Interviewing Women: a Contradiction in Terms. Pp. 30–61 in *Doing Feminist Research* edited by Helen Roberts Boston: Routledge and Kegan Paul.

Olney, James. 1972. *Metaphors of Self: The Meaning of Autobiography.* Princeton: Princeton UP.

————. *Autobiography: Essays Theoretical and Critical.* 1980. Princeton: Princeton UP.

Olsen, Tillie. 1978. *Silences.* New York: Delacorte Press.

Ortner, Sherry. 1975. Is Female to Male as Nature is to Culture? In *Women, Culture and Society* edited by Michele Zimbalist Rosaldo and Louise Lamphere. Stanford: Stanford University Press.

Pachter, Marc. Ed. 1979. *Telling Lives: The Biographer's Art.* Washington, D.C.: New Republic Books.

Reinharz, Shulamit. 1984. *On Becoming a Social Scientist.* New Brunswick, N.J.: Transaction, Inc.

Rich, Adrienne. 1979. Women and Honor: Some Notes on Lying. In *On Lies, Secrets and Silence* edited by Rich New York: W.W. Norton.

Rose, Phyllis. 1983. *Parallel Lives: Five Victorian Marriages,* New York: A Knopf.

Rowe, Mary. 1975. Micro-inequities and Unequal Opportunity: Saturn's Rings II. Cambridge, MA: unpublished paper, MIT.

Russ, Joanna. 1983. *How to Suppress Women's Writing.* Austin: University of Texas Press.

Ryan, Leo. 1971. *Blaming the Victim.* New York: Pantheon Books.

Sayre, Robert F. 1964. *The Examined Self: Benjamin Franklin, Henry Adams, Henry James.* Princeton: Princeton University Press.

Schibanoff, Susan. 1983. Early Women Writers: In-scribing, or, Reading the Fine Print. *Women's Studies Int. Forum* 6, (5): 475–89.

Schurr, Edwin M. 1984. *Labeling Women Deviant: Gender, Stigma and Social Control.* New York: Random House.

Showalter, Elaine, Ed. 1985. *The New Feminist Criticism: Essays on Women, Literature and Theory.* New York: Pantheon Books.

Smith, Dorothy. 1979. "A Sociology for Women." In *The Prism of Sex: Essays in the Sociology of Knowledge,* edited by Sherman, Julia and F. Beck. Madison: University of Wisconsin Press.

Smith, Hilda. 1976. "Feminism and the Methodology of Women's History." Pp. 368–84 in *Liberating Women's History* edited by Carroll, Berenice. Urbana: Univ. of Illinois Press.

Smith, Sidonie. 1987. *A Poetics of Women's Autobiography: Marginality and the Fiction of Self-Representation.* Bloomington: University of Indiana Press forthcoming.

Spacks, Patricia Meyer. 1980. "Selves in Hiding." Pp. 112–133 in Jelinek, Ed.

Spengemann, W.C., and Lundquist, L.R. 1965. "Autobiography and the American Myth." *American Quarterly* 17: 501–519.

Sprinker, Michael. 1980. "Fictions of the Self." In Olney, *Op. Cit.:* 322ff.

Stanton, Domna. 1984. Autogynography: Is the Subject Different? In *The Female Autograph* edited by Domna Stanton. New York: New Literary Forum.

Sterling, Dorothy, Ed. 1984. *We Are Your Sisters: Black Women in the Nineteenth Century.* New York: W.W. Norton & Co.

Washington, Mary Helen. 1986. Lecture delivered at Syracuse University.

Westkott, Marcia. 1979. "Feminist Criticism of the Social Sciences." *Harvard Educational Review* 49, 4: 422–30.

———. 1986. *The Feminist Legacy of Karen Horney.* New Haven: Yale University Press.

Wood, Ann D. 1971. The "Scribbling Women" and Fanny Fern: Why Women Wrote. *American Quarterly:* 3–24.

Woolf, Virginia. 1976. *Moments of Being.* New York: Harcourt Brace Jovanovich.

———. May 1925. The Lives of the Obscure. *Dial:* 381–82.

CONTINUITIES AND DISCONTINUITIES IN ELDERLY WOMEN'S LIVES:

AN ANALYSIS OF FOUR FAMILY CAREERS

Katherine R. Allen

INTRODUCTION

Women's lives are organized mainly around their families. Regardless of marital and parental status, the family in which one is born and the family one establishes at marriage are critical to the structure of life events and to the subjective perception of those events. In this article, the family careers of four women are examined in depth. The subjects are from the same birth cohort and socioeconomic status, but they have led different lives in accordance with their marital and parental careers. The first woman married, had children, and is still married. The second woman was married, bore children, divorced, remarried, and di-

Current Perspectives on Aging and the Life Cycle,
Volume 3, pages 225–242.
Copyright © 1989 by JAI Press Inc.
All rights of reproduction in any form reserved.
ISBN 0-89232-739-1

vorced her second husband. The third women was married, had children, and was widowed. The fourth woman remained single and childless her entire life. These four family careers, represented by a married, divorced, widowed, and single woman, respectively, are typical of the wider population of women aged 65 and over, consisting of 36.5 percent who are married, 3.2 percent who are divorced, 52.2 percent who are widowed, and 8.1 percent who are lifelong singles (Glick, 1979).

THEORETICAL ORIENTATION

The life course perspective (Elder, 1977; 1981; Hareven, 1977; 1982) guided this inquiry into the variant family careers of white, working class women from the 1910 birth cohort. The concepts of timing, sequencing, and transitions were used in the analysis of the women's lives. The analysis of the life histories investigated here proceeded from the assumption that the present cannot be understood without an appreciation of the cumulative nature of past experience. The life course approach offered an opportunity to view diversity in women's lives from a perspective of variation rather than deviance. Typically, the family life cycle model is used to examine women's life history, but this concept considers only normative stages in the marital and parental careers. The present study included the experiences of women who did not marry or bear children in order to more accurately portray the extent of women's involvement in their families over time (Allen and Pickett, 1987). A key finding from the original study was that never-married women *maintained* the family by taking care of ancestors, parents, and siblings, and serving as surrogate mothers for their siblings' children. Widows *extended* their families through descending generations of children and grandchildren.

The themes identified in these four life histories reflect three levels of experience. Using Runyan's (1984) scheme of three levels of generality in the study of lives, the analysis of life histories encompasses the search for universals in human experience (i.e., what is true of all people); group differences due to sex, race, class, culture, historical period, etc., (i.e., what is true of groups); and individual characteristics (i.e., what is true of particular individuals). The purpose of the present study was to examine the experiences of the women in general, the experiences of women in each of the marital groups, and the unique experiences of each individual woman (Allen, 1989).

RESEARCH DESIGN

The larger study from which these four life histories are taken consisted of a sample of 104 women born around 1910. The major purpose of the larger study was to examine the family careers of never-married women, and to compare their life histories to a matched sample of widows—women who had experienced the traditional life course of marriage, motherhood, widowhood, and grand-motherhood (Glick, 1977; Uhlenberg, 1974). The focus was to understand varia-tion in female family careers from the perspective of the women who had experi-enced normative events and those who had not.

Data collection proceeded in three stages: 104 women were interviewed ini-tially to collect demographic information and to highlight their major life events and transitions. At that time, if they fit the requirements for the full study, they were invited to participate in the remaining two interviews. If they were not eligible, the interview process ended. The second phase involved an in-depth interview, which was tape-recorded and transcribed verbatim. The women were asked a series of open-ended, semi-structured questions and statements regarding their daily activities, their family and friends, perceptions about their life events, and comparisons of themselves to others. Finally, in a third interview, a detailed life history was completed on their residential history, occupational history, health history, family of orientation, and family of procreation.

Of the 104 women initially interviewed, only 30 were eligible for the com-parison of never-married women and widows. These 30 women completed ap-proximately five hours of interviews in three sessions. This sample is described elsewhere (Allen and Pickett, 1987), but a brief summary follows. The in-depth group consisted of white, working class women from the 1910 birth cohort. Their actual birth years ranged from 1907 to 1914. They held positions such as homemaker, hairdresser, secretary, factory worker, and nurse's aide. They were selected by contacting senior citizen clubs and centers in a metropolitan area of upstate New York. They lived in their own homes or apartments at the time of the interviews which were conducted in 1984.

In addition, the life histories of several other women were completed for the purpose of comparing the four family careers that are discussed in this article. A woman who was still married and a divorced woman also completed the entire interview process. These two women match the women in the in-depth sample of 30 in every way except for marital career. In other words, they were also white, working class women who lived independently and were in their seventies at the time of the interviews. For the discussion of the four life histories in the present article, one never-married woman and one widow were selected from the in-depth group of 30 to include with the married woman and the divorced woman noted above.

What follows is a summary of each of the four women's life histories, as organized around their families. The narratives are highlighted by direct quota-tions from the interview transcripts, enabling much of the life histories to be told by the women themselves, as Geiger (1986), Glassner (1980), and others have advocated. Names and minor details about each life have been changed to protect anonymity. Continuities and discontinuities in their lives are shown in each vignette and then discussed following the presentation of each story.

Mae: A Married Woman

Mae was born in 1913, the daughter of Italian immigrants. She was from a large family, and when she was six, her parents divorced. The children were separated, and she and her sisters went to live in an orphanage for girls in the same community where her parents and brothers lived. Her experience of going to the orphanage was a common one for women in the in-depth sample. Half of the subjects or their siblings lived in an orphanage for some portion of their childhood. Recourse to charity was common for members of this sample, as I have described in greater depth elsewhere (Allen, 1989).

After seven years, Mae left the orphanage to live with her mother again, who had since remarried. By the age of 21, she married the man with whom she recently celebrated her 50th wedding anniversary. They had three children, born during the Great Depression, and five grandchildren. She was unlike most of the women in this working class sample in that she did not work outside the home following her marriage. Specifically, among the 30 women in the in-depth group, only one widow and one never-married woman were homemakers for their entire adult lives.

Mae described her lengthy marriage as comfortable, and her husband as her closest companion. They had very separate roles as husband and wife:

> When my children were growing up, going back that far, I didn't get to see anyone much, because I was home all the time. I took care of my children, and I didn't believe in going out and leaving the children. In those days they didn't do like they do now, go to work and those things. My husband wouldn't let me work anyway . . . He said if I go to work, he'd quit. But he's a darn good man, very active, won't sit. In 50 years, I don't think he's missed four days of work.

Like other women in the sample, Mae gave me a tour of her home. It was immaculate, and every object was neatly placed. It was striking to see the separate spaces allotted in the home to her things and to her husband's things. Their separate roles as husband and wife were expressed in the demarcation of his space and her space. She showed me her husband's extensive workbench, pulled out all the drawers, and lingered over each of his tools. She opened her cupboards to reveal her collection of dishes. Her home had the quality of a shrine, but the homages to people were missing. In contrast to the homes of other women, there were no pictures of her children, siblings, parents, or grandchildren on the walls. Her perfectly ordered home stood in direct contradiction to the painful disruption of her childhood family, as if to reveal the unresolved theme in her life: intact home, disrupted family.

As she described the routine she follows in her life today, she slipped back into memories of her childhood. She linked the regimentation of her present life back to the orphanage and the discipline imposed by the nuns who became her surrogate mothers:

Everything I know I think they taught me, because I wasn't very old when I went there. And everything was regimentation, you know. The girls got up at 6:00, everybody knelt by their beds and said their prayers, everybody marched down to the breakfast table, everybody said their grace on time. Then after that, everybody had a certain duty to perform. And after that was over, we went to class, and then after class, we were told what to do; we'd go to the library or things like that.

The early discontinuity in her life—separation from parents and brothers when she was just beginning school and having to live at the "asylum, as it was called back then"—was the memory around which her life history was organized.

That was just the most outstanding thing in my life, was going to that home, and all the things that happened. It was the biggest thing that happened to me. . . . It makes me blue at times, it saddens me. Oh, it's all gone, it's all over with now, why even bring up the subject. That's why I hate to talk about it . . . and the only time I really talk about it is if somebody else is talking about it. Oh, there are days when you feel not as good as others and my husband will say, "What's the matter?", and I'll say, "Oh, I feel blue today, I'm thinking." But that's why I like to get out as much as I can and do activities. I love to get out and forget it. . . . To me they're wonderful memories in a way, because I was happy while I was there. But the reason I was there wasn't a good one.

Regarding her parents' divorce and the disruption to her young life, she noted:

We lost out on a lot of love. Mom and dad loved us, but they didn't give us the attention that we would have liked. But when I got to the orphanage, they showed us a lot of attention, and they taught us everything I learned.

A common experience of these working class women was to blame one of their parents for the problems in their families of orientation. For Mae, it was her mother's fault more than her father's. After discussing the importance of her parents in her life, she noted that her father had been more important to her than her mother:

It just seems like yesterday. I think he was hurt more by her than mother was. If I can say this, he used to come and see us at the orphanage and he'd say, "Mae, there's a tear on every sidewalk coming here."

The theme of mother being more to blame for the family disruption than father was echoed in the care she provided her parents and in the places where they died:

Dad used to spend alot of time with us. We'd go down and pick him up; he had a home down by the school, one of the homes that they took down. We'd go down there and pick him up and bring him here. Saturday and Sunday he used to spend with us all the time. In fact, he was here when he died. He died in my back room here. I took care of him for, oh, he was only ill about five months or so, and I took care of him. Then, I took care of mother before she passed away, although, she had to go to the hospital.

Her mother's efforts to obtain custody of the children did not soften Mae's perception of her mother's previous irresponsibility:

> Mother fought for us. See, when we were to come out of the orphanage, mother and dad had a big court trial, and we had to go to court to see who would get us. Of course, my dad didn't have anyone to take care of us. That's why he didn't get us. Mother got us because she was working, but she had some lady come and take care of us. We used to go to her house every night after school, and they made sure we were there. So this is how mother got us.

A recent reunion underscored the importance of the nuns for mediating the disruption of her parents' divorce. On a visit to a city several hundred miles away, the bus tour stopped at a home for retired nuns, and Mae realized that she might see one of the nuns from the orphanage.

> So I walked into the chapel, and there stood a nun from the orphanage, and I said, "Sister Josephine". It had been 40 or 50 years ago since I saw her last. I just put my arms around her and said, "Oh, my mother." I was so glad. I just couldn't believe it. I just cried like a baby. When it was all over with, I was so filled up I went back to the bus and I cried and cried. . . . I recognized her, and I said, "you haven't changed a bit."

She carried with her the memory of the disruption and reflected that it gave her the determination to stay with her husband:

> I wouldn't want anything to happen. I wouldn't want it to be the same as my parents were. Through thick and thin, if I wasn't happy, I'd still make it go because I wouldn't want my children separated from me.

Toward the end of the interview, as the life review process was winding down, more reflective questions were asked. The women revealed a different kind of memory. Here, they summarized their lives in capsule form, and an overall assessment was offered. In response to a question about whether she had ever thought about what her life would be like if she had not married, Mae replied:

> Oh, I suppose those thoughts have run in my mind when my children were too much to handle, or if I get a little bit depressed. I'll say, "why did I do this, or why did I do that?" Don't we all say that? But when it's all said and done, its a different story, and I'm glad I had them. . . . If you have two or three children, it makes you feel better. You get around 70 years old, and you don't have any children, you're just tossed in a nursing home. So, if I had to do it again, I'd still have children, I really would.

In response to a question asking her to describe the shape of her life, she replied:

> When you have three babies, you can't do anything. Your home is where you are, you just can't move out of there, cause you've gotta take care of them. Then, when they go to school, you've got all that free time, that's a different time. And then when they go to high school, it's different again, and when they graduate, it's different again. Now, everything is different

for me. But you know, I can remember when I had my three babies, and they were all constantly under foot, all the time, and I used to get so tired, then I'd stand by the window, and I'd see an elderly lady walking up the street, this is the God's honest truth, just as casual and carefree and I'd say to myself, "I can't wait till that day." Now I've come to that day and I feel free, I feel rested, I've got nothing on my mind. . . . I don't have anything to worry about because I've been very fortunate. I've had my health, my husband had his, my three children so far are very healthy and they've had a good life and I think I'm very fortunate.

Thus, two kinds of memories were offered: there was a core memory woven throughout the interview process, and there was an overall assessment of one's life in retrospect. For Mae, the core of her life review was the early discontinuity in having to live in an orphanage and being separated from her parents and brothers. But, the overall assessment of her life, as reflected upon from the perspective of having lived 70 years, was one of acceptance and peace about the choices she made to marry, have children, and stay with her husband.

Dora: A Divorced Woman

Dora was born in 1910 to parents of English ancestry. She did not complete grammar school due to poor health and frequent hospitalizations in childhood. It was her mother's second marriage; her mother had eight other children by her first husband, but Dora was her father's only child. Her parents' divorce coincided with Dora's first marriage at 17.

With her first husband, Dora had four children, one of whom died recently. She has approximately 30 grandchildren and great-grandchildren. When she was about 40, she left her husband for a "drifter", but they lived together for about 20 years. He was physically and mentally abusive to her, and she eventually left him. They were married in the last few years of their life together. Like other women in the sample, she has known a great deal of tragedy, recalling divorce, poverty, and abuse in her childhood family, and similar events as an adult. Old age is a time of relative peace, however. She lives in a senior citizen apartment building, with her days routinized by visiting with her children and friends, taking care of her two room apartment, and occasionally participating in senior citizen events.

Her perspective on aging reflects a certain sense of wonder and disappointment about how quickly time has passed:

You get old too soon and smart too late. You think your life is going to last forever, but it don't. It is the shortest part of your life to be young. . . . My life isn't too exciting now, but it could be worse. . . . You have to take the good with the bad. That's what life is all about. . . . Life is funny, you think it'll last forever, but it's a very short time. If you lived to be a certain age, you don't feel any different now than you feel when you was younger. You wonder where it all went. . . . I feel young inside, but your body gets old outside. . . . If you ever notice older people, course you're young, but their voice is the same as say when they were in their teens. That was a mystery to me, how young your voice can be and how old you get outside.

A major focus of her life history was the transition from her first marriage to her second. She said she never really loved her first husband, but she married him because there was no one else around. He was a distant cousin. They had four children and lived in poverty in a rural area in upstate New York, but she wanted something more for herself:

> We were too poor, you know, raising four children, and there was no work up north to speak of. Just exist. . . . We had food, but not a very large space to live in. Course we had no electricity. We didn't have the luxuries of a washing machine. A scrub board you had. You had to carry the water from the well. You couldn't turn the faucet on. . . . My first husband was a good man, he treated me good, and he always worked hard . . . but my life with him was dull, and I wanted more excitement.

She met her second husband in a bar, and they ran off together to a nearby city:

> I had to work to support him. I only made $25 per week, and he never worked. He drank up all my money, and spent it in bars on other women. He slept with other women and ran around on me and beat me up regular. . . . I was really disappointed. A lot of abuse and I couldn't take it. . . . Then I just left him and came home. My son said "I told you so" when it happened, and he has never forgiven me for it. . . . I feel ashamed to tell you I lived with a man before I got married to him. . . . I wish I could go back and had stayed with my first husband. I feel young people should never marry young.

Thus, the biggest disappointment in her life was being in love with her second husband and not having it work out as she planned. She noted that she was "looking for love in the wrong places, like the country and western song". However, like Mae, in her overall assessment of her life, she expressed more acceptance of her past:

> My life is more peaceful after I became by myself. I don't have the trauma, things on my head. It seems like now, I'm tired of thinking about the past. How bad it was, you know. Course, there's a lot of people who had a tragic time, I'm not the only one. You come in the world fighting and you go out fighting. I mean, life is like that, you fight to exist.

The regret and disappointment in her married life was mediated by the presence of her children:

> When you have your children, it's the best part of your life. It's something God give you, he trusts you to do the right thing with them. . . . I think to have children, that is the most gifted thing in your life. I think my life would be very empty if I hadn't had children. . . . You can have a husband, but your children, it's something that's a gift to you.

As echoed in the lives of most of the women in the sample, married or single, children were important whether they were one's own or a sibling's.

Willa: A Widowed Woman

Willa lost her mother as a child, quit school after the eighth grade, married early, had a baby while still a teenager, had two more children, never worked outside the home, had grandchildren, lost her husband, adjusted to widowhood, and at the time of the interview, felt for the first time that she was living her own life. She was born in 1912, and her parents were immigrants from Wales. Her father was a coal miner. Her home was filled with pictures of her descendants— mostly all over the refrigerator, an image that mirrored her association of nourishment with home and family.

She got married because it was the thing to do, and she had children for the same reason:

> I just wanted to leave school. My father said I could, if I wasn't going to be happy in school. But everybody did it then. All my friends did, too, In fact, the girlfriend I ran around with, her and I both quit at the same time, and we went together to the continuation school, until I was 16. And then I got out, and I worked a little while. I didn't have no job or education for a job or anything, and my father wouldn't let me work in a factory or a mill, because the women were too rough in them places. That's what he told me so I listened. . . . They used bad language. Hey, we couldn't even have a deck of cards in the house. He called them "devil cards", and he was so strict. Both of my parents were.

Like other women in the sample, Willa had her share of tragedies: her mother's death, the death of four of her six siblings in the same year, her husband's sudden death, and her subsequent depression of two years duration. She has had numerous operations. She said she responded to each illness with depression, but has always managed to pull herself out of them.

In spite of her many losses, she said that women like herself, who had a good marriage and children in old age, have their memories to fall back on, their own lives to lead now, and do not need to remarry. She explains:

> Because they already had that life. The ones who are looking for a husband are without kids or they had a bad marriage. Old men aren't the best prospects for a widow to look for because they are likely to be trouble and to get sick on you.

The transition to single status following her lengthy marriage was expressed this way:

> Oh, when I had a husband, we used to bowl together. But, he was more of a homebody. He loved to stay home and watch his sports on television. And we traveled a lot, too, But now, this is my life. I had to organize my life, and adjust my life for myself. Before I had him, and we were both involved and did different things, some together. We were always together. . . . If he wanted to stay home, I would do just what he wanted to do. Usually, we went out to dinner from time to time, but he wasn't one to go out to eat. He would rather have me cook for him. I liked to go out every once in a while, and I still don't like to go out too

much to eat. I'd rather cook home. But I do it now just to get out and be with the girls. . . . I think I am doing more now and enjoying life more now, than I had in a long time. Since my husband passed away, it is a very different life. You can come and go, and if I go shopping, I don't have to worry about my husband being home alone, but now I go, and close the door, and there is nobody to worry about . . . You really do have an altogether different life than when your husband is alive. You're not pressed for time to get home and to get meals on time. Or be here when he comes home. My husband was one who wanted me here when he come home. If I didn't have dinner ready, before I was shopping, I would always manage to have something on the stove, because the minute he'd come in, he'd check the stove to see if I had dinner cooking. Oh, he was a good husband. I really miss him.

Although her husband had been dead for over 10 years, she still thought of him as a major force around which she organized her life. They had been married for more than 50 years when he died.

To get adjusted. . . . You just have that feeling that its a shock. You feel that they are coming home. At night especially, when it is time for him to come home, you wait for him. . . . They were good years. I would do them over again with the same man.

Like other widows who felt they had successful marriages, she acknowledged the importance of her children, but she believed that the marriage relationship is the cornerstone of a family:

I always was putting him ahead of the kids. I figure you should for a good marriage. I keep telling my kids that, too, but with this woman's lib, well! Ah! I go down to my daughter's for a vacation, and I'll try to wait on her husband, and she'll say, "Now mother don't, you will spoil him like you spoiled Daddy". But I love to wait on other people, you know. She don't wait on him. He's great. He does most of the housework and the washing. I can't see that. Years ago, we didn't do that. We did it all. Everything. Even my daughter, she said, "Mother, its your fault I don't know how to cook because you would never let me cook." But see, I never worked, and that means a lot. Now, she has got her little girls trained. Hey, they start supper and clean their room and everything. I like it. I think it is great. I can see where my fault was where I didn't teach my kids. I wished I did. I did it all by myself, even hang up their clothes. But see that's what a mother does when she is home. If I had to work, I would have trained my kids to do more.

Yet, she claimed she was less emotional than her husband about their children:

My husband was pretty devoted to his family. As each one got married, he took it so hard. I didn't because I was with them from the time they were born until they left. But if he got lonesome for them, he would say, "call the kids" and get me to call them up to come visit. I used to say, "Oh for crying out loud, can't we be alone for a while?" I wanted to be alone, and he wanted the kids back again.

The most important people in her life have been men: her father, husband, and son, although she spends her days with her women friends, and frequently refers to her daughter. Perhaps her fondest memories of male relatives is related to her mother's death when she was age 10.

My mother died when I was 10 years old, so I don't know too much about her. Only that she was around until I was 10, and she was a good mother. I remember she was an old-fashioned mother. She used to wear a long apron. She did all of her cooking and stayed at home raising seven children. My father was great. He took right over when my mother died. There was four of us still at home . . . Of course, when I was growing up, I always thought my father was too strict. After I got married, and had children, I realized why. . . . He took care of us. He brought me up good. He made me go to church all the time. . . . We were very religious. Yeah, we had a good upbringing.

And about her son,

My daughter always told me that I never let my oldest son go. I used to say, "Oh, why don't he come home?" See, he was in the service. That's the one I depend on now. My daughter would say, "Mom you've got to cut your apron strings on him". She said, "You've had him in your apron strings long enough, let him go". My daughter is very outspoken. She'll tell me what she thinks.

The progressive movement toward more and more openness in her life, going from a severe upbringing to a marriage controlled by her husband to her current openness with her children and grandchildren, was evident in her characterization of dating rituals in her girlhood versus her response to her grandson's cohabitation:

At night, we stayed home and popped corn and played checkers, and did our homework. We went to bed early, too. . . . We used to meet at each other's houses, maybe four of us, and we'd have a party. And then they only had those victrolas, which you had to get up and change the record, and we used to be yelling at each other to change it. Especially if you had a boyfriend. . . . There was no petting going on like today. Maybe steal a kiss once in a while. Or hold hands.

But today, regarding her grandson's behavior, she explained:

I say, "Hey, you kids, nothing surprises me anymore." I tell them they can do what they want. You get educated with the television. . . . My husband was more out than I was. I was the real pious one. . . . I think I changed with the times. And with the kids. You see them living together today. Oh my God, if I heard of my grandchildren living together without getting married! Hey, I wouldn't believe it. Not my grandchildren! Maybe somebody else's. But they do, I mean my grandson did, before he was married. . . He didn't come right out and tell me, but I just knew it. I said that when he got married, well, thank God, you're making it legal . . . I thought it was terrible at first. Hey, how did I know they weren't living together when they went to college? Nobody told me. They kept things away from me. But now, it don't bother me. Nothing bothers me.

Typical of the widows in the in-depth sample, Willa knew very little about her family of orientation. Rather, she was prolific in her answers about her own descendants. For example, her own mother's death was not perceived as great a loss as her husband's death or the near fatal car accident of her daughter-in-law on the eve of her son's wedding:

See, I was only 10 years old when my mother died. It wasn't as bad as these other things were. I was just that young, that I didn't, well, I did miss her, but I didn't remember it much. Very vaguely.

Willa's life history was like the other women in that she had experienced a great deal of loss and change. She had no regrets about the past, however, because she felt she had lived through and experienced every possible aspect of a woman's life. She described her life in normative terms, in relation to what other women her age and in similar circumstances had also done.

Nan: A Never-Married Woman

Nan was born in 1912, six months after her father's death. Her parents were born in Italy, and they came to America and had four children in four years. The circumstances of her birth, in combination with her gender, birth order, and her mother's immigrant status, laid the groundwork for what was to become a lifelong pattern of caring for others. As the youngest daughter in a family where the sons went on to become a priest and a businessman, and the other daughter went to live with the mother's childless sister, Nan was elected to become her mother's lifelong companion and main source of emotional and financial support. Nan is a devout Catholic, having spent much of her life connected to the church. She graduated from a Catholic high school. For many years, she was a housekeeper in the rectory where her brother was a priest, and she lives in a church sponsored residence for the elderly at the present time.

Her life has been in service to others. Her training in being a caregiver began with the expectation that she live with and support her mother, at the expense of leading an independent adult life:

I had to take her into consideration. No I couldn't do anything. She had to be my prime concern. . . . I don't think I was college material. I might have gone to business school, had we been able to afford it. . . . I wanted to get to work, because I knew we needed the money. So there was just no question about it. . . . It was my responsibility because my older brother was married, and my other brother was in school, so I was elected.

In discussing how she spends her days, now that she lives alone for the first time in her life (her mother died 10 years ago), she noted that she does things for others, just as she has done all her life:

Well, like today, I had to do some shopping. We have a woman here who has a bad heart condition. She just came home from the hospital. She needs something from the grocery store, so I did some shopping for her. And, I came back, and then my aunt called, and I talked to my aunt, and she asked me if I would come over, and I had to do some shopping for her. I went over there and helped her get dinner, but I got home here about 3:00 or 3:30, and I just sat down and relaxed and read the paper. And that's it.

She and her mother had a strong attachment with each other that was forged by a family tragedy:

> My mother was left a widow before I was born. I had two brothers and a sister. My sister is dead. . . . We were never that close. She lived with my aunt. After my father died, and I was born, my sister was taken very sick, and she went into convulsions, and my aunt, to help my mother, took care of her. And they lived close by. Then, my aunt and her husband and family decided to move to open a grocery store, and they took my sister with them. At the time, it was just after my father died and my mother was sort of at loose ends, and she had a house full of boarders she had to take care of, to raise us. . . . My sister used to come and visit us. She called my aunt "mother". She said she had two mothers. . . . Of course, at the time, it was just one of those circumstances. But, afterwards, my mother realized what she had done, and so she told my aunt, "I want her back, there has never been any agreement that you can have her". She just said, "I'll take her and take care of her", and my mother agreed. Well, when that took place, my aunt said "I won't give her up, she's like my own," and her husband refused, too. Oh, and I wanted to bring up the fact that my mother and my aunt married two brothers, so we were really double related.

> . . . So, she was their own. And, so, my mother really always felt bad. But, she couldn't press the point without making hard feelings. And when my sister got married, she got married at my aunt's house. I never had that really close feeling towards her. Not as a sister.

Nan and her mother's bond is more evident in the context of the severance of this early relationship. An additional factor was her mother's attitude about remarriage:

> She was a very sincere and very hard working person. And of course, we were her whole life. . . . There was just nothing else. My father was very determined that the boys would get a good education and that was her thought, too. . . . My brothers were very very smart. I didn't inherit the brains in the family. But they got scholarships and were on debating teams and in athletics . . . When they had the opportunity to go to college, that made her very happy. And she worked hard all her life. And she never remarried. She didn't want to because she said she was afraid of who she'd get. You know, someone might be abusive to us. I was deathly afraid that she would because people would say "Oh, maybe one of these days, you're going to have a stepfather."

Nan idealized her mother, even though she realized that her mother became a "professional widow":

> She wore black for 12 years after my father died. We hated that black. Oh, it was so depressing. Oh, just terrible. And we kept at her and at her and finally one day, we kept saying to her, "please won't you take it off", and finally she took it off. And in later years she said, "That was such a foolish thing for me to do. I don't want you to wear black when I die", and when all of her sisters died, she didn't wear black. She had enough of it. That isn't going to bring them back.

Nan described her life in terms of "we". Presently, she refers to "we" in terms of her brother. In discussing her life with her mother, she described events

through her mother's attitudes about them. For example, when her young
nephew died, she expressed her grief through her mother's experience:

> Oh, we loved him so deeply, especially my mother. She never got over it because he was the
> first one and the first born and the first grandchild. And the first boy, and she just revered
> him.

The only regret Nan has about not marrying was that she did not have her own
children. Regarding a husband, she echoed her mother's sentiments:

> Maybe I'm a little happier (than married women). I never had to contend with, you know, you
> can never tell what you're getting. You can get a lemon.

On the other hand, she felt she may be better off than some of the women she
knows who have their own children, because sometimes they have a strained
relationship with them. Whereas, she is very close to her nieces and nephews and
has been involved in their lives since their births.

Contrary to the stereotype that never-married women are bereft of family
connections, Nan's life reveals how coupled she was with her mother and now
with her brother. Like Mae, the married woman, Nan has never perceived of
herself as alone. Willa and Dora, however, both formerly married women today,
consider their later years as a time of independence and living on their own. By
juxtaposing these four life histories, it is clear that knowledge of a woman's
marital or parental status does not accurately capture the complexity of her life
history and family connections. These life histories reveal that women's lives are
organized around their families, but not necessarily in normative ways.

DISCUSSION

Too often, women's experiences have been flattened by an emphasis on "the
woman" or "the monolithic family", without attention to the variations that
exist among women and families (Allen and Pickett, 1987; Thorne, 1982). The
choices and circumstances that shape individual lives have been obscured by
collapsing diverse experiences into normative types. Life history analysis
focuses on the particular rather than the general in order to protect the integrity of
an individual's experience from distortion and premature generalization (Geiger,
1986; Runyan, 1984). The present research began with four women representing
different marital statuses, and revealed the limitations of marital status as a
singular description of their family experiences. Rather, as Runyan suggests,
group differences is just one way in which lives may be understood. Two other
levels of generality are universals in human experience as well as characteristics
that are unique to a life.

The goal of life history analysis, then, is to describe and understand commonalities in life course experience without sacrificing the integrity of the individual lives. The four life histories presented here reveal that all the women's lives are centered and organized around their families, even though the timing, occurrence, and sequencing of events in their lives differed in dramatic ways. The data suggest three central themes, or tentative generalizations, which acknowledge the discontinuities in each life history but ultimately reveal the commonalities the women shared.

One experience shared by these women was hardship in the early years of life by virtue of their working class status. Economic stress and family history often collided, leading to family disruption from the death, divorce, or illness of parents and recourse to public charity on a temporary or permanent basis. Early hardship was played out in different ways over the life course of these four women, however. For Mae, early disruption led to continuity in her marital career. She married one man and stayed with him throughout adulthood. For Nan, early disruption, through her father's death and the separation of her sister, also prompted a continuous relationship, though not in the typical way. Rather, she maintained a primary relationship with her mother. For Dora, her parents did not divorce until she was 17, but she did experience a disruption in her early family relationships by being hospitalized so often. Unlike the first two women, Dora married and divorced two men; now, she is spending her later years relatively alone. Willa, too, experienced an early disruption through the loss of her mother, but did not perceive that loss as significant when she reflected upon it during the interview. The loss of her husband was a significant transition for her, taking two years to overcome. But having done so, she created a new life for herself. Thus, all the women experienced an early loss or disruption in their relationships with the family of orientation. They did not respond uniformly to the disruption.

Another common experience was that all the women were involved with children, either their own or their siblings' children. A major finding of the wider study (Allen and Pickett, 1987) was that never-married women were very involved in the lives of younger family members. In the life histories analyzed here, marital and parental status did not differentiate the women in terms of their connection to younger people. Even without being a biological parent, Nan was still deeply involved with her younger kin. Having peers was a very important connection, as well, but the presence of children was different from any other relationship. Without the arbitrary boundaries of marital and parental status differences, the similarity across these four lives in terms of the actual and symbolic value of children can be captured. This quality of caring about and for younger people, of being generative in Erikson's (1963) sense, would not be visible if the focus was only on the women as biologically reproductive.

Finally, to return to the central theme, women organize their lives around their families. All the women shared the experience of weaving family relationships, events, and transitions throughout their personal life histories. The particular events actually differed, either as individuals or as members of variant marital groups, but the universal experience here was the way in which individual and family careers were entwined. Regardless of marital or parental status, the women chose to tell their stories in relation to others. Life was organized around family, as reflected in their homes, in their metaphors, and in the description of those they cared for, felt betrayed by, forgave, loved, and shared their lives.

Hareven (1982) points out that individual and family careers are often in conflict, and such discontinuity can be seen in the four lives presented here. For example, Nan relinquished her claim to independence early in life as the family's need for her to care for her mother took precedence over her own career goals. For Dora, leaving her first husband and children to live with another man revealed a choice for herself. She took the risk of losing all for "love", and became resolved to that choice when her second marriage proved disappointing. But, in the long run, the summary statements these women made about their lives reveal how they negotiated the two demands, how they managed to juggle their own lives and their families without giving up completely one or the other.

The tension between individual assertion and family need was also expressed in the character of the recollections. The women revealed two types of memory. The first was a core memory woven throughout the interview which was an unresolved issue or life theme that they returned to again and again, with emphasis and emotion. They also offered an overall assessment of their lives, which revealed an acceptance, however, tentative, about where they had been, and how today, "this is my life now". The connection between individual and family history is complex, and these life histories offer a glimpse into how the women lived within or rebelled against that tension.

ACKNOWLEDGMENTS

Special thanks to David Unruh and Jane F. Gilgun for their helpful comments in preparing this article.

REFERENCES

Allen, K.R. (1989). *Single Women/Family Ties: Life Histories of Older Women*. Newbury Park, CA: Sage Publications.

Allen, K.R., and Pickett, R.S. 1987. "Forgotten Streams in the Family Life Course: Utilization of Qualitative Retrospective Interviews in the Analysis of Lifelong Single Women's Family Careers."*Journal of Marriage and the Family* 49: 517–526.

Elder, G.H., Jr. 1977. "Family History and the Life Course." *Journal of Family History* 2: 279–304.

Elder, G.H., Jr. 1981. "History and the Family: The Discovery of Complexity." *Journal of Marriage and the Family* 43: 489–519.

Erikson, E.H. 1963. *Childhood and Society* (2nd ed.). New York: W.W.Norton.

Geiger, S.N.G. 1986. "Women's Life Histories: Method and Content." *Signs: Journal of Women in Culture and Society* 11: 334–351.

Glassner, B. 1980. Role Loss and Manic Depression. Pp. 265–282 in *Research in the Interweave of Social Roles: Women and Men,* edited by H. Z. Lopata Greenwich, CT: JAI Press.

Glick, P.C. 1977. "Updating The Life Cycle of The Family." *Journal of Marriage and the Family,* 39: 5–13.

Glick, P.C. 1979. "The Future Marital Status and Living Arrangements of the Elderly." *The Gerontologist* 19: 301–309.

Hareven, T.K. 1977. "Family Time and Historical Time." *Daedalus,* 106: 57–70.

Hareven, T.K. 1982. The Life Course and Aging in Historical Perspective. Pp. 1–26 in *Aging and Life Course Transitions* edited by T.K. Hareven and K.J. Adams. New York: Guilford.

Runyan, W.M. 1984. *Life Histories and Psychobiography.* New York: Oxford University Press.

Thorne, B. 1982. Feminist Rethinking of the Family: An Overview. Pp. 1–24 in *Rethinking the Family: Some Feminist Questions,* edited by B. Thorne and M. Yalom. New York: Longman.

Uhlenberg, P. 1974. "Cohort Variations in Family Life Cycle Experiences of U.S. Females." *Journal of Marriage and the Family,* 36: 284–292.

EMOTION WORK AND EMOTIVE DISCOURSE IN THE ALZHEIMER'S DISEASE EXPERIENCE

Jaber F. Gubrium

ABSTRACT

The sociology of emotions has almost come theoretically full circle. Existing programs span so-called positivist and constructionist views. This paper critically reviews the programs, arguing that they either overly objectify or overly subjectify emotional experience. Evidence of the craft-like quality of emotion work, drawn from the Alzheimer's disease experience, suggests that (1) concrete feelings can be sociologically addressed, (2) native concern is more analytically astute than the programs portray, (3) affective privacy and uniqueness are foundationally social, and (4) a distinctly emotional discourse is natively called upon to communicate real feelings. Only a fully social sociology of emotions addresses the evidence.

Current Perspectives on Aging and the Life Cycle,
Volume 3, pages 243–268.
Copyright © 1989 by JAI Press Inc.
All rights of reproduction in any form reserved.
ISBN 0-89232-739-1

INTRODUCTION

Humphrey Bogart's facial expressions teach us something about emotional experience. There are two telling scenes in *The African Queen,* one in which Charlie (Bogart) begins to realize he's falling in love with Rosie (Katharine Hepburn) and the other where they are about to be married and then hung by the Germans. In both scenes, we see the inimitable Bogie visage shift, in seconds, from expressions of surprise and dismay to bliss, fear, and amusement. As is the film, the facial expressions are both serious and comedic. Expressions and frame play on each other in such a way that the emotions conveyed by the lead players both manage what we expect in the film and produce our overall mood. The scenes teach us that we are both passive and active in experiencing emotion, accordingly altering our orientation as we consider, attend to, fix and refix what we are feeling while, in the same process, learning what it is that affects us so. They teach us, too, that feelings more or less rapidly transform each other as we encounter their passing expressions. We tacitly work at, and through, what we feel, with ourselves and others, who altogether reflexively forge and shape the body's assigned collaboration with the proceedings.

I shan't argue here that the feelings of everyday life are cinematic contrivances. They are at once more serious, reflexive, and less scripted. What I do propose is that recent sociological programs for understanding emotions commit one of two analytic errors, either overly objectifying or overly subjectifying its experience. The outcomes, respectively, gloss over the everyday problematic of feeling or trivialize its intersubjectivity. In what follows, I shall (1) review the programs accordingly, (2) consider evidence for the craft-like quality of emotion work drawn from a study of Alzheimer's disease (senile dementia) patients and those concerned for their well-being, and (3) offer concluding considerations for understanding emotional experience from a fully social standpoint.

THE PROGRAMS

Since 1978, four sociological approaches to, or programs for, studying emotions have appeared. Their theoretical roots are grounded in positivist sociology, symbolic interactionism, and phenomenology.

Kemper's Positivist Program

In a brief paper, Kemper (1978a) first outlined what would soon appear in expanded form (Kemper 1978b) as an ostensible "social interactional theory of emotions," still later (Kemper 1981) classified by the author as a positivist approach. The broad outline laid out in that first paper, detailed in his book, conveys the essential message; Kemper's 1981 paper is a defense.

Declaring the importance of the organic roots of emotions, Kemper (1978a, pp. 36–37) happily reports that there is a significant sociological side to the physiology. Presented in awesome detail in his book (1978b), Kemper argues that power and status can be meaningfully linked with the psychophysiology of feeling.

The sympathetic and parasympathetic branches of the nervous system guide the body's emotional excitement, the first one associated with "negative" feelings of arousal such as fear and anger, the second one related to "positive" pleasure and satisfaction. Each branch is physically ramified with peripheral organs and moderates the other. As such, excitation is not organ-specific but tends to be generalized throughout the body, the system working to prevent overexcitation.

As far as negative emotional excitation is concerned, Kemper points out that a large body of physiological data shows that there is a neurochemical basis for the difference between anger and fear. What is "generally agreed" (Kemper 1978a, p. 37) to be anger, a negative confrontational state, is underpinned by the release of the neurochemical norepinephrine (noradrenaline). That believed to be fear or anxiety, on the other hand, is experienced when epinephrine (adrenaline) is released, producing a state of flight or negative withdrawal. As Kemper (1978a, p. 37) emphasizes, *"no other hormones or neurochemicals have been found to be so specifically related to particular emotions."* What he does not dwell upon is the question of whether it is anger or fear as such, or overly activated approach and avoidance states (whatever their motivation), or even particular cultures of affective interpretation, that are at stake. The ontological issue of what those "particular emotions" are that have been found to be so specifically related to neurochemicals, if they are even to be taken as emotions, is assumed to be settled—taken to be a matter of general agreement. The concerted resolution of the issue would require an investigation of the social organization of what came to be "generally agreed" upon as a particular emotion or bodily state, not the technical rigor that entered into it. It would necessitate our setting aside the acceptance of established bodies of experimental research for a movement into the laboratory itself to peruse affective categorization and coding (see Latour and Woolgar 1979; Knorr-Cetina and Mulkay 1983).

Kemper goes on to argue that the emotional/physiological combinations isolated underpin select social relational conditions, relations of power being linked to fear and anxiety, status relations being the social conditions for anger. Personal power deficits or excess power in others is fearful; loss of customary or deserved status is maddening. As such, Kemper concludes, we can study the sociological conditioning of these emotions.

Kemper also refers to what he calls the "positive emotions," those that produce feelings of satisfaction or well-being. These pertain to the parasympathetic nervous system, where the two forms of epinephrine are absent. The major neurotransmitter in the parasympathetic system is acetylcholine, a substance associated with positive feelings, moods, or states of mind. As far as social relations are concerned, Kemper suggests that the parasympathetic nervous system is predominant "when both power and status are sufficient and no specially compelling emotions are felt" (1978a, p. 37). Thus a specificity theory of psychophysiology grounds a sociology of feeling.

Kemper notes that the specificity theory has important opponents, particularly in studies that show a common physiological basis for all the emotions, emotional differentiation being articulated by social observations and comparisons. As folk do, Schachter and Singer (1962), for two, would seem to argue that "getting psyched up"—whether fearfully, lovingly, competitively, euphorically—is a matter of "getting that old adrenaline flowing." Kemper curiously concludes that it doesn't matter one way or the other whether specificity theory is correct. What must be investigated is the extent of association in the variabilities of social relational and psychophysiological systems, the enticing elegance of which, if found, would justify the effort.

But it does matter. If emotions are to be treated, "by general agreement [as] phenomena of the autonomic nervous system" (Kemper 1978a, p. 30), their variety would necessitate a specificity theory. If, on the other hand, emotions are essentially social phenomena in which the body is assigned specific states in various degrees, specificity theory is unnecessary baggage for a fully social understanding.

The sociological theory that Kemper lays out and diagrams could very well do without its physiological specifications. As such, though, it loses some of its positive punch and determination. According to Kemper (1978a, p. 32), "an extremely large class of human emotions results from real, anticipated, imagined, or recollected outcomes of social relations." He points out that the social relationships should be expected to affect emotion. It is clear from this and other passages that the causal direction is from social relationships to emotions, social relationships virtually "triggering" emotion, by inference, triggering the neurochemical basis for the subsequent perception of feelings. Now this a rather strong argument, one which Kemper backs away from in denying the need for a specificity theory in a "strictly" sociological theory. Bereft of specificity theory and its critical neurophysiological link, what we are left with is a theory that attempts to draw together conditions of power and status, on the one hand, with *judgments* about associated bodily states, on the other. Kemper develops the connection by means of a taxonomy of social relationships which, according to him, "can constitute a systematic definition of the situational matrix which produces emotion" (p. 32).

Losing its organic bearing (but maintaining it implicitly), Kemper's positivist program draws upon a different form of determination. In this form, it is more fully subject to sociological critique. Indeed, it is open to verification by such detailed facts of emotional expression as those shown in the quick turnings of Bogie's face.

As a strictly sociological theory, Kemper's program ties the configuration of experienced emotions to what he calls the "social relational matrix" of power and status. Anyone's social relations are comprised of particular amounts of power over others as well as select funds of honor or status. Kemper is careful to point out that the relations are tied to particular dyadic settings (p. 33). The actor attributes responsibility to himself or to others for deficient or excess power or status. Based on the two dimensions–status/power and responsibility—Kemper figures that particular combinations are associated with specific emotional states. For example, guilt results when someone senses that he has used excess power against another. When the agent responsible is someone other than oneself, the resulting emotion is said to be a form of megalomania. Shame is tied to status and so on. The so-called positive emotions are dealt with at greater length in Kemper's book (1978b), especially love.

It is evident that what Kemper means by social relational matrices is something far removed from what is apparent in the title of his book, a social interactional theory. While he states that the relations from which he draws emotional conclusions are located in dyadic settings, the settings seem to congeal in his analysis, becoming quite settled. As argued, a particular kind of dyadic setting has a specific matrix of power/status and agency that, in turn, produces the expected emotion. Kemper's vision of social interaction presents its actors as eager emotional tools of its relational structure (see Gouldner 1970). Interactions are simply the automata of structured conditions which, albeit "situationally" varied and definitional, serve to more or less transmit power/status and agency into bodily feeling (see Kemper 1982, p. 359). While the definition of agency is presumably the actor's prerogative, the overall social relational matrix and its logic are not, being social forms and forces within which the action and interaction take place, playing out their appropriate feelings.

The social relational matrix is not a cognitive structure, not a configuration of collective representations. Its externality is definitely not cultural (Kemper 1981, p. 344). As Kemper quips, "Would that emotional life were as plastic as the social constructionists maintain! In that case there would rarely be dull parties" (p. 344). Rather it is of existing *structures* of power and status that Kemper speaks. Yet in the following paragraph, Kemper describes a curious mix, "the ineluctable fact [of] social relations of power and status, [that is] sometimes crystallized into an enduring structure, sometimes fluid and evanescent, determine out feelings, [referring] to real emotions, not forced smiles . . ." (344).

Now this kind of emotion theory is all to the good if we enter into its empirical affairs in a certain way. Yes, as Kemper states or implies in many passages, there is fluidity in social structure. Actors do define and attend to situated variety in their daily lives, tinker with agency, and take norms (feeling rules) into account. But, when all of that is said and done—that being the perennial interpretive vicissitudes of everyday life—it is "the social relations that constitute the existing social structure . . . that evoke our emotions" (p. 344). And these are real emotions, not appearances, rather concrete psychophysiological states of feeling. As such, complex arrays of everyday practice are granted and quickly retrieved, replaced by fixed distillates: existing social structure and real emotions, sociophysiology. In his own defense, Kemper (1981) nods time and again to the concerns of the so-called social constructionists, to what is admittedly all, no doubt, empirically true. But he abruptly, though politely, trivializes the concerns. He does grant that the two approaches—the positivist and constructionist—are not so much antagonistic as complementary, but the peace offering is clearly one-sided. At no time is the array of everyday concerns with emotion and feeling conceived as productive of its own enduring explanatory structures; rather the reverse and thus the complementarity. The empirical emotional order addressed by Kemper is a when-all-is-said-and-done matrix of social and physiological objects, an order that, in the final analysis, fully backs its theory.

Kemper's program overly objectifies emotional experience. The social interaction presented is never strong enough, itself, to lay out relational matrices of power/status and agency as game plans for the interpretation and practice of feelings. Kemper's actors speak of emotions and have feelings; they don't do things with words (see Wittgenstein 1953). They certainly don't "do" feelings, nor their explanation (see Garfinkel 1967). Yet, as I shall show later, serious attention to the craft of everyday concern with feelings can bring relational matrices to their knees. When the bewildered and overly burdened wife of a senile husband whom she lovingly cares for pleads for an explanation for, and definition of, what she achingly feels, she invites objective offerings to frame and evoke her emotions. What rewards her may or may not be a Kemper-like understanding and realization, because Kemper's is only one vision of her emotional experience.

Hochschild's Management View

Hochschild (1979, p. 552) agrees that what she refers to as "two possible approaches to the social ordering of emotive experience" and what Kemper later calls the social constructionist and positivist approaches, are compatible. Indeed, she points out that her program "relies on some accumulation of knowledge garnered from [the other]" (p. 552). However, Hochschild reserves a more significant place for what Kemper, by and large, treats as the automata of emotional relations: interpretive contingencies.

Against a background of concern with the theoretical linkages between personality and social structure, Hochschild presents an emotional world that resembles Kemper's on a number of points. She places the concrete stuff of feelings in the body, in physiological states, the expressions of which are affected and conditioned by feeling and framing rules, and emotion work, respectively. The convergence with Kemper on the theoretical status of feeling is limited, though not incompatible. In contrast to Kemper, who treats the distinction between real bodily feelings and their expression as trivial, Hochschild highlights the difference. She wants to explore how people manage the expression of feeling against public expectations, feeling rules. As such, there are two orders of feeling, to paraphrase Goffman (1959, p. 2), given feelings and feelings given off. Given feelings, as it were, are psychophysiological states of emotion of importance to Kemper, the concrete bodily affects known to those who experience them, the personal side of emotion. Feelings given off are the expressions of feeling presented by persons, more or less aligned with what is expected, the public side of emotion. True feelings may not converge with what the person who possesses them appears to feel.

While there is a difference in emphasis on the place of bodily feelings, given feelings, in their scheme of things, Kemper and Hochschild both take given feelings as meaningful, objective states of affective being. Their texts reference them as things, readily cataloguing them by name: guilt, shame, joy, love, despair, and so on. For both, these "things" are objects to be referenced and analytically located in their respective theoretic frames. On that, there is little or no disagreement: given feelings are ontologically secure.

The objective treatment is challenged by practical concerns with their essential reality. Detailed below, the Alzheimer's disease experience shows that everyday concern with ostensible "feelings" is, at times, a project without an apparent bodily object, one which, it may be claimed, doesn't deal with the body at all. Both Kemper and Hochschild ignore the ongoing issue of given feelings as a practical reality, ignoring its everyday problematic. But this is less important for Hochschild's analytic aims since she, in contrast to Kemper, focuses on the connection between appearances (feelings given off) and circumstantial proprieties (feeling rules). Still, there are times when even the management of feelings conjures questions of authenticity and reality, not simply their veneer. On these occasions, Hochschild's otherwise useful gloss is lacking. A great deal of reality work on the status of that thing—feeling—is engaged prior to, and as part of, the emotion work considered by Hochschild. It cannot always be assumed what the given feelings are (unless, of course, they are literally treated as given) or, in practical fact, whether they even exist at all as meaningful, nameable, bodily entities.

Naming, speaking, goes a long way in solving the ontological question under consideration, especially when the names used are common expressions, part of a culture of related object references. We all readily speak of our feelings, the concurrent state of our bodies. The facility belies the equal ease with which the proposed quest for affective reality is engaged. That side of emotion work is undeveloped in Hochschild's otherwise handy phrase, the side that turns emotion work from management into craft, from a resource by which to analyze the world of feelings to a topic in its own right, one both processual and objective.

Feeling rules are the explicit or implicit norms referenced in public considerations of personal feelings. Feeling rules are tacitly present in certain admonitions:

"You should be ashamed of yourself." "You have no right to feel so jealous when we agreed to an open marriage." "You ought to be grateful considering all I've done for you." (Hochschild 1983, p. 58). As Hochschild states, it is taken for granted "that there are rules or norms according to which feelings may be judged appropriate to accompanying events" (p. 59).

It is not always clear what the feeling rules are that apply on particular occasions, which brings us to what Hochschild calls "framing rules," the interpretive side (the "bottom side") of social structure, the third and final component of her program. Framing rules are the varied means by which "we ascribe definitions or meanings to situations" (Hochschild 1979, p. 566). Hochschild provides an example. When a person is fired, the situation might be defined as yet another case of capitalist abuse or the outcome of personal failure. The two rules for framing the dismissal suggest corresponding feeling rules. As Hochschild states, "According to one feeling rule, one can be legitimately angry at the boss or company; according to another, one cannot" (p. 566). Feeling rules are applied according to how the situation is framed, which, in turn, produces proper appearances.

Now there is something altogether disingenuous about all of this, not in the sense that affective management does not happen but in its portrayal as the general nature of emotion work. The work of dealing with feelings in everyday life is not chiefly a matter of social repair, by appearance or otherwise. It is important to make theoretical provision for the production of feeling and framing rules as well. While folk sometimes behave as if they were structural puppets, they are not always so. They at least participate in the development of the definitions and expectations that are sometimes their emotional undoing. Indeed, they craftily engage the whole affair, from the work of realizing given feelings to the construction of fitting frames and, reciprocally, the managed indulgence of the products of their labors. Hochschild builds her program around the respectable side of craft.

As Kemper's program envisions the link between feeling and social structure as *factual,* Hochschild draws the connection as *managerial.* The terms are as metaphorical as they are analytic. As we have seen, for Kemper, the world of emotions is to be definitively, objectively delineated and surveyed for resultant covariances. Kemper's is a strait-laced affective world of sociophysiological and structural, not cultural, facts, having no truck with willy-nilly rules. Hochschild's program is informed by the very commercial symbols she finds in her student and Delta Airlines data. Hers is a managed world of feeling, one she skillfully teases out of the flight personnel she studies, showing us diverse details of the commercialization of human feeling. It is a lament, yes, but one that decries the by-product of the feeling work engaged, not appreciative of the full range of good and bad reasons for why things are as they are in the worlds of feeling.

While all programs and visions are metaphorical, leaving us with no choice but to accept their actualization of our views of things, it is important that they be clarified and compared. Clarification and comparison reveal the ways in which the empirical potential of our data is delimited. So it is with Hochschild's, Kemper's, and others' programs. The forbidding feature of Hochschild's program, in contrast to Kemper's, is that the management metaphor tends to hide the reality of the very objective—feeling—of central concern to all. While Kemper's program is overly objectivistic, at the very least, it brings us face-to-face with feelings as objects to analytically deconstruct and experientially display, as the craft in folk's emotion work does.

Shott's Symbolic Interactionist Analysis

Shott's (1979) program is an apparent application of Blumer's (1969) interpretation of symbolic interaction to the world of feelings. Her analysis is built on the practical nexus of the concrete bodily realities of emotional experience and their interpretation. As she puts it, "Hence, two elements—physiological arousal and cognitive labeling—are necessary components of the actor's experience of emotion" (p. 1318). Emotional experience is a process of continual socialization, not in necessarily reproducing social structures but productive of more or less structured understandings.

In line with Blumer's view of the persistent emergence of social life and social forms, the ongoing reciprocity of reality and definition, Shott states that what is felt is not determined by cultures and feeling rules, only guided by them. Following Hochschild, cultures and feeling rules are taken as ontologically secure in their own right, to be applied according to the wits of the actors who encounter or discern them while, at the same time, being experientially challenged by the physiological arousals of their bodies. Likewise, the latter do not, alone, constitute emotions. Accepting the Schachter and Singer (1962) view—the nonspecificity position—Shott notes, "What is required, in addition, is the belief that some emotion is the most appropriate explanation for a state of arousal" (p. 1322). Shott's program thus focuses on what she calls the "interplay of impulse, definition, and socialization."

Curiously, Shott illustrates the interplay by means of what turns out to be a rather structurally deterministic portrayal of social control and role taking, applied to "role-taking emotions," an area of emotional experience in which "the sociological relevance of sentiments is most evident." Presumably, sociological, as opposed to another discipline's, relevance is displayed deterministically, something for which Shott alleges symbolic interactionist theory is quite useful. For example, she addresses the emotions of guilt, shame, and embarrassment as facilitating social control because they "check and punish deviant behavior" (p. 1325). On the positive side, the emotions of pride and vanity serve to keep social matters on an even keel, "often rewarding normative and moral conduct," as does empathy, "which moves us to feel joy or grief at another's situation and thereby ties us, at least momentarily, to that person'' (p. 1326). Concluding her illustrations, Shott writes, "All of the role-taking emotions, then, further social control by encouraging self-control; they are largely responsible for the fact that a great deal of people's behavior accords with social norms even when no external rewards or punishments are evident" (p. 1329).

Are we to conclude from this that what in Blumer's original formulation was a program for addressing the symbolic dialectic of realities and understandings in human exchanges, has become an apologetic for the status quo? Shott is obviously aware of the problem as she parenthetically writes, with respect to symbolic interactionists' common focus on the self-control side of social control, ". . . sometimes so much that coercive elements of social control have been unduly minimized" (p. 1329). It seems that what we are left with is an ironic vision of self-controlled emotional expression directed by existing social forms, a kind of voluntary affective determinism, allegedly the side of emotional experience especially appropriate for sociological analysis. With the theoretical co-optation of the actor and his feelings, as proper objects in a world of controlled emotional fact, what started out as a balanced treatment of the dialectics of emotional experience comes very close to Kemper's positivist sociological program.

Denzin's Phenomenological Program

In Denzin's (1984) program, we cross over the theoretical horizon for an approach to emotional experience in which subjectivity is focal. The plentiful attention to social interaction only nibbles at what is taken to be the essential and unique subjectivity of feeling—hardly a "social" phenomenology. For Denzin, "emotions are self-feeling" (p. 3). It is lived emotionality, which in the phenomenological lexicon means that its experience simply is, not categorical, not meaningfully obtained, not objectively encounterable, rather felt in the flow of its unique passages.

While Denzin does not deny the importance of the body for the understanding of emotion, he rejects the organismic view (Darwin 1955; James and Lange 1922) that emotions are the by-product of the automatic triggering of bodily change caused by the perception of exciting conditions. Rather, emotions are conceived as embodied experiences, in which the body does not "call out interpretations" but, rather, is the locus of tactile adjustment to emotional consciousness (p. 30). Referring to Sartre (1962), Denzin states that the bodily disturbances that present themselves during emotion are, in their own right, rather ordinary, resembling such feelings as fever and indigestion. What distinguishes a bodily disturbance as emotional is the assignment of emotional significance. The body is an instrument of the experience, not the reverse.

Because emotion is self-feeling, its affective experience is possessed by the person and that person alone. Essentially, emotion is something privately felt, privately lived, only infringed upon by bodies and others. In the context of discussing what he calls ''spurious emotionality,'' Denzin stresses the privacy of self-feeling. Quoting Scheler (1970, p. 33), Denzin writes, ''A man's *bodily* consciousness, like the individual essence of his *personality, is his* and *his alone*'' (p. 154). Denzin adds, ''Pure or complete understanding of another's experiences, feelings, and thoughts is an impossibility'' (p. 154). Spurious emotionality is the mistaken condition, called ''egocentrism'' by some, when one's own feelings are taken for others. No one can be authentic to emotion except the very one who personally, genuinely experiences it, deeply and privately (p. 159).

As Denzin entitles the fourth and fifth chapters of his book, emotion has individual and social margins. The distinction, however, is lopsided, with authenticity clearly an individually private state of emotional affairs. The social facets of emotion—emotionality—are ''temporally and relationally rooted in the social situation'' (p. 52), the side of emotion of concern to Goffman (1971), Gross and Stone (1964), and Hochschild (1979), among other students of emotive ritual. Denzin's use of the term ''ritual'' to describe the inauthentic, organized social minutiae of emotionality is quite telling in revealing his approach, worth quoting at length for its metaphorical hints.

> Deep emotionality is buried, hidden, clouded, and distorted within the tiny, fleeting, ritual/emotional/cognitive exchanges that make up the daily round. Individuals come into and out of each other's presence, exchanging greetings, smiles, apologies, sympathies, and small talk about the weather and the economy and setting times for lunch, while they make minor requests of each other. In these small actions and others like them, individuals keep their emotionality from themselves and from one another. The consequence, of course, is that, to the degree that they adhere to the public ritual order, they fail to reveal their deeper meaning to themselves. A certain kind of emptiness is felt. (p. 159)

In the final analysis, it is evident that Denzin sees social experience as getting in the way of authentic self-feeling, the substitution of emotionality for emotion, the affective version of the individual-versus-society distinction, cloaked in personal sentimentality. As Denzin writes, ''Adherence to surface rules of conduct (Goffman 1971) increases the likelihood that spurious emotional intersubjectivity and emotional understanding will be produced. . . . The deep meanings of the person are not displayed or felt in ordinary, everyday discourse'' (p. 159).

What are we to make of this? Emotion is said to exist, to be something which Denzin urges we get back to—to the thing itself, as phenomenologists oblige us to do. Yet, we also are told that the reality under consideration is private and unique. There is something very strange about this, for if emotion is private and unique, how are we, let alone the person who experiences it, ever to know it, to know he or she has felt it? Must we not all use recognizable categories, shared linguistic contrivances to communicate it to each other and to ourselves? Aren't the resources and structures of communication applied by the person in relation to self, in principle, the same as those used in relation to others? If they are to be categorically distinct, privacy must be meaningless, bereft of meaning, an empty discursive class.

That is the key. Denzin wants to provide a means for recognizing essentially private feeling as an experiential category, but he lacks a language for doing so. In fact, there is no such language, for a private language is inconceivable, its conceivability itself being social, intersubjective. It is a self-defeating program for even privacy is a social category, as is authenticity—communicable and understandable states of being. There it is again: sociality. Denzin refuses to take seriously the extent to which sociality penetrates experience. Indeed, "penetrate" itself is an inappropriate usage, for it presents the very image Denzin offers us.

The passage quoted earlier provides the metaphor that informs Denzin's program. Like J. Alfred Prufrock's, the passage, too, is a lovesong, one bemoaning the loss of authenticity, an existential being mired in the coma of social ritual. Denzin's sentimentality twists it into a plea for personal understanding, the focus of which is hurt, misunderstood feelings, a signal gesture of popular culture. Because it provides no understandable solution for the riddle of emotion, it must wallow in its desires, perennially hurting and misunderstood, a hallmark of country music (Denzin 1984, p. 77). It grants the social side, the ritual, of emotional life, yet it understandably avoids its analysis, preferring to cast aspersion on its categorically presented details. The upshot of this is that intersubjectivity is trivialized. To paraphrase Ma Bell, no one seems able to reach out and touch anyone.

EMOTION WORK AS CRAFT

I propose a different metaphor for the details of emotional experience. The metaphor is work, but not the vision of it that informs Hochschild's analysis and usage. Her program is rooted in managerial activity, adjustment, the technology of social repair. I shall treat emotion as craft, whose skillful activation and activities serve to plan, produce, appreciate, and indulge the felt products of its labor, emotional labor. As such, emotion work is as much reality-productive as it is reality-reactive: feelings are as much "done up" as they are "served up."

As craft, emotion work is both analytic and active. Not only do we, as sociological analysts of the world of emotion have our programs for interpreting it, but, varied as they are, emotional folk do too. Indeed, they occasionally remind each other of that as they defend or present alternative visions of emotional experience, visions that, in broad outline, resemble the programs just reviewed. Emotion workers not only display emotional data for sociological consumption, they produce the very frameworks that make "their" data meaningful to themselves, whether conceived as the triggered automata of shifting relational matrices, the ostensibly incommunicable agony of private hurt, or the like. As craft, emotion work is also active. Leaving off the analytic side of feeling, folk proceed to feel or not feel according to their understandings, or, in reverse order, indulge understandings befitting what they take to be their real feelings. In its practice, the craft of emotion work is a multi-layered process of concern with, conduct through, and usage of, both the realities and the realizations of feeling.

The Alzheimer's Disease Study

Data for the reconsideration of emotion work are drawn from a field study conducted among those concerned with the care or custody of Alzheimer's disease (senile dementia) patients (Gubrium, 1986). Once a "silent epidemic," Alzheimer's has become what many say is the "disease of the century." Slogans of the growing Alzheimer's Disease and Related Disorders Association (ADRDA), the phrases signal the expanding awareness of a devastating illness that is only now being distinguished from normal aging. There is no prevention or cure for Alzheimer's disease, only a process of progressive degeneration of the brain and cognitive impairment (Katzman, Terry, Bick 1978; Katzman 1983; Reisberg 1981; Wells 1977). The disease presents an inscrutable burden of care for family members. Caregivers are often said to be the other victims of the disease. For them, it bodes personal and financial ruin as the disease's "36 hour" daily travails take their toll (Mace, Rabins 1981).

The study made use of a variety of data. Research and caregiving literature was content-analyzed for experiential themes and understandings, especially the proceedings of Alzheimer's disease conferences and the newsletters of ADRDA chapters. Participant observation was conducted for several months in a small day hospital for the care of Alzheimer's disease patients. The day hospital's support group for caregivers, mostly spouses or adult children, was observed, as were ADRDA-sponsored support groups in two cities. Unstructured interviews were conducted with 16 caregivers, in their own homes, and with the professional staff of the day hospital studied.

Data were interpreted to reveal and distinguish details of the emotion work that enters into attempts to understand the disease experience. Focal were the concerns of caregivers, articulated against the patient as victim, on the one side, and the service providers who attempt to ease and aid those who, day in and day out, face the 36 hour day, on the other. Taken as socially organized, the disease's distinct everyday reality was not only considered to be a result of its organic markers and behavioral effects, but also an artifact of the practical ways it was perceived as a disease entity, as were the feelings that entered into its experience. The data raised a number of questions about existing programs for the sociological study of emotions.

THE QUESTION OF ANALYTIC INNOCENCE

By and large, existing programs take it that those who experience emotions do so with little or no theoretical understanding of their feelings. At best, folk understandings are treated as reactive or adjustive, where cognitive factors merely serve as communicative conduits for sociophysiological arousal (Kemper) or facilitate the management of existing bodily states and rules (Hochschild). As those who are emotional stop to consider their feelings, they are innocent of analytic interest. The earnest inquiries and explanations of caregivers in the support groups studied suggest that folk are not lacking in the theoretical understanding of feelings. Their analyses don't differ from sociological ones as much in level of understanding as in the service to which they are put.

Consider the following series of exchanges among caregivers participating in a support group, where the issue being discussed is the impact of a change in environment on the feelings of the Alzheimer's patient. As argued by some behavioral scientists, participants believe that the very ill are especially vulnerable to environmental changes. For example, relocation from one nursing home to another is likely to be bodily and emotionally traumatic. The conventional wisdom is to maintain familiar environments lest stress levels be dangerously raised.

One of the participants, the wife of a patient, stated that when her husband first entered the day hospital, he became very anxious, being an unfamiliar environment with strange sights and sounds that "really set him off so that he couldn't get anything at all together." Responding in a way strikingly reminiscent of Kemper's (1978a, p. 35) description of the social relational dynamics of anxiety, a caregiving daughter remarked:

> I know just what you mean, Fern. My mother reacts the exact same way. That's the way they seem to get when there's a change. They come here and, somehow, they know they've come down a lot, like they're here because they can't make a go of it. They feel vulnerable. Well . . . they might not know it in so many words, but you just know they feel it. And it really scares them. You know what I mean? That's really hard on a person like my mother who used to come on strong and usually get her way.

Compare this with Kemper's description:

> Anxiety: When there is an imbalance in the power relationship between actors, the one with relatively less power is vulnerable to the encroachments of the other, and the anticipation that other will use power is the core of anxiety.

A period of discussion followed, elaborating the, until then, implicit theory. As it unfolded, attention shifted from particular patients' feelings to what *they*— all similarly affected persons—must feel. Likewise, feelings, as such, became focal as participants zeroed in on the general question of the causes of anxiety. In fact, at one point, a participant analytically reflected:

You know, that's an interesting view of it, that the changes and losses make them fearful. All along, I just thought it was just Richard [her patient] but I can see now how all that happening so fast-like, and not being anything like you used to be, just triggers the fear. I'd be fearful, too, if that happened to me. Gosh, it builds fear in me when I even think of putting myself in Richard's shoes. It's an interesting theory when you think about it.

Several participants joined into the appreciation of the now interesting theory. Its application was expanded. One after another, sometimes simultaneously, each drew data for the emotional impact of environmental changes and the loss of accustomed social effectiveness out of their respective patient's disease experience, presenting it as proof of the soundness of the theory being entertained. Each bit of information offered served to support the way feelings in general seemed to operate. Theory and data became mutual documents of their practical validity (see Garfinkel 1967). Well into the discussion, several participants reported that it made them feel better to know how things work to produce people's moods.

In the course of the appreciation, one of the participants interjected:

It's not that! It's not that at all. That's not what you're seeing. They're not frightened. I think they're just stubborn. I really can't say that Carol [his demented wife] is frightened of the van [which takes her and others to the day hospital] or the program here or even all that's happening to her. None of them are. She just plain doesn't want to come here sometimes and gets all hot about it. She knows what's happening . . . and so does everyone of them really; they feel it anyway. I know Carol. She'll act like she's frightened sometimes just to make me feel sorry and think she's afraid and all shook up inside. Heck no. It's just a big act that feeds into the way they want us to think they feel.

Another participant, the adult daughter of a patient, added:

I don't think they're afraid either. My mother enjoys the van ride. She's always talking about the nice people and how nice the driver is. I think, though, that she forgets that and doesn't want to go sometimes. It can be stubbornness, not fright.

Carol's husband then responded:

Look, they manage to get their way. They feel a certain way or they see that you think they feel a certain way—they sense it—and, bingo, they dig in the heels and start resisting. Or they use the way *you* feel to their advantage. Look, we're every one of us just like that sometimes, aren't we?

That familiar environment stuff only goes so far. We've been living in a beautiful condominium for quite a while now and Carol's still upset about it. She wants to get out all the time and, well, you've heard it before, does she get hot and hoppin'. She wants to go back and live on Oxford Street, where she used to 47 years ago, before we got married. That's more familiar, as you say, that's more familiar even than the house on Jersey that we left and sold three years ago. Familiar's to everyone's liking, ain't it though.

Sometimes I get the distinct feeling that they're making all the rules and pulling all the strings. Carol will create a scene, like she's frightened, just to get me to react, and that's the way she gets her way. She doesn't give a damn what people think. She's always been like that, way before she started the Alzheimer's. As you all know, Carol's got a mind of her own. And, if you think about it, there's a little bit of that in all of them.

Finally, a confused newcomer to the support group remarked:

> Well, then, what's it going to be? I've heard both sides of it and I can see some of both in
> Harry [her husband]. Well, it's all supposition anyway, isn't it? How do we really know what
> anyone feels deep down?

With this, a lengthy discussion developed, periodically quite heated, over the question of access to personal feelings, debated not only in regard to the patients but to each other's privacy, a concern to which we shall return shortly. In terms of the foregoing bits of data as they apply to the question of analytic innocence, it is evident that there is a sometimes explicit but regularly implicit theorizing about the cause, dynamics, and consequences of feelings, everyone's feelings, nonpatients included. What those feelings are, is not simply a matter of their careful investigation. They seem to rise and fall, in fact, with the theoretical frame applied to them. The rather complex series of exchanges from which the preceding comments were extracted showed that not only did the figure and ground of each feeling and theory oscillate in the discernment and documentation of emotional understanding, but each shift served to alter the empirical nature of the data under consideration. In this particular session, support group participants moved from a Kemper-like view of emotions and the actual feelings under consideration, to a view and facts resembling Hochschild's program and data, eventually becoming more constructionist than managerial about feeling and framing rules. (Carol makes her own rules, as do all folk at times.) The confused newcomer finally suggested that all might as well give up on the whole question since privacy is, after all is said and done, precisely that, unknowable to others and thus futilely pursued, reminding us of Denzin's phenomenological view of emotion as self-feeling.

It cannot be said that, being disease facts, the concrete feelings and theories of feeling under consideration by the participants are not relevant critiques of existing sociological programs. It is evident in support group proceedings that it is not only the feelings of the demented that are at stake but also those of the other victim of the disease, the caregiver. In part, caregivers organize their own feelings about the disease in accordance with their current understandings of how the disease affects the thoughts and emotions of the patient. The understandings, in turn, are verified by means of what caregivers and others interpret patients' emotional status to be, further warranted by mutual confirmation of the emotional facts under consideration. The organization of caregivers' emotion work makes it clear that feelings, whether those of the patient or caregivers themselves, are articulated by conditions far more socially inclusive than their separate medical status would suggest. The feelings of the diseased are as much conditioned by the everyday related activities of its "second victims" and concerned others as those same feelings are caused by social forces in the patients' lives as such, as are the feelings of the nondiseased affected by their patients and each other. Caregivers not only debate the abstract merits and demerits of particular "theories" of emotional life, but, in discerning the status of patients' and their own emotions, they apply them to interconnected arrays of feeling. Typical in this regard is a telling comment made by the daughter of an Alzheimer's patient:

> I'd be less worried and fearful for her [her mother, the patient] if I only knew that, in her own mind, she was comfortable and happy. God knows, I try to make her happy. You'd be surprised how much my own feelings are affected by hers, even though things for her are really different—like I'm not the one that's afraid to say things in public. Me? I don't care a hoot, really, what people think. I'm a good deal more secure than Mother. I really feel sorry for her sometimes. I can't tell you how embarrassed she really gets and I sure get embarrassed for her too.

THE QUESTION OF PRIVACY

This brings us to the analytic status of concrete feelings as such. Kemper presents the varied combinations of power, status, and responsible agents that cause particular emotions. Hochschild outlines the broad normative and situational imperatives that actors take into account in managing affective expression. Shott offers examples of the personal relevance of structured sentiment, controlled by role-taken emotions. Denzin categorizes forms and processes of spurious emotionality, types of inauthentic expression of self-feeling. The mode of analysis is social psychological—from social conditions to individual feelings. The connection is envisioned without serious consideration of the analytic consequences of a variety of rather obvious, fully social complications.

Recall the ostensible individual feelings at stake in the preceding extract. A daughter readily ties her "own" feelings to her mother's. According to the daughter, she feels whatever way her mother feels. Yet, we know, too, that the daughter's social relational matrix is not the same as the mother's, being more secure. Still, their feelings are the same and flow together. While the two are individually differentiated by the daughter, they coincide with respect to emotional agency. As such, the daughter does not completely own her feelings but shares disease-related ones with her mother, the daughter's emotions being vicarious reflections of what the daughter takes to her mother's feelings.

Hochschild (1973) once described the vicarious daily life of a community of grandmothers whom she studied in the San Francisco Bay area. She found that the grandmothers, who lived together in a housing project called Merrill Court, shared a fund of activity not evident in the actual goings-on of their own daily lives. What they spoke of, took pride in, were disappointed over, and anxiously awaited, among other joys and sorrows, extended to activities and events in their children's and grandchildren's lives. Calling it "altruistic surrender," Hochschild reminded us that the experiences of everyday life cannot be understood by accountings of individual activity, that what one experientially owns, claims, and exchanges are as much symbolic (communal) realities as they are concrete entities. So it is with what might similarly be called "affective surrender."

Affective surrender varies in degree. There are some who tie only certain feelings to those of the patient. For example, a caregiver might explain that her patient only affects her when the patient seems to be depressed, which depresses ("spreads to") the caregiver as well. Otherwise, she reports, their affective lives are relatively independent. Other caregivers readily note, in contrast, that they have no life of their own to speak of, that they are not only burdened by the chores of the proverbial 36 hour day but that their feelings, too, are completely taken up by those of the patient. When the patient is happy, they're happy; when the patient is depressed, they're depressed; and so on. Indeed, some say that such is a hallmark of love. As one rather annoyed support group participant didactically shouted at another who had fondly surrendered herself to the affective life of her demented husband, "My dear, let me point something out. You're not only an old man's darling, you're an old man's slave in mind and in heart!"

The statement suggests something else about the individuality of feeling. Individuality is also taught. It is commonplace in support group proceedings for participants to offer judgments about the degree to which select caregivers own their own feelings, keeping them separate from the patient while remaining sympathetic to the latter's plight. Caregivers repeatedly remind each other of the prevailing wisdom that they must make a life of their own, taking time out for themselves, developing their own feelings as separate individuals, away from the burdens of daily care. As Roth (1963) once informed us of the comparative and shared quality of individual recovery timetables in tubercolosis, so caregivers teach us and each other how to have and keep individual sentiments, altogether displaying the social construction of privacy.

A fair portion of support group proceedings is devoted to the process of individualizing sentiment, as caregivers often put it, "getting caregivers to accept the fact that they're not the patient and have a life of their own." Ironically, the process is, at the same time, fully social. Not only is individualization a collective enterprise, but what the caregiver is taught to be—a unique individual in his or her own right with particular feelings—is something collectively recognized. Whether one declares or affirms that one has private feelings, known to oneself alone, or is informed or told the same, the communication is nonetheless social in its realization. For what it is that is communicated, to oneself alone or to others, is something tellable and, being so, procedurally public. Even the private assertion of privacy can only be publicly meaningful, for as one *addresses* another openly, one must *address* oneself inaudibly. Anything less remains meaningless, experientially empty.

In this connection, Denzin's program is dubious. He wants to establish a meaningful, private ground for essential self-feeling, separate from its public encumbrances. Presumably, the substance of private self-feeling can be known to oneself alone, as indeed, caregivers and others themselves sometimes claim. But the unique knowledge of self-feeling must come from somewhere, assuredly for Denzin, not the body alone. It is here that serious consideration of the practical organization of references to self-feeling is rewarding. Whether they are private ruminations or part of conversation, their structure is the same: one to another. Likewise, whether ruminations or open conversation, their content must be recognizable to be meaningful, again to oneself or others. As such, if it is private and unique at all, self-feeling is, through and through, socially recognized as being so.

As others do, caregivers have a working knowledge of the social nature of private feelings, addressed with a wary eye to learning about them. Indeed, one of the major goals of the ADRDA in this connection is educational: offering support, information, and counsel to those burdened by the care of the Alzheimer's patient. Anyone faced with the gradual mental demise of a loved one, to the point where a spouse or children are no longer recognized, suffers great emotional stress. Individual caregivers often are in a quandary over what to do and how to feel about it. The ADRDA teaches a way out.

Two kinds of working question are frequently raised by caregivers and concerned others in regard to the emotional status of those affected by the disease. Newcomers or not, they repeatedly ask if the Alzheimer's patient has any particular feelings about his experience, frequently raising the same question of themselves. Typical in this regard was a tearful series of questions posed by a caregiver as she reflected on "state of mind" over what was happening to her beloved husband:

> You know, it's kinda hard to face up to it, seeing his mind just going like it is. He's just not the same Harold that he used to be. He was always sharp as a tack. And you'd never find a kinder and more loving husband. [Weeps] Oh, I'm sorry, but I just don't know what to do or think. My state of mind now . . . I'm empty. I don't know what I feel. I don't even know if I feel anything. I'm just numb. And what does he feel, the poor guy? Do you think he knows? Does he feel anything at all? Maybe you can tell me. Am I different or something? I'm really frightened of not knowing how to feel. Should I be feeling something that I'm not? Or not feel the way I seem to be? God knows, I must be a sight! Excuse me, please. [Leaves to regain her composure]

The second kind of question concerns the form and shape of feelings. Assuming they do have feelings, caregivers ask for help in sorting them out, in discerning which ones they are and their particular status in their lives. Typically, the questions are raised in considerations of the patient's feelings. When a support group participant or facilitator asks, "Where are you at now?" or "I hear you saying that you feel . . . ", the query addresses the status of the deep feelings taken to be affected by the gradual mental demise of a loved one and the burdens of care. When the participant addressed responds that he doesn't know how to feel or where he's at, he informs his audience that he, who ostensibly has the best access to his emotions, is in doubt as to what they are.

Whether feelings are supplied or elaborated, it is taken for granted that some of the concerned know the felt side of the disease better than others. Significantly, those considered to be in the know are not necessarily, indeed are often not, the ones whose feelings are under consideration. Participants assume that feelings can be made and, once there, can be shaped for the holder's benefit, on behalf of his or her mental health. This, of course, doesn't mean that participants don't argue over who, in fact, knows best in particular cases. For example, responding to an explanation of what her real feelings were, a patient's daughter once objected, "Look, I know better than anyone here how I really feel deep down inside." On the other hand, commenting on a similar kind of explanation of his true feelings, offered by another participant, a caregiver might answer that the explanation sheds a different light on how he feels about his personal state of affairs, feelings of which he hadn't been aware. The social distribution of access to private feelings is a principle that, in practice, is applied in accordance with prevailing interpretations of relative individual access.

Whether patient, caregiver, or concerned others, each person's feeling are understood to fall within or between recognizable emotional categories, which, in one outline, resemble the stages of Kübler-Ross' (1964) view of the dying process. The growing public culture of the disease, evident in ADRDA chapter newsletters, educational forums, widely distributed informational and promotional brochures, and in the broadcast media, carries the available stock of exemplars for the disease's emotional life. All concerned make use of the public culture in discerning what or who the "really" feel like, how they once felt, and how they expect to "really" feel in time to come, presenting their feelings in terms of more or less well-known figures whose experiences have been documented and shared. Paraphrasing Mills' (1959, 1963) description of the significant link between public issues and private troubles, the deeply personal and unique feelings of the concerned are construed by means of recognizable understandings, which, in turn, are collective representations of privacy, symbolic realities all (Durkheim 1973). Not only do caregivers and concerned others speak for the patient's feelings, the former's feelings are likewise spoken for, by themselves and others.

The social production or discernment of deep feelings, not appearances, has implications for Hochschild's and Denzin's programs. Though, of course, Hochschild does not deny the existence of deep feelings, she concentrates her program on the alignment of their expression with public expectations. What her program ignores is the production and understanding of what caregivers call "real" feelings, not their public veneer or pretense. Emotion work extends as much to what is believed to be the heart and soul of feeling as it does to "the managed heart." As for Denzin's program, it is evident that a principle of discernment underpins the search for self-feeling, a rule that is fully social in form—collective, representational, reflexive. The principle lends reason to the pursuit of emotional privacy, which, in its very application, socially realizes its unique individuality (see Durkheim 1947, 1954). It is clear that the form and shape of unique self-feeling are idiosyncracies assigned.

EMOTIVE DISCOURSE

In and about the consideration of feelings, the question of emotional communication frequently arises. The issue is whether or not it is possible to convey feelings without using verbal language. It is an important one to those concerned because, sooner or later, the Alzheimer's disease patient loses the ability to communicate in words and sentences. This does not mean that the patient can't communicate, rather that a vehicle appropriate to the transmission of feelings is needed. It is important, too, because caregivers and concerned others believe the feelings at stake, including their own, to be so deep as to defy the power of plain words to convey them. The related craft of emotion work concerns the provision and application of emotive discourse as a means of discerning real self-feeling.

The possibility of emotive discourse is an implicit critique of existing sociological programs. By and large, the programs leave aside the pursuit of concrete emotional experience in its own right, rather being mainly concerned with their social causation, management, control, or disingenuousness. Denzin, of course, is the exception, but he secures emotion (self-feeling) in such a way as to leave it unanalyzable. Yet, it is clear in the earnest talk and interest of patients, caregivers, and concerned others that emotion work extends to the pursuit of concrete emotional experience. In discursive application, the pursuit is productive of feelings that are admittedly indescribable, exposing a portion of emotional life that existing sociological programs take as foundational but do not analyze. The possibility of emotive discourse allows us to zero in on what typically is studiously avoided.

To those concerned, real feelings connect with the body—they are felt—yet they are, at the same time, significant, mysterious, but meaningful states of being. As Denzin (1984) aptly puts it, they are "embodied" states connected with significant events of everyday life, the mysterious underside of life's highs and lows. Caregivers often speak of real feelings as *hidden behind* their surface appearances, behind gestures, activity, and talk. Routine forms of communication suit appearances; they cannot convey real feelings. It is that characteristic combination of qualities—significance, mystery, meaning—that requires the contrivances of emotive discourse.

As far as the Alzheimer's disease experience is concerned, real feelings are hidden in two places: in the real person behind the disease's patient and in the depths of the caregiver's disease-related experience. Time and again, caregivers and concerned others are reminded not to confuse what the patient says and does with what he actually feels. Varied personal accounts of the disease experience, reported in ADRDA chapter newsletters, point out the lessons to be learned from the distinction. In one account, a wife wrote that she had had a very loving and trusting relationship with her husband, who, upon contracting what was later diagnosed as Alzheimer's, became abusive and cunning toward her. At first, she was shocked and confused by the strange behavior. She reported that when she was told that it was the disease that was causing his crude talk and accusations, not the real, loving husband, she tried to ignore everything he said and did. She now knew that the accurate expression of her husband's true and deep feelings was prevented by the disease, which had gradually eroded her husband's intelligence. One lesson to be learned from this and similar experiences is that reactions to the patient by all concerned must be such that the disease, not the person behind it, is blamed for symptoms and the burdens of care. A second lesson is that some means other than verbal communication must be used to make contact with the patient's real feelings.

Real feelings are also hidden in the depths of caregivers' disease-related experience. Caregivers, family members, and concerned others, it is said, are likely to hide what they really feel about the patient's conduct, degeneration, and the burden of care. Indeed, they often hide their actual feelings from themselves, called "denial." Caregivers' real feelings are envisioned as a series of cognitive and affective responses from denial to anger, bargaining, depression, and acceptance. In the Alzheimer's experience, while the denial of real feelings may occur at any time, it is a normal adjustment only in the early period of care.

The term "denial" is regularly used by both denying caregivers and others to describe an undesireable condition: the reactive management of feelings. Caregivers' true feelings are kept hidden, publicly overridden by what are mistakenly believed to be appropriate responses to the burden of caring for a loved one. Anger, fleeting hate, and guilt are suppressed for expressions of love and devotion, a scenario nicely addressed by Hochschild (1979, 1983). In contrast to her portrayal, though, in the Alzheimer's disease movement, there is a thrust toward tactful authenticity in this regard. In their own ways, caregivers, supporting facilitators, and service providers know very well the elements of Hochschild's scheme: feeling and framing rules; emotion management. An important aim of support groups is to teach caregivers to be, in effect, craftier with their emotions, to see through emotional veneers and penetrate the depths of real feeling, to cast side emotional management for the appropriate expression of affective realities—for the benefit of all concerned. Denial, then, is the emotional state where self-feelings are known better by persons other than those whose feelings are under consideration, an emotional condition hopefully recognized, in time, by the holders of the feelings themselves.

The distinction between hidden feelings and appearances sets a course of action. All concerned are to look beyond the obvious to the real. To that end, a variety of discovery rules has become part of the disease's public culture. In contrast to feelings rules, which inform one of how one should (appear to) feel in certain situations, discovery rules are recipes for discerning how those under consideration actually feel. The most formalized of these are contained in a widely quoted and copied article by Bartol (1979). Called "nonverbal communication," the rules are a means of reading the real feelings of the patient. Caregivers and service providers are urged to "actively listen" for clues to real feelings. By means of examples, Bartol shows how to see through surface expressions and gestures, to what the patient really feels as opposed to what is actually being done or said. Facsimiles of Bartol's approach appear throughout the Alzheimer's disease literature, mainly in handbooks for caregivers and family members and in ADRDA chapter newsletters. Informal schemes of various kinds are widely applied and shared in and out of support groups and chapter meetings.

Emotional language takes two forms. One is tonal and tactile; the other poetic. The concerned inform each other that the only way to convey one's true feelings to the Alzheimer's disease patient is by means of touching, bodily affection, and the pitch of one's voice. Proof of their communicative effectiveness is provided by examples of "what a difference a little patience made" or "what a long way a simple touch of the hand goes." Exceptions are proof of effectiveness, too, for the times when touching and tone don't work are those when either the reception isn't just right, the patient is not affectionate by nature, or the right form of affection hasn't been applied. In such cases, caregivers are urged to try other tactile or tonal modes. In the matter of conveying real feelings, the craft of affective communication has limitless technical resources.

Sometimes, the object of communication is to convey either the patient's or the caregiver's true feelings to persons other than the patient. Because of their cognitive deficit, patients are unable to verbally inform others of how they feel. Because the emotions that lie buried in the caregiver's experience are so formidable—significant, mysterious, yet meaningful—plain words cannot convey what he or she really feels. Caregivers point out that words cannot tell what it's like, that it's impossible to put into words.

On one occasion, a wife had been openly, emotionally, describing the deep feelings that entered into her devoted care for her loving but demented husband, how the feelings oscillated, collided with each other, and fleetingly flitted through her mind and heart. She attempted to convey the subtle and flowing stream of her emotive experience, her self-feeling, but she could not. She admitted that what she felt she couldn't even describe to herself, as she added, "not in so many words." She then reached for her purse and pulled out a small piece of paper, saying,

> Maybe I can relay my feelings in another way. Let me get this out. It's a poem that I clipped out of the newspaper. I believe it was written by a woman in St. Louis, who had taken care of a husband bedridden and nonverbal for years. I think it tells it exactly the way it is, what I can't put in so many words.

She then read the poem, intending thereby to communicate her feelings. The reading ended, the listeners agreed that, yes, that was the way it felt, that the poem had said all there was to be said and had put it very well. With words, the poem conveyed what words could not do.

There are many poems in and about the disease's public culture. Some are said to be written by patients themselves, some by caregivers on behalf of patients, and others by caregivers to communicate their own feelings. Their contents would, no doubt, be judged crude by some. Certain poems contradict what others present as the true feelings of the disease experience. Still, as conveyed and received, the poetry of the disease is taken as concrete emotional expression, doing something that prose cannot. What is important about poetry as emotional discourse is that it is poetic, more what poetry does than what it says, a means of making meaningful something that is otherwise altogether too mysterious for words.

CONCLUSION

The beauty of Bogie's face is that it not only manages to express feeling, at times full tilt, but openly tinkers with self-awareness and appearance. He shows us how to manage a variety of fleeting emotions and conveys to us that he is in full control of his conduct, as if to wink an eye at us from behind the facework. The craft of emotion work resembles Bogie's agency. It is fully engaged in managing self-feeling against audience (public) expectations, yet never sinks its resolve in appearances, keeping fully attentive to the distinction between appearances and realities, between the emotional expression and real feeling.

The feelings involved in the Alzheimer's disease experience toe the margin of appearance and reality. While there is ongoing concern with the expression of feeling, with tacit feeling rules, and how to align or not to align feelings with them, there is, at the same time, a persistent, even urgent, desire to get to the real emotions being felt. The desire is not just a lurking hope for affective insight, but a concerted search. As caregivers, service providers, facilitators, and concerned others gather in support groups, they attend to both the management and discovery of feelings. While there is the occasional claim that the latter can only be a strictly private matter, there is nonetheless an open, sometimes quite studied, pursuit of privacy. The experiential rewards of their affective deliberations inform us of several shortcomings in existing sociological programs for the analysis of feeling.

First, it is evident that while feelings are taken to be concrete realities—real feelings they are as much realized as real. Alzheimer's folk do not dance about social relational forces, the objects of their affective worlds captively luring them into emotive embodiment. Theirs is an affective world at once made, reproduced, and responded to. Their collective deliberations present them with a problematic, the point of which is to define and interpret the world of feeling under consideration so as to decisively act within it. They seem to be more impatient with appearances and feeling rules than the latter would suggest. Their affective cause implores them to turn to the discovery of real feelings lest a submission to appearance do them in. Their related actions and interactions tell us they are actively forthright in articulating both the conditions and shape of feelings, amusedly and insightfully smiling at those who expertly try to tell them otherwise.

Second, Alzheimer's folk rush in where sociologists fear to tread. Bridled by the daily trials of caregiving, they nonetheless confront the concrete feelings involved. They desire to know whether or not they have feelings and, if so, what they are and whether they are unique in that regard. Their attention to self-feeling is anything but strictly private, their streams of emotional consciousness under continual scrutiny, shared, interpreted, and understood. Their world of feelings is hardly a brooding, private domain, even though that state of being is firmly recognized.

Third, as Bogie's face also tells us, the world of feeling is not a catalog of discrete emotions. Feelings drift into one another; they flit about—affective rainbows. They transform each other such that the interpretation of new feelings lead to redefinitions of what earlier feelings actually were. Affective experience thus undoes the matrix of secure variables laid upon it, seemingly not holding still long enough in practice to serve as empirical warrants, certainly not long enough, in principle, for the positivist program.

Fourth, the Alzheimer's disease experience shows that the intersubjectivity of feeling is anything but marginal to the discovery of self-feeling. Whether it is internal or external conversation, intersubjectivity serves the discovery of self-feeling. Indeed, it even underpins a distinct language: emotional discourse. That language is perhaps the most visible evidence of the attempt to address deep, concrete feelings on their own terms—things significant, mysterious, yet meaningful.

REFERENCES

Bartol, Mari Anne. 1979. "Nonverbal Communication in Patients with Alzheimer's Disease." *Journal of Gerontological Nursing* 5:21–31.
Blumer, Herbert. 1969. *Symbolic Interactionism.* Englewood Cliffs, NJ: Prentice-Hall.
Darwin, Charles. 1955. *The Expression of Emotions in Man and Animals.* New York: Philosophical Library.
Denzin, Norman K. 1984. *On Understanding Emotion.* San Francisco: Jossey-Bass.
Durkheim, Emile. 1947. *The Division of Labor in Society.* Glencoe, IL: Free Press.
_____. 1954. *The Elementary Forms of the Religious Life.* Glencoe, IL: Free Press.
_____. 1973. *On Morality and Society.* Edited with an introduction by Robert N. Bellah. Chicago: University of Chicago Press.
Garfinkel, Harold. 1967. *Studies in Ethnomethodology.* Englewood Cliffs, NJ: Prentice-Hall.
Goffman, Erving. 1959. *The Presentation of Self in Everyday Life.* New York: Doubleday.
_____. 1971. *Relations in Public.* New York: Basic.

Gouldner, Alvin W. 1970. *The Coming Crisis in Western Sociology.* New York: Avon.

Gross, Edward, and Gregory P. Stone. 1964. "Embarrassment and the Analysis of Role Requirements." *American Journal of Sociology* 70:1–15.

Gubrium, Jaber F. 1986. *Oldtimers and Alzheimer's: The Descriptive Organization of Senility.* Greenwich, CT: JAI Press.

Hochschild, Arlie Russell. 1973. *The Unexpected Community.* Englewood Cliffs, NJ: Prentice-Hall.

―――. 1979. "Emotion Work, Feeling Rules, and Social Structure." *American Journal of Sociology* 85:551–575.

―――. 1983. *The Managed Heart: Commercialization of Human Feeling.* Berkeley, CA: University of California Press.

James, William, and Carl B. Lange. 1922. *The Emotions.* Baltimore: Williams & Wilkins.

Katzman, Robert, ed. 1983. *Banbury Report 15: Biological Aspects of Alzheimer's Disease.* Cold Spring Harbor, NY: Cold Spring Harbor Laboratory.

Katzman, Robert, Robert D. Terry, and Katherine L. Bick, eds. 1978. *Alzheimer's Disease: Senile Dementia and Related Disorders.* New York: Raven Press.

Kemper, Theodore D. 1978a. "Toward a Sociology of Emotions: Some Problems and Some Solutions." *The American Sociologist* 13:30–41.

―――. 1978b. *A Social Interactional Theory of Emotions.* New York: Wiley-Interscience.

―――. 1981. "Social Constructionist and Positivist Approaches to the Emotions." *American Journal of Sociology* 87:336–362.

Knorr-Cetina, Karin D., and Michael Mulkay, eds. 1983. *Science Observed.* Beverly Hills, CA: Sage.

Kübler-Ross, Elisabeth. 1969. *On Death and Dying.* New York: Macmillan.

LaTour, Bruno, and Steve Woolgar. 1979. *Laboratory Life: The Social Construction of Scientific Facts.* Beverly Hills, CA: Sage.

Mace, Nancy L., and Peter V. Rabins. 1981. *The 36-Hour Day.* Baltimore: Johns Hopkins University Press.

Mills, C. Wright. 1959. *The Sociological Imagination.* New York: Oxford University Press.

―――. 1963. *Power, Politics and People.* Edited with an introduction by Irving Louis Horowitz. New York: Ballantine.

Reisberg, Barry. 1981. *Brain Failure.* New York: Free Press.

Reisberg, Barry, ed. 1983. *Alzheimer's Disease: The Standard Reference.* New York: Free Press.

Roth, Julius A. 1963. *Timetables.* Indianapolis: Bobbs-Merrill.

Sartre, Jean-Paul. 1962. *Sketch for a Theory of the Emotions.* London: Methuen.

Schachter, Stanley, and Jerome E. Singer. 1962. "Cognitive, Social, and Physiological Determinants of Emotional State." *Psychological Review* 69:379–399.

Scheler, Max. 1970. *The Nature of Sympathy.* Hamden, CN: Archon Books.

Shott, Susan. 1979. "Emotion and Social Life: A Symbolic Interactionist Analysis." *American Journal of Sociology* 84:1317–1334.

Wells, Charles E., ed. 1977. *Dementia.* Philadelphia, PA: F. A. Davis.

Wittgenstein, Ludwig. 1953. *Philosophical Investigations.* London: Basil Blackwell & Mott.

THE BEREAVED PARENT's POSITION;

ASPECTS OF LIFE REVIEW AND

SELF-FULFILLMENT*

Meira Weiss

Our challenge is not changing life into death, but vice-versa. We feel that death should be changed into life . . . as it is written: In your blood shall you live!

- a bereaved mother's words, Israel Television, eve of Memorial Day for the Fallen, 1980

INTRODUCTION

The purpose of the study on which this paper was based was to examine the reactions of parents after the loss of a son, and to interpret the significance of this behavior. The questions of uniformity and process in this behavior were investigated. I believe that such information may be useful to professionals, particularly clinical psychologists, in their understanding and treatment of bereaved parents.

Current Perspectives on Aging and the Life Cycle,
Volume 3, pages 269–280.
Copyright © 1989 by JAI Press Inc.
All rights of reproduction in any form reserved.
ISBN 0-89232-739-1

The possibilities and indicators of rehabilitation or adaptation to bereavement suggested by various researchers (Eliot, 1955a, 1955b; Gorer, 1967, Eliot, 1932; Hill, 1955)[1] reflect the investigators' value judgement; unclear terms and circular thinking are used. These studies involve an underlying assumption of a unilinear adaptation process, a perception that the bereaved must go through a given set of predetermined stages.

A few researchers, as well as some philosophers and poets, mention other aspects of the behavior of the person who has encountered death, such as painful recollection of past experiences, recognition of liberties formerly denied or forfeit of the right to self-fulfillment, and a new uncompromising demand for immediate fulfillment of individual aspirations (Buber, 1948; Tolstoy, 1960; Campbell, 1969, p. 211).[2] It is this line of thinking which inspired the present study.

In order to avoid the shortcomings of the value judgements inherent in previous studies, I requested that the respondents define the bereaved's recognition of past failures and their perceptions of self-fulfillment. The results show that in spite of frequent behavioral uniformity among the various respondents, the messages conveyed in those behaviors may be different for each respondent and even for the same respondent in different contexts.

Furthermore, this work introduces the voluntary aspect of the behavior of the bereaved. It is shown that the loss leaves the bereaved with a zone of action in which he is free to direct his behavior toward attainment of his various aims. Here we see that the various modes of behavior chosen often express like messages.

The study differs from others in the field in its investigative focus as well. Most investigators explain the bereaved's behavior as a reflection of suffering over the death of the beloved. Yet, social scientists are unable to observe pain per se[3], and it seems to me that there is nothing inherent to suffering which imposes certain social behaviors.[4] It is rather the social context that directs behavior. Here, then, the focus is on the expression of bereavement in various social contexts.

The study revealed the tendency of bereaved parents to use their loss as a new starting point in their life. It seems that the social circumstances related to the loss encourage the individual to examine his relations in various realms, and to seek fulfillment of opportunities perceived as previously blocked.

The present article describes and analyzes the bereaved's efforts to review past experiences, recognize failures, and strive for self-fulfillment, in two areas: husband-wife relations and work relations.[5]

RESEARCH METHODS

The fieldwork on which this paper was based applied the participant observation method. As the available research on bereavement provided no relevant hypotheses for a statistical survey along the lines described above, this work was based on thorough examination of a small number of respondents, with the hope of deriving new insights into this subject. Perhaps its findings can now be used as the basis for more widespread statistical research.

The fieldwork was conducted on three couples of bereaved parents, who lost their sons in the Yom Kippur War. During my voluntary service on behalf of the Ministry of Defense, I visited about a hundred bereaved families during the first week after they were notified of their sons' death. I stayed in contact with ten families for about half a year. I kept in close contact with the three families who serve as my respondents,[6]

- *Fania and Yolek,* who immigrated from Poland in the 1930s, live in a prestigeous area within a big city in the North. The area is homogeneous in terms of social status. Both husband and wife are employed as high-ranking public officials.
- *Cuna and Yetti* immigrated from Rumania ten years ago, but still view themselves as "new immigrants." They live in an ethnically-mixed neighborhood, populated mainly by workers. The family maintains almost no social contact with the community. Cuna is a high-ranking public official; Yetti is a non-professional worker in a large factory.
- *Abigail and Yochanan* are of Yemenite origin. They live in a large urban neighborhood. Yochanan is a low-ranking public official, and Abigail has occasionally worked as a cleaning woman. She returned to this work about a year after the loss of their son.

These couples were chosen according to three sets of criteria. First, they share some common features: For all couples, this is their first marriage, all have other children, they have all been married for about thirty years, and their ages range from fifty to sixty. All are urban families who generally define themselves as non-religious. Second, none have any pathological history, or other stigma that might divert the research direction or its conclusions. None of these families were treated by a social agency before the loss; they have no history of mental illness. Their economic situation is satisfactory. All functioned normally before the calamity and appear to do so after it as well. The third consideration was the establishment of a deep relationship between myself and them to the extent that I was accepted as part of the family. Life went on as usual in my presence.

Observation of the families began in October 1973 and continued until 1978. In this period I visited the families during various hours of the day, I stayed with them at meal-times, when they hosted guests, when they visited government offices, in memorials for their sons, in organized meetings with other bereaved parents, etc.

Following is a discussion of two aspects of the behavior I observed during this period, which characterize the process undergone by bereaved parents.

The Bereaved's Quest for Self-Fulfillment
in Husband-Wife Relations

The respondents' behavior indicates a reexamination of past marital relations and sex-roles. Reviewing their marriage, the bereaved become aware of a waste of their potential within their family life; they believe that they have given up the opportunity for a meaningful life and acted like marionettes, for the sake of family stability. From now on, the bereaved do not want to reflect their spouse's values, but rather wish to be themselves.[7] For this they are willing to destroy the pretensious and false life upon which, according to them, their marriage has been based, and to establish a new life-style with their spouse. The price they previously would have paid for such behavior (family instability) no longer serves as a deterent. And so, after the loss, the bereaved turns to his or her spouse to convince (or force) him or her to recognize the legitimacy of his past claims for self expression and fulfillment of desires. This is done mainly by means of reference to issues of suitable behavior after the disaster, which are central to the overt behavior of the bereaved.[8]

This interpretation is based on an examination of responses to these issues, not only in terms of their content, but also in their underlying significance, as revealed by ongoing observation and in-depth analysis of the subjects' words and actions. The study-claim is exemplified in an analysis of Abigail's behavior.

As other respondents, Abigail expresses a sense of failure to fulfill herself as an independent person within the family. She now rebels against men's world in general, and her husband in particular, for suppressing her and robbing her freedom. While up to now she has outwardly resigned to her fate as a woman, a fate of sacrifice for her children and husband, now she revolts: "Why," she cries, "should I sacrifice myself for my husband and daughters? Why? I have my own life. I have already sacrificed enough. I am not going to do anything I don't want to." And so, with a sense of urgency, a feeling that this is the last chance to fulfill her wishes, Abigail begins to struggle for her freedom.

First, she perceives that she could have real freedom if her husband, Yochanan, had fallen instead of her son. Yochanan's death would have freed her from the burden of living with "a boring and unsuccessful" man, and would have given her respect and freedom. She believes that her independence lies in "getting rid" of her husband. "He is the first thing I want to get rid of," says Abigail ten days after her son's death.

Her efforts to "get rid of him" concentrate on degrading him as a man For instance, Abigail knows that her husband ascribes great importance to having her listen to "his boasting of imaginary success" at work and in the army, and to her readiness to have sexual relations with him. Therefore, she disregards him exactly in those realms. She ceases all contact. She demonstratively stops talking to him. She puts his meals on the table without uttering a word, and leaves the kitchen. In her intentional humiliation and negligence of Yochanan, Abigail rejects his self-understood right to her. She feels that her main power lies in denying him her recognition; she believes that by giving him attention she will lose her chance to feel free.

Abigail generally attributes this behavior, particularly her refusal to have sex, to her mourning over the son. But there are many indications of her strong intent to avoid "wasting" her life in the future. A thorough follow-up of her arguments reveals this. "I do not want to have sexual relations," she explained excitedly and firmly in our first meeting, "what is it, my son dies and all is as usual? Should I enjoy myself? . . . And also, nothing interests me since. There is nothing to live for. And my husband understands nothing. He never did. I do not care about him. Let him go to other women. I don't want him. I am not sacrificing myself anymore for him or the girls. Enough. It's over. I have my own life. Let it be clear to everyone that I'm not doing anything I don't want to . . ." Abigail repeats this declaration word for word on several occasions; at all times the order of reasons presented for her behavior is identical. At first, the son's death, or "being nervous with her husband" are raised as the reason for abstention from sexual relations. Next comes the sentence, "There is nothing to live for", which means, it emerges, there is nobody to live for, since no one's life justifies sacrifice of her own. Towards the end of her presentation of reasons, Abigail's actual intent is clear. She has firmly decided to free herself from all bonds with her husband. Self-examination after the loss has given Abigail an opportunity to seek an alternative style of life with her husband. Now, after her son's death, she makes public her sense of past failures and tries to attain what, according to her, she was denied before.

A second example of the same process is demonstrated by Polish-born Yolek, about sixty years old. Shortly after his son's death, Yolek secludes himself and painfully examines past years. He now recognizes his wife Fania's dominance over the couple's relationship, and choice of friends: "My role was just to approve."

A conflict arises as to the proper behavior after the loss of their son. Since the loss, Yolek wishes to visit the cemetery often, and wants to meet other bereaved parents, who, according to him, suffer as much. His wife, on the other hand, clings to matters which entail neither suffering nor social contacts with those who remind her of her grief.

However, it seems that Yolek's insistence on these issues is actually rooted in their contradiction to his wife's views, and in the essential change involved in disagreeing with her. He is striving for a better position in husband-wife relations. For it does not seem likely that Yolek is seeking an individual style of mourning. If this were so, he could fulfill this desire during the hours in which he and his wife are not together. He knows that many of the bereaved parents he likes visit the graves during work hours. Yet, he does not go alone; he insists on the necessity of going together with Fania.

Moreover, his indulgence in suffering is neither absolute nor consistent. He does not abstain from all activities which give him pleasure. He strongly insists on having sexual relations.[9] It appears that Yolek's real wish is to receive his wife's approval for his special way. Solitary visits to the cemetery would have not gained him this recognition. His wife's consent to act according to his wishes is interpreted as legitimization of his decision both concerning the mourning pattern and household management.

Similarly, it seems that Yolek clings to contact with the other bereaved not because of their importance to him, but mainly because they constitute an alternative to those friends who were not desirable to him prior to the loss.[10]

As time passes, the struggle becomes overt and direct, to the point that after three years, Yolek declares an overall rebellion against his wife and his oppression: "Now I want to be an egoist. I feel my friends and my family treat me like a puppet . . . being danced as they wish . . . but they forget that a puppet has open eyes, and one day it also starts talking, and not only dancing. . ."

The bereaveds' behavior described above exemplify a general process which occurred among all respondents. They express distresses which appeared prior to the loss, but now, with the change of circumstances, they believe, they have a chance to express their dissatisfaction.

The Bereaved's Quest for Means to Fulfillment at Work

The bereaved also tend to review their past occupational achievements. While doing so, they often recognize their failure to express individuality and their subjection to "vocational masks." This submission, according to them, reflected societal expectations, and made self-fulfillment in work harder to attain. The positions they hold now seem to be degrading and inappropriate, denying their individual potential and skill. Now that the status quo has been changed, they resolve to express their individuality in work. This is done, for example, by finding a new occupation fitting their skills and areas of interest, or by investing less energy in a job intended for livelihood only, and devoting more effort to the development of personal interests.

According to Abigail's own report of the past, she was generally submissive to representatives of authority, and did not publicly manifest her resentment of them. After her son's death, Abigail overtly fights the world which exploited her in the past. This is exemplified in her treatment of her supervisor in work, as well as in her treatment of me.

Abigail makes her work supervisor the target of her war against men, who, according to her, are responsible for the world's wrongs. She manages to turn the supervisor into a subordinate, she mocks him and enjoys seeing him humiliated and depressed. Abigail also believes that her new liberty should be expressed in blatant disregard for her work, non-compliance with regulations, and the demand for honorary and monetary compensations from work.

Abigail sends me daily, for three months, to different employment places; in the end, she rejects all offers presented by me. She feels she can punish me for her exploitation by those I represent. While once she felt dependent on me, and those like me, she now demands I serve her.

Yochanan also uses his son's death as a power resource. This is demonstrated in his relations with co-workers and his supervisors.

In 1940, Yochanan enlisted in the British Army, and became a cook for a regiment, the officers of which now hold high positions in the army and the government. Here Yochanan seems to have developed a wish to attain success without hard work, admiration for external signs of success and a wish to imitate the status symbols of the high-ranked.

Upon his discharge, Yochanan became a sanitary supervisor in the municipality. "From the moment I got there, they decided I was gullable," Yochanan recalls, "and they gave me the hardest and the most humiliating tasks . . . also, I wasn't promoted for years . . . I delicately hinted about it, but they ignored me." As during his military service, Yochanan did not identify with his employer, but was proud of the title "municipality employee," which he bestowed upon himself. He was willing to suffer numerous humiliations in return for the rare moments in which he received approval.

After his son's death, Yochanan reviews his life, and his work relations in particular; the function of his reminiscing changes. While in the past it helped him to tolerate humiliation, he now draws on his memories as legitimization for refusal to submit, for a new sense of strength. Yochanan perceives that he has fulfilled the ideal masculine image (shaped by his recollections of the British Army) through his son's death.[11] Now, as "a bereaved father and veteran of the British Army" he feels the right to object to being humiliated by his boss, and to not being promoted and being assigned detested jobs. He now demands just treatment and that which he believes was withheld from him in the past.

Yetti also expresses bitterness over past injustice and deprivation in her work relations. She in unwilling to continue her silence in these matters. I would attribute the change in Yetti's behavior mainly to her new identity as an Israeli. While up to now she abstained from demanding what was coming to her because of the language deficiency, and her label as "a Rumanian," now the opportunity arises to feel equal (and sometimes better than others). Her son's death establishes her Israeli identity which then gets stronger. Her resentment of previous employers is expressed in a determination to get promoted, to gain consideration, and respectable work. However, Yetti's general behavior at work includes rejection of all suggested jobs, and incitement of quarrels. This can be understood as a wish to punish the employers for depressing her past desires for advancement in work.

Like Abigail and Yochanan, Yetti explains the change in her behavior by nervousness, increased weakness since the loss, and claims that her employers are insensitive. There is, of course, an objective basis to most of these claims; such loss may diminish the capacity of the bereaved for suffering, their tolerance level and their physical strength; and some of the claims against their employers can be verified. However, the complaints about exhaustion and nervousness emerge only in certain contexts and not in others, in which they might just as well be relevant.

In all the cases described above, the work situation was used as an arena of establishing control, demonstrating resistance to prior oppression, demanding recognition, obtaining rewards and punishing those who prevented the fulfillment of these wishes in the past.

DISCUSSION AND CONCLUSIONS

Can we note a behavioral uniformity among bereaved parents? This examination shows that all respondents demonstrate a process, the components of which are similar: a review of life prior to the death, including examination of relations with significant others, explicit recognition of the waste of their potential before the loss, and a demand to fulfill it immediately.[12] It shows that the death raises specific issues, the reference to which imposes reexamination of past self-identity. However, the bereaved do not test the most basic assumptions upon which their life had been based. The parents do not shatter the family institution and work, and are not interested in actually destroying the existing social order. They only challenge their position in it.

It further emerges that this process of reexamination has no firm deterministic development. The parent is left with room for maneuvering. The difference among respondents is expressed in several areas. Some parents scatter their efforts over all arenas (Abigail, for example), while others limit their activities to a specific realm (Yetti, for instance). The behavior takes various forms (Yetti and Abigail curse, while Fania uses more camouflaged strategies). The chronology of the elements of the process differ from subject to subject, as does the starting point of its development.

These differences may be attributed to various factors, such as sex (particularly in the family realm), cultural background (particularly in the form the behavior takes) and personal life history. The similarity that was found among the respondents lies in the messages underlying the respondents' behavior. While "symptoms" and explicit sayings may often vary, an examination of the significance of this behavior points to the common denominator—the process.

This study raises further questions. For instance, could the respondents' behavior be attributed to factors other than bereavement, such as (a) the overall atmosphere in Israeli society after the Yom Kippur War,(b) the role which I fulfilled, or (c) the the fact that all were secular, economically middle-class citizens.

There is some evidence that these factors are not significant. A content analysis of discussions and articles on bereavement from the last decade testifies to their similarity to the present information.[13] A personal encounter with other bereaved parents indicates that at some point every bereaved person finds someone with whom to share his thoughts, who fills similar role to that which I performed. My experience in meeting religious bereaved parents shatters the assumption that they automatically accept routine and defined relationship patterns.[14]

It seems, moreover, that the need to make basic changes in one's life systems is common to other bereaved persons. In the cases studied, however, we do see the importance of the realistic possibility of improvement, as provided to those bereaved in war by the Defense Ministry and other agencies. It seems that this aid is significant in the individual's efforts to affect such changes.

A factor which still remains to be investigated is the economic factor. While all the respondents in this study were relatively well-established, it may be argued that those involved in daily struggles for existence may not be involved with issues of self-fulfillment.

This analysis may be the basis for further research on bereavement. First, a more systematic examination should be made of the behavior of bereaved parents who differ according to the above-mentioned factors (particularly economic status). It would also be interesting to examine the applicability of these findings to the behavior of other bereaved (war widows and orphans, or those otherwise bereaved).

ACKNOWLEDGEMENTS

In the analysis of the material I have been helped by my instructors and friends. In particular I am thankful to Professor S. Deshen and Professor M. Shokeid, who have greatly contributed to the development of my thinking. I thank Professor I Marx, Dr. A. Aviel, Dr. N. Rubin, Dr. P. Palgi, Dr. M. Schwartz and Dr. I. Sheffer for their interest and the great encouragement I have received throughout the research stages.

NOTES

1. See, for example, Brough (1975, p. 499); Goldberg (1973); Lindman (1944); Hinton (1971, pp. 167–174).

2. See also Jonesco (1963, 1967) and Lamont (1973) who refer mainly to people for whom death has become the motivating power in life (such as the dying the old and philosophers), and whose behavior resembles that of the bereaved.

3. Students of social behavior lack the tools to fully observe the life of the bereaved. They only have observational expressions for suffering as manifested in various social contexts. We cannot comprehend the bereaved's sensations, thoughts, longings and dreams when he is alone. We can only know how he expresses his pain in the presence of his spouse, children, relatives, neighbors, etc.

4. If suffering were the only explanation for and conditioner of the bereaved's social behaviors, how could the behavioral variations be explained? Why is a person's suffering shown, sometimes simultaneously, by different expressions? Why does yearning yield different expressions in different

social contexts Why does a certain expression emerge under certain circumstances and another in different contexts? The sensation of suffering, as such, cannot explain these variations.

5. The original research included two additional areas: contact with high-status people and relationships with relatives and neighbors.

6. Names and some characteristics are fictitious.

7. These expressions, which recur in various forms among my respondents, emerge in Feifel's reports (1959 XII, p. 124), (1963, pp. 10–11), and Kubler-Ross (1970, pp. 51–81).

8. For example: How should the contact with the fallen son be continued; what is the nature of the contact with other bereaved parents, or with old friends; should pain and suffering be openly expressed or concealed; should sexual relations be maintained?

9. Other respondents also claim self-expression in their struggle over having sexual relations. For example, since the loss, Yochanan does not agree to submit to his wife's refusal to have sex with him. The importance of this demand is beyond mere fulfillment of a sex urge, since alternatives that would satisfy his sexual drives are rejected by him. Yochanan uses the issue of sex to insist on a basic right which, he claims, was withheld from him—the right due to a self-respecting male. From Yochanan and his wife's descriptions, we see a sort of battlefield, in which the issue of sex will determine the outcome of the general struggle between the couple.

10. This subject is elaborated on in the original study in the chapter dealing with efforts to improve status. Some similarities do exist between groups of bereaved parents and "factions" (Shokeid, 1968). Consequent to the disaster, a human reservoir, which can be manipulated according to the desired aims, is formed. Bereavement, which seemingly is a common denominator between the people is important because of the legitimacy it grants to the formation of a group functional for the fulfillment of interests.

11. In the original study (Weiss, 1978), in the chapter dealing with the use of bereavement for status improvement, it was shown that, according to Yochanan, his son's death should raise his status and should bring him back to the officers' society. This is because the son's death proves his affiliation with the family of military men.

12. The bereaved's "personal growth" as a result of his suffering appears in Frankel (1970, pp. 92, 100, 136) and Hendin (1973, p. 183).

13. In these contexts one can detect very similar behavioral expressions to those of our respondents. These were voiced in a very different general atmosphere than that which accompanied the last war. Also, the resemblance in the kibbutz members' thoughts expressed before and after the war is remarkable.

14. In these meetings, such expressions were voiced against God, as, "Why have you chosen us?"

REFERENCES

Berger, P.L. 1963. *Invitation to Sociology.* Doubleday & Company, Inc., Garden City, New York.

Bott, E. 1957. *Family and Social Network.* London, Tavistok, pp. 218–219.

Brough, W. 1975. "Mourning: Normal and Abnormal." *Update Publications,* 10, (5): 499–506.

Buber, M. 1948. Tales of Hasidim: The Later Masters. New York, Shocken, pp. 173, 268.

Butler, N.R. 1968. "The Life Review: An Interpretation of Reminiscence in the Aged," in *Middle Age and Aging,* edited by Neugarten, B.L. University of Chicago, pp. 486–496.

Campbell, E.Q. 1969. "Death as a Social Practice," Pp. 209–230 in *Perspective on Death,* edited by Listen, O.M. New York: Abingdon Press.

Camus, A. 1956. *The Rebel.* New York: Vintage Books.

Clayton, P. et al. 1968. "A Study of Normal Bereavement," *American Journal of Psychiatry* 125, (2): 168–178.

Deshen, S. 1989. *The Mellah Society: Jewish Community Life in Sherisian Morocco.* University of Chicago Press.

Eliot, T.D. 1932. "The Bereaved Family," Pp. 184–190 in Annals of the American Academy of Political and Social Sciences, 160.

Eliot, T.D. 1955a. "Adjustment to Bereavement," Pp. 170–171 in *Social Problems in America,* edited by Lee, B.E. Holt.

Eliot, T.D. 1955b. "War Bereavement and their Recovery," Pp. 339–346 in *Sourcebook in Marriage and the Family,* edited by Sussman, M.B. Houghton-Mifflin.

Engel, G.L. 1964. "Grief and Grieving," *America Journal of Nursing,* 64:93–98.

Feifel, H. 1959. "Introduction" and "Attitudes Towards Death in Some Normal and Mentally Ill Populations," Pp. XI–XVI, 114–130 in his (ed.), *The Meaning of Death,* New York, McGraw-Hill.

Feilel, H. 1963. "Death" Pp. 8–21 *Taboo Topics,* edited by Farberow, N. New York: Atherton Press.

Frankl, E.V. 1970. *Man's Search for Meaning.* Tel-Aviv, Israel: Dvir Co., Ltd.

Goldberg, S.B. 1973. "Family Tasks and Reactions in the Crisis of Death," *Social Casework* 54 7:398–405.

Gorer, G. 1967. *Death, Grief and Mourning.* Garden City, N.Y.: Doubleday Books.

Hendin, D. 1973. "Grief and Bereavement," Pp. 164–184 in his *Death as a Fact of Life,* New York: Norton.

Hill, R. 1955. "Social Stresses of the Family," P. 303 in *Sourcebook in Marriage and the Family,* edited by Sussman, M.B. Houghton-Mifflin.

Hinton, J. 1971. *Dying.* Baltimore: Penguin Books.

Homans, G.C. 1967. "Fundamental Social Processes," Pp. 27–28 *Sociology: An Introduction,* edited by Smelser, N.J. New York, John Wiley & Sons.

Ionesco, E. 1963. *Exit the King.* New York, Grove Press.

Ionesco, E. 1967. *Journal en Miettes,* Editions Mercure de France.

Kalish, R.A. 1969. "The Effects of Death upon the Family," Pp. 79–107 *Death and Dying: Current Issues in the Treatment of the Dying Person,* edited by Pearson, L. Cleveland, Case Western Reserve Press.

Kubler Ross, E. 1970. *On Death and Dying.* New York, MacMillan Company.

Lamont, R. 1973. "The Double Apprenticeship: Life and the Process of Dying," Pp. 198–224 in *The Phenomenon of Death,* edited by Wyschogrod, E. New York: Harper & Row.

Leshan, L. and Leshan, E. 1973. "Psychotherapy and the Patient with a Limited Life Span," Pp. 3–13 in *The Phenomenon of Death,* edited by Wyschogrod, E. New York: Harper & Row.

Liebow, E. 1967. *Tally's Corner.* Boston, Little Brown & Company.

Lifton, R.J. 1973. "On Death and Death Symbolism: The Hiroshima Disaster," Pp. 77–102 in *The Phenomenon of Death,* edited by Wyschogrod, E. New York: Harper & Row.

Lindemann, E. 1944. "Symtomatology and Management of Acute Grief," *American Journal of Psychiatry,* 101:141–148.

Palgi, P. 1974. Mavet, Evel Uschol Bachevra Haisraelit Beitot Milchama (Death, Mourning and Bereavement in Israeli Society at Wartime), Jerusalem: Ministry of Health.

Rubin, N. 1978. Defussei Haevel Hayehodiim Beeretz Israel Bitkufat Hamishna Vehatalmud (Jewish Mourning Patterns in Israel During the Talmud and Mishna Period, Ph.D. Dissertation, Bar-Ilan University.

Shokeid (Minkovitz), M. 1968. "Immigration and Factionalism: An Analysis of Factions in Rural Israeli Communities," *British Journal of Sociology,* 19:385–405.

Tolstoy, L. 1960. "The Death of Ivan Ilych." In *The Death of Ivan Ilych and Other Stories.* New York: New American Library.

Weiss, M. 1978. Ma'amad Hahoré Hashakool: Haibatim shel Skirat Haim M'ered Vehagshama Azmit (The Bereaved Parent's Position: Aspects of Life's Review, Rebellion, and Self-Fulfillment). M.A. Thesis, Tel-Aviv University.

METAPHORS, AGING AND THE LIFE-LINE INTERVIEW METHOD

Johannes J.F. Schroots and Corine A. ten Kate

INTRODUCTION

The purpose of this chapter is to introduce a new method in research on aging and the life course, the so-called *Life-line Interview Method* (LIM). This assessment method is a powerful instrument for eliciting biographical information, especially from the elderly. As one of the older interviewees said at the end of a LIM-session: "I never told this before to one single person in my life. I feel entirely exhausted and I really don't know what else to say!" This reaction is not only typical of the emotional involvement of interviewees during a LIM-session, but also reflects the power of metaphors when people are asked to describe their life.

Essentially, the LIM has been developed on the basis of several studies in metaphors of aging and the individual course of life (Birren and Schroots, 1980a; 1980c; Schroots, 1982). Before discussing these studies and their implications, it

Current Perspectives on Aging and the Life Cycle,
Volume 3, pages 281–298.
Copyright © 1989 by JAI Press Inc.
All rights of reproduction in any form reserved.
ISBN 0-89232-739-1

seems necessary to present a synopsis of the role of *metaphor* in modern science. Once having been sensitized to this role, we will next turn to the task of discussing some old and new metaphors of aging and the life-course, their implications for research methods, and—finally—their functions in developing the LIM. This paper will be completed, eventually, by a comprehensive description of the Life-line Interview Method, illustrated with a case study.

Science and Metaphor

Until recently metaphors had a bad reputation in science. Leatherdale (1974), for example, summarizes no less than fifteen possible objections to the metaphorical view of science, most of them raised by scientists who favor a logical positivistic approach. An important feature of positivism is the notion that literal language is required for a clear, unambiguous, and objective description or characterization of reality. Metaphors—as a species of figurative language—are in this view a kind of anomaly of language: they violate the language rules and they stand in the way of a general theory of reference of meaning.

However, in recent years claims have been made that science is in an essential way metaphorical and characteristically employs metaphors (e.g., Cassirer, 1946; Richards, 1936). These claims are based on the *constructivist* view of reality in science which holds that reality or the objective world is mentally constructed on the basis of the constraining influences of the individual's knowledge and language (Sapir, 1921; Whorf, 1956).

Although many theories in science give the impression that they are constructed solely under certain formal rules which exclude metaphorical reasoning, recent research evidence suggests that theorists typically do not operate under these rules but rather are guided by an *implicit* metaphor in "discovering" new phenomena (Valle and Von Eckartsberg, 1981). This research shows that this is not only so for a number of social policy and personality theories (Mehrabian, 1968; Schön, 1979), but also for physical and biological sciences (Boyd, 1979), in fact, for all sciences (Gentner and Grudin, 1985; Ortony, 1979; Schroots, 1986). We ought to become aware of these implicit metaphors by making them explicit, not only because they provide us with a deeper understanding of the existing theory, but also because they tend to generate or create a whole body of theoretical problems and solutions.

Paraphrasing Schön (op.cit.) one might say that the framing of scientific problems often depends on certain pervasive, tacit metaphors underlying theories which generate problem setting and point to the directions of problem solving. In one of the prevailing views of social services, for example, the main problem is diagnosed as "fragmentation" while the remedy is prescribed as "coordination". But services seen as fragmented might be seen, alternatively, as "autonomous". Fragmented services became problematic when they are seen as the shattering of a prior integration. Fragmented services are seen to be something like a vase that once was whole and now is broken.

When we have made explicit the implicit metaphor, we are in a good position to elaborate the assumptions which arise from it, and to examine their appropriateness and consequences. In the above example the obvious implication is that fragmentation is bad and that coordination is good. But this sense of obviousness depends very much on the metaphor's remaining tacit. After we have made explicit the metaphor which generates the problem, we may ask, for example, whether the services appropriate to the present situation are just those which used to be integrated, and whether there may not be benefits as well as costs associated with the lack of integration. The notion of a generative, tacit, implicit, theory-constitutive, or root metaphor as they are called alternately, becomes then an interpretive tool for the criticial analysis of scientific theories and/or social policies.

Metaphors of Aging

Having discussed the metaphorical basis of science, we now will examine some metaphors of aging as they are still used in the sciences and society. In the text to follow terms that have particular metaphorical significance are set in italics.

Since the relative and absolute number of older people is growing in modern society, and established institutions like the family, church and state apparently do not know how to deal with growing numbers of older people, aging is more and more defined as a "problem". The way this "problem" of society is defined and solutions are discussed, reflects some implicit metaphors. In general, it might be said that aging is considered to be one or another form of *disease* (Lasch, 1979). Such a view is expressed by different professions, which look at aging as a biological, psychological and/or social disease.

According to Birren and Renner (1979, p. 4), definitions of biological aging primarily involve concepts of deterioration or *decrement,* i.e. "decreasing behavioral, psychological and biochemical adaptation to internal and external environment challenges" (Ordy, 1975, p. 17). The behavioral sciences, especially psychology, often favor the view that aging is the slow *decrease* of mental capacity and adaptability, which may result in feelings of incompetence and worthlessness (Craik, 1982). Finally, a common social viewpoint embraces aging as the increasing social *disorganization* of society, in which social and technological change and cultural conflict lead to a breakdown of social rules and restriction of opportunities for the elderly (Schroots, 1984).

From the above, one might conclude that theories of aging are heavily influenced by medical-biological conceptions of human life. Birren and Schroots (1980a) pointed out that development is often compared with incremental processes like biological growth, and that aging stands for decremental processes like deterioration. The old fashioned calendars of life with people of all ages arranged in order of age on a platform, are an excellent illustration of this implicit "increment-decrement" or *hill* metaphor of the nature of human life (consider, for example, the saying "over the hill" to express that one has passed the apex of his/her capacities or abilities).

Psychological and social phenomena, however, do not necessarily follow the same course in life as biological phenomena. Clayton and Birren (1980), for example, pointed out that the psychological attribute "wisdom" traditionally represents a progressive aspect of change in adulthood and the later years of life. This conception of wisdom suggests that the hill metaphor is not an universally valid one to embrace all of the processes of development and aging at the biological, psychological and social levels of organization of man. For this reason, Birren and Schroots (1980b; 1984) tried to develop new metaphors of aging that would help integrate current views and stimulate further thought and research according to the principle of "catachresis" (Boyd, 1979). That is, new metaphors are developed to introduce theoretical terminology where none previously existed so that they become—almost imperceptible—constitutive of the theories or conceptions they express.

To emphasize the various conceptions of aging, Birren and Schroots proposed that aging be viewed as consisting of three metaphors. The process of biological aging, which results in increasing vulnerability and a higher probability of dying, was identified as *senescing*. Concurrent with senescing, individuals show *eldering;* that is, the processes of role change and behavior in mature adults in a direction toward those expected and displayed by older individuals in a society. To these two metaphors, one should add psychological aging, called *geronting*. This is defined as the processes of optimizing self-regulation and independence of environmental variations in the presence of some decreasing capacities and resources which the mature individual may experience.

In concluding this section, we want to notice that the deliberate use of the separate terms of senescing, geronting and eldering, would help to differentiate our view of the processes of aging.

Metaphors of Life

One of the oldest metaphors of life in history is that of the *Tree of life,* situated in the center of the earthly paradise, a symbol of the immortality of man. The tree as metaphor of life needs hardly explanation: in the literature one can find numerous analogies between the annual cycle of growing, budding, blooming, shedding leaves, and dying on the one hand, and the individual life cycle on the other hand. A recent example of this cyclical nature is given in Erickson's *The Life Cycle Completed* (1982).

At the turn of the century psychology had discovered this botanical metaphor for the description of mental processes (Vygotsky, 1978). Gesell (1928), for example, described child development in terms of plant growth, which should be well kept by a gardener or teacher, hence the concept of kindergarten. Since that time the tree metaphor has found many applications in psychology. Well known in modern cognitive psychology is the so called *decision tree,* which represents —via a series of branching points—the potential and/or actual choices a person makes or might make to reach a decision. The *branching tree* as metaphor of life is another example in which the branching points symbolize the events, experiences, and happenings that significantly affect the direction of individual lives.

Although the tree as a metaphor of life is very powerful, there is one serious flaw: the tree is basically a spatial metaphor which only vaguely suggests some kind of temporality (cf. Gentner and Grudin, 1985; Lakoff and Johnson, 1980). Therefore, the *flowing river* is a more useful metaphor of life, because flowing implies time as well as change. Waddington (1957) has given some scientific prestige to this spatio-temporal metaphor of life by introducing the concept of *chreode* (the pathway of desire or necessity) or *canalized pathway of change.* This concept refers to the developmental trajectory of a living system as it crosses the metaphorical *epigenetic landscape.* In this metaphor the path followed by the developing organism as it rolls downwards corresponds to the developmental history of the organism. As development proceeds there is a branching series of alternative paths represented by valleys. These correspond to the potential pathways of development of the organism.

Waddington's metaphorical chreode and landscape can not only be applied to living systems and developing organisms, but also to subsystems or parts of the organism, e.g. organs, cells, tissues. Thus, in the course of the developing organism, the three-dimensional landscape changes to a multi-dimensional landscape of increasing *complexity,* traversed by a branching series of canalized pathways of change. However, this conception of increasing complexity refers only to the spatial organization of (behavioral) development in linear time (chronological age). The metaphor of *branching, multidimensional* development becomes more complicated when one takes into account that development— generally speaking—proceeds nonlinearly over the life span. That is to say, behavioral change does not take place at the same rate in each age-interval; also, it may vary in rate from one canalized pathway to another, i.e. polychronic change (cf. Hall, 1983). The resulting temporal variability is less at younger ages and increases as development proceeds and as the number of pathways increase with age. From this it follows that as development proceeds the organization of behavior becomes increasingly complex, not only spatially but also temporally.

Implications

By now it is apparent that the authors believe that metaphor is implicit in all conceptions of aging and the life course and that an important role of metaphor also exists in theory development. There is also the continual task of discussing the research implications of old and new metaphors in order to further develop research programs, methods and techniques.

In an earlier section we suggested that the deliberate use of the metaphorical terms of senescing, geronting and eldering would help to differentiate our view of the processes of aging. This suggestion is based on the developmental axioma of increasing variability with age, or the increasing uniqueness of the individual over the lifespan; an axioma which is often specified with the phrase that older adults become more distinctly themselves and less like each other than at any other point in the lifespan (cf. Gatz, Pearson and Weicker, in press). To emphasize the importance of increasing variability of all kinds of functions and processes at biological, behavioral and social levels with age, Schroots (in press) named this life-span phenomenon *individuation,* a term originally developed by C. G. Jung, and in our view an accurate, single descriptor of aging as a finite series of transformations in an organism toward patterns (biological, behavioral, social) of increasing uniqueness. As the traditional psychometric approach with its universal and group-specific generalizations fails to describe increased uniqueness and to make generalizations applying to specific individuals (Runyan, 1982), the implication of individuation is obvious: new methods and techniques need to be developed, tailored to the unique qualities of older individuals.

A similar implication follows from Waddington's metaphor of branching, multidimensional, polychronic change, i.e., linear models of behavioral development with their stages, phases, transitions, seasons and passages (Erickson, 1963; Gould, 1978; Levinson, 1978; Lowenthal, Thurner and Chiriboga, 1975; Maas and Kuypers, 1974; Sheehy, 1977) are inadequate descriptions of the human life span. As the branches with their varying rates of development grow in number with age, it will become increasingly difficult to find a single, spatio-temporal denominator for the developing and aging organism. This increasing spatio-temporal variability with age also explains why a single variable such as biological age (Birren and Cunningham, 1985) is an inadequate predictor of aging.

Recently, Schroots (1988) suggested a possible solution of the measurement problem of increased *complexity* with age by introducing the new geometrical concept of *fractal,* which is a model of complex structures arising from order and randomness, with a dimensionality between one and two (Nicholis, 1984). Although lack of space does not allow us to further develop this line of thought, fractal geometry—combined with Prigogine's nonequilibrium systems theory (1979) and its main concept of *bifurcation* (branching point)—might well be the new tool for the description and analysis of increasingly complex changes over the individual's life span.

While the search for advanced tools on the basis of new metaphors of aging and the life course is still in a theoretical phase, some other implications of the use of old metaphors should be mentioned here. Essentially, these implications—of the 'hill' metaphor, for example—can be reduced to the very simple notion of (unidimensional) age *differences,* e.g. between younger and older individuals, instead of (multidimensional) age *changes* over the life span. On the basis of age differences (which cannot explain the aging process) rather than age changes, traditional methods and techniques like questionnaires have been developed and continue to be developed. However, in spite of their obvious merits, questionnaires do have some serious drawbacks, especially with the elderly (cf. Johnson, 1982).

First, the use of pre-determined questions in structured or semistructured interviews not only presumes that the investigator knows all the relevant questions, but also rules out the possibility of collecting new information which is *unique* for the person being interviewed.

Second, due to a very strong tradition of neo-behaviorism, the dimension of *time* as represented in the life history of an individual has been completely left out in psychological inquiries. In discussing the concept of life satisfaction Back (1981) refers to this serious omission as follows:

> Social scientists . . . . have tended to measure satisfaction in current time, partly because it is technically easier to do so. It is easier to make sure that the respondent is concentrating on current feelings than on the whole life span, perhaps even including expectations for the future. Thus most interviewing scales on life satisfaction consist mainly on questions about current feelings . . . . What is needed, therefore, is a convenient measure of assessing the whole life. (p. 1)

Back (1981) and Johnson (1976) are not the only advocates of the reintroduction of a temporal framework within which the behavior of the respondents ought to be interpreted. Many others are also of the opinion that it is necessary to use the life history of the individual as the major framework for their inquiries (Chiriboga, 1978; Cooper et al., 1981; Freeman, 1984; Rakowsky and Hickey, 1981; Reese and Smyer, 1983; Rosenmayr, 1981). Raynor and Entin (1982) even declare that the motivation of individuals cannot be understood when the temporal dimension has been omitted.

In addition, questionnaires seldom take into account the *affective* tones and feelings of personal life events. Questions are asked and interpreted at the cognitive level of organization of behavior, even to the extent that respondents are asked to check a list of words, indicative of personal feelings and emotions. It seems to be extremely difficult to gather information about the affective qualities of events by means of semi-structured interviews and questionnaires. Apparently, these instruments appeal to the more cognitively and verbally oriented qualities of the individual's behavioral organization.

From the foregoing it follows that the development of new methods, instruments and techniques—particularly in the field of aging and the life course—asks for special research-efforts and extra attention.

Life-line Interview Method

Rationale

Essentially, the Life-line Interview Method (LIM) has been developed on the basis of *metaphors* which people use to describe their life histories and expectations for the future. Restating briefly what we put before in earlier sections, metaphor seems to be the key to understanding both the methodological and creative aspects of scientific discovery and progress. Metaphor is an important way by which we create new meanings, making sense and alternate sense where there is little or none before. Metaphor, in short, is a means of entering the unknown through the gateway of the known. The metaphor allows us to map what we know onto what we vaguely know and gives rise to new hypotheses and integration.

When older persons are asked to describe their life, they frequently use metaphors like the *river* or *footpath* (cf. Vischer, 1961). The river symbolizes the stream of life, and the footpath stands for the journey one makes from birth to death, when one alternately crosses the mountains and valleys of life. Both metaphors enclose the forementioned temporal dimension, but only the 'footpath' metaphor refers explicitly to the dimension of affect. For example, when people say "I'm feeling *up*" or "I'm really *low* these days", they are using a spatial metaphor, i.e., hilly country, to express the positive and negative feelings they had in life.

From the 'footpath' metaphor to the LIM is only one step as soon as one realizes that the graphical, two-dimensional representation of a footpath—with *time* on the horizontal dimension and *affect* on the vertical dimension—symbolizes the course of human life. For in a LIM-session a person is asked to place perceptions of his or her life visually in a temporal framework by drawing his or her life-line. Thus, with the help of this method one can elicit biographical information about important life-events in a non-verbal, visual way. Consequently, one avoids the faults of more cognitively-oriented techniques like questionnaires, open interviews and (auto)biographies which appeal primarily to the verbal capacities of the individual. Before discussing the LIM in more detail, we will describe briefly the course of a typical LIM-session.

Administration

The interviewer first introduces the general plan of the session by saying that he is interested in the human life course with its ups and downs, level periods rises and declines, etc., which are all completely different from one person to another. He then explains that he would like to hear his subject's life story in a special way, which he will illustrate by giving some typical examples. However, before doing this, he hands a piece of paper that looks very much like a blank-grid, consisting of a bottom and top line, two solid and one dotted vertical line (see figure 1, p. 292). The interviewer then says "You will notice that the bottom and top line represent chronological age and that this vertical line with the birth dot in the middle represents the time of your birth (0 Yr). The vertical line just beyond the middle represents your age, please indicate your age at the bottom line." The interviewer continues in saying "Look, this is your life space. Can you now visualize how your life could be put into a graph, starting at this birth dot? First I will give you some examples of the life-lines other people have drawn." These examples are further illustrated with explanations like "The higher the life-line, the happier this particular person feels himself" or "This dip shows that this person felt very depressed at this age." The subject is then asked to draw his or her life-line from birth dot to the present age-line.

As soon as the life-line has been drawn, the subject is asked to label each peak and each dip by chronological age and to tell what happened at a certain moment or during an indicated period. At the same time the interviewer makes a verbatim report of what the subject sees as the most important events in his or her life.

After the subject's past (and present) have been visualized and described in detail the future can be explored in the same manner. Starting from the point where the life-line has stopped, the subject is asked to continue the line until the dotted age-line of expected death is reached. Then the whole procedure of explaining the future life-line and making a verbatim report is repeated.

When the LIM-session is over, the final result is a visual and verbal life history of the individual and his or her visual and verbal future as well, in terms of what he or she thinks/feels has been or will be important in life.

Commentary

Starting-point for the development of the LIM was the basic idea that this method allows a person to place perceptions of his or her life *visually* in a *temporal* framework. This idea turned out to be very important for several reasons.

Firstly, as most people are familiar with the graphical representation of time by a straight line, they don't need much thinking before drawing their life-line. However, in actual practice most interviewees realize only after the drawing of the life-line fully what they are revealing to the interviewer. Some of them even go so far that they try to 'correct' the original line. All this is a positive sign of the LIM's claim to elicit biographical information at the *affective* level of the behavioral organization of the individual.

Another, positive sign of this claim is that some persons become emotionally very upset because of the continual confrontation with their life-line during the interview. Normally speaking, one can deny or just ignore what has been said a few minutes ago. However, when the life-line has been drawn, denial or ignoration doesn't help the person who wants to escape the sometimes painful truth of his or her life, because the visualized truth is there all the time. As a matter of fact, this means that the LIM needs protection and may be used by professional workers only.

A second, innovative aspect of the LIM is its *self-pacing* quality. The older person draws his life-line and tells his personal story in his own pace. So, this method not only meets the objections we raised with regard to modern question-naires, but also synchronizes perfectly well with the tempo of individuation of each individual person. In practice, the non-directive atmosphere of the inter-view facilitates that someone's real opinions, beliefs and attitudes are expressed.

Last, but not least, we would like to mention the methodologically-innovative quality of *self-structuring*. One of the unsolved problems with regard to the content analysis of open, free flowing interviews or verbatim reports concerns the categorization of unstructured interview-data in such a way that they can be analyzed statistically and significantly. With the LIM a method is presented in which first of all the person him or herself categorizes and structures the data in terms of number of events, age and affect. This inherent structure has a validity of its own for it represents the facts of an individual's life as he or she sees it.

CASE STUDY

Content Analysis

Although the LIM produces verbal and graphical data as well, we intend to analyze the verbatim report of a typical case study only. On the basis of even a brief review of relevant literature, it might be expected that numerous statistical and methodological problems emerge in the analysis of verbal data. However, contrary to this expectation, qualitative data analysis methods are rarely reported in detail (cf. Miles and Huberman, 1984). This observation applies also to various methods of content analysis. Krippendorf (1980) and Rust (1983), for instance, have written extensively about theoretical and practical issues, but spent only a few words on explicit, systematic methods for drawing conclusions which can be tested and repeated, respectively.

Content analysis is a multipurpose research method in which the content of verbal data—e.g. dialogue, text, report—serves as the basis of inference. A rather broad, but widely used definition is given by Holsti (1968): "Content analysis is any technique for making inferences by systematically and objectively identifying specified characteristics of messages" (p. 601). This definition refers to Lasswell's classical 'formula' of communication from 1952, i.e., "who says what to whom, how and with what effect". These questions can be answered by content analysis in an objective and systematic way. Objectivity is arrived at via interrater agreement, and the analysis is carried out by applying certain rules systematically and consistently.

Nevertheless, there are some serious methodological problems. Psychometric concepts like 'reliability' and 'validity' need to be redefined as they are used in a different context of primarily qualitative analysis (Campbell, 1979; Ericsonn and Simon, 1980; Runyan, 1984). Reliability, for instance, is here referred to as the clear description of methods used to collect and analyze data (LeCompte and Goetz, 1982). As for a definition of validity, we have stated earlier that the verbatim report has a validity of its own: it represents subjective truth (Kohli, 1981).

The content analysis of the case to be presented in the next subsection, is based on Cuilenburg's graph theory (1985; 1986), which can be traced back to the pioneering work of Osgood (1956). According to this theory, each sentence of the verbatim report is analyzed in terms of 'subject', 'meaning object' and 'connective terms'. The 'subject' is defined, as the first person (P) or the 'I' of each sentence. 'Meaning objects' (MO) can vary from people, objects and animals to concepts and ideas; in brief, to all objects which the person considers to be meaningfull (cf. Bromley, 1977, 1978). 'Connective terms' are all other words in a sentence which show an affective relation between the person and its meaning object, affective relations being viewed—generally speaking—as positive (+), negative (−) or neutral (0).

For example, in the next case Mrs. K. starts her life-story with saying: "I had a very happy childhood". In terms of Cuilenburg's theory, this very first sentence of the verbatim report would be analyzed, then, into P/+/MO, in which P = 'I', + = 'had a very happy', and MO = 'childhood'.

The final result of this type of content analysis is a series of affective relations between P and its Meaning objects, which might be analyzed further in various ways, dependent on the purpose of the (case) study. In the study to follow, we decided to do a so-called 'branching point analysis', in which the branching points (turning points or transitions) of the subject's life-line are assessed. Branching points being defined as those changes in the life of the individual, which direct the life-path distinctly, and which are separated in time from each other by one or more affective events, i.e., meaning objects.

Case of Mrs. K.

In an early detection pilot-study of the dementia syndrome, one of the subjects, Mrs. K., an 81 yr old widow, appeared to be rather depressed. On the Geriatric Depression Scale (Brink et al., 1982) she scored 15, which suggests a moderate depressive state with 80% sensitivity and 100% specificity. As we are interested in the antecedents and prospects of current behavior, the LIM was administered subsequently. In figure 1 Mrs. K.'s life-line is shown, followed by the verbatim report.

Verbatim Report of Mrs. K.

Past (O). I had a very happy childhood, that's for sure. When I was 16, I had to leave school (1). I suppose I was fortunate in being able to stay on even for that long, but I would have

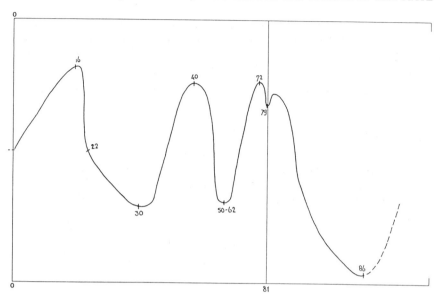

Figure 1. Mrs. K.'s life-line, drawn in the LIM grid (dimensions: 300 x 185 mm), with chronological age (yrs.) on the horizontal dimension and affect on the vertical dimension.

liked to have been a teacher like my eldest sister. She was allowed to continue her studies because the family didn't consider her suitable enough to help at home as I, being more practical, was made to do. My mother wasn't very strong, she had a weak heart, and therefore needed help. Because of her bad health, we were advised to move to the country side. I continued helping at home, until I was 22, when I married (2). A year and a half later my husband contracted ulcers. Those days weren't too good. Well, we did have a lovely family (I had eight children of whom I am very fond), but my husband was often ill. Because of this we had financial difficulties (3). When I was 35 the war started and the problems became worse. It can be considered a miracle that we all came through it so well. Inspite of having to eat bulbs (etc.) we did manage to have a good meal a day. Two of my children were born during the war. Now I suppose you'll be thinking: "how could she have let it come so far in a time like that''; but they weren't as clever then as they are nowadays. Anyway, we survived. After the war (4) my husband started to work again, you could consider this a highlight, but his health deteriorated. He had at least 10 operations (5) and when I was 62 he was taken into a nursing home. This totally confused him and two years later he died. One year later my eldest son died. Dreadful. I often thought "How will I survive?" But I've learnt from life: "You'll always manage somehow''. My son died of lung cancer, he was only 42 years old. Shortly afterwards one of my grandchildren died, and also a daughter in law, she had a sudden fatal attack, whilest doing shopping. That time I went through a depressive period. Of course, I've had a good life, but I've been through some very hard times. You have to be grateful for life, after all there is no choice. I was 72 (6) when I moved to this appartment and things improved. I did a lot of traveling: Scandinavia, Russia, Israel, Egypt, Africa, etc. I really enjoyed visiting those countries, so why shouldn't I? Two years ago (7) I had dizzy spells, so I had to go to hospital for a month, but they couldn't find anything. A few months ago I started to suffer from something again. Now I'm really old. I find it difficult to accept this. I still want to do such a lot, but I'm not able to do anything. I have difficulties in walking and I'm afraid to travel. I have become so dependent on others. Life seems meaningless, except perhaps for my children who still come and see me, but they don't really need me anymore. I try to be interested in their lives, but it's extremely difficult and very trying.

Future. (I) I haven't much faith in the future. I'm constantly afraid of my failing health. I also think of the end of the world. When I look around me and see what is happening to people: for money they seem to do everything. Apart from that, all they do is complain. And all those unemployed. . . No, I don't want to sound like a pessimist, but I hope that I, nor my children, will have to live through it all over again (II).

In explanation of figure 1, it should be added that Mrs. K. did not expect to become older than 86 years. Unasked for, she continued drawing her life-line after expected death via a series of dots.

In summarized form, table 1 presents the results of the content analysis of the verbatim report in terms of age (years), branching points (number in parentheses), events (meaning objects) and affect (affective relation positive, negative of neutral).

Discussion

Table 1 is self-evident and needs no explanation. Nevertheless, by way of illustration, a picture of Mrs. K.'s 'branching tree' is presented in figure 2, with chronological age on the horizontal dimension of the LIM grid and number of affective events (positive, negative or neutral) on the vertical dimension. This tree is composed of branching points (numbered 0 to 7 for the past, and I to II for the future) on the one hand, and affective events on the other (see table 1). The continuous, solid line or the 'trunk' of the tree is the mean number of affective events for each period between two consecutive branching points. The dotted lines or the 'branches' of the tree are the extrapolations of the various solid lines in time. Thus, the branching tree is composed of consecutive trunk-segments or relatively stable periods of affective events, as well as moments of instability or branching points from which branches of potential life-paths originate, i.e., not realized possibilities in an individual life.

In discussing Mrs. K.'s branching tree the question might be asked 'What are the antecedents and prospects of her current, depressed state of behavior?' As is well-known, 'illness' or 'being an older person' does not trigger a depressive state automatically (branching point 7). Many people are sick and old, and still do not feel depressed. So, objectively speaking, there is no reason for Mrs. K. to feel depressed. However, from a subjective point of view, there is a different situation, as Mrs. K. finds it difficult to be dependent on others and—even worse—not needed anymore by her children. From this we might conclude that the feeling of (in)dependency, combined with the (un)ability of taking care of relatives, are the main sources of her depression. The question arises whether there is any evidence in the pattern of Mrs. K.'s branching tree that the combina-

Table 1.   Summarized Analysis of Mrs. K.'s Verbatim Report

Age	Branching point	Event	Affect
		Past	
0	birth (0)		
0–16		childhood	+
		school	+
		teacher	+
		sister	+
		being practical	+
		mother	−
16	leaving school (1)		
16–22		move	−
		help at home	−
22	marriage (2)		
22–30		marriage	0
		husband	−
		children	+
		finanical problems	−
30	crisis/war (3)		
30–40		crisis/war	−
40	end of war (4)		
40–50		husband	+
50	illness husband (5)		
50–72		husband	−
		son	−
		grandchild	−
		daughter-in-law	−
72	move (6)		
72–79		move	+
		traveling	+
79	illness (7)		
79–81		being ill	−
		being old	−
		Future	
81	future (I)		
81–86		no faith in future	−
		failing health	−
		black pessimism	−
86	death (II)		

tion of dependency and unability to care for other persons would cause a depression.

In principle, one might distinguish three types of combinations: (a) independence and ability to care, (b) independence and unability to care, and (c) dependence and unability to care. Inspection of Mrs. K.'s branching tree shows that combination (a) with branching points 1 and 5 has caused a depression several times. On the other hand, combination (b) with branching point 6 does not result

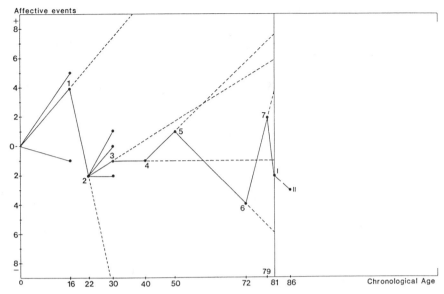

Figure 2. Mrs. K.'s branching tree, projected in the LIM grid, with chronological age (yrs.) on the horizontal dimension and number of affective events (positive, negative or neutral) on the vertical dimension.

in a depressive state, as one might expect on the basis of Mrs. K.'s affection for her children. For, at branching point 6, Mrs. K. decided to move to a new place and do some traveling afterwards, even though relatives were lacking to take care of during that period. Thus, after inspection of the branching pattern and on further consideration, we may conclude that the primary source of Mrs. K.'s depression is not as much the unability of taking care of relatives, but the dependency of other people.

Given the nature of this dependency, i.e. being ill, it seems improbable that Mrs. K., who is 81 years old, will do much better in the future. As such, her future life-line, which goes down almost immediately, makes a rather realistic impression, unfortunately. Any interventions, therefore, should be aimed at the reconciliation with expected death.

SUMMARY AND CONCLUSION

In the foregoing some old and new metaphors of aging and the life-course have been discussed at length, as well as their implications for the development of new instruments, methods and techniques. Two metaphors of life, i.e., the branching tree and the footpath, turned out to be of major importance for the development of the Life-line Interview Method. The branching points of the 'tree' metaphor symbolize the events, experiences, and happenings that significantly affect the direction of individual lives, while the 'footpath' metaphor stands for the journey one makes from birth to death, when one alternately crosses the mountains (positive affect) and valleys (negative affect) of life.

The development and administration of the LIM has been primarily based on the 'footpath' metaphor, visualized into a two-dimensional graph with time on the horizontal dimension and affect on the vertical dimension, while the graph itself symbolizes the course of human life (events, experiences, etc.). With the help of this method one can elicit biographical information about important life-events in a non-verbal, visual way.

A case study has been presented to illustrate the methodologically-innovative quality of self-structuring of the LIM. That is to say, the interviewee him of herself categorizes and structures the visualized data of his or her life (in terms of events, age and affect) in such a way that the following verbatim report can be analyzed accordingly. In the final analysis, the 'branching tree' metaphor plays the role of visualization of the course of human life in terms of branching points or turning points.

It might be concluded that the LIM can serve as a diagnostic and process-facilitating tool, especially with the elderly, not only because of the quality of self-structuring, but also—and may be even more important—because of the self-pacing quality, which allows the older person to set his or her own pace in giving biographical information.

REFERENCES

Back, K.W. 1981. *Types of Life Course and Gerontology.* Paper presented at the 12th Int. Congress of Gerontology. Hamburg, July 17, 1981.

Birren, J.E. and V.J. Renner 1977. Research on the Psychology of Aging: Principles and Experimentation. Pp. 3–38 in: *Handbook of the Psychology of Aging,* edited by J.E. Birren and K.W. Schaie New York: Van Nostrand Reinhold.

Birren, J.E. and J.J.F. Schroots 1980a. *A Psychological Theory of the Organization of Behavior with Age.* Paper Presented at the International Congress of Psychology, Leipzig, July 6–11.

Birren, J.E. and J.J.F. Schroots 1980b. A Psychological Point of View toward Human Aging and Adaptibility. Pp. 43–54 in *Adaptibility and aging.* Proceedings of the 9th International Conference of Social Gerontology, Quebec Canada, August 27–29.

Birren, J.E. and J.J.F. Schroots 1980c. *Aging, from Cell to Society; A Search for New Metaphors.* Paper Presented for the WHO Global Program for the Care of the Aged, Mexico City, October.

Birren, J.E. and J.J.F. Schroots 1984. Steps to an Ontogenetic Psychology. *Academic Psychology Bulletin,* 6:177–90.

Birren, J.E. and Cunningham, W.R. 1985. Research on the Psychology of Aging: Principles, Concepts and Theory. Pp. 3–34 in *Handbook of the psychology of aging* edited by J.E. Birren and K.W. Schaie; *2nd edition.* New York: Van Nostrand Reinhold.

Boyd, R. 1979. Metaphor and Theory Change: What is "Metaphor" a metaphor for? Pp. 356–408 in *Metaphor and thought* edited by A. Ortony. Cambridge: Cambridge University Press.

Brink, T.L., J.A. Yesavage, L. Owen, P.H. Heersema, M. Adey and T.L. Rose 1982. Screening Tests for Geriatric Depression. *Clinical Gerontologist,* 1:37–43.

Bromley, D.B. 1977. *Personality Description in Ordinary Language.* London: John Wiley & Sons.
Bromley, D.B. 1978. Natural Language and the Development of the Self. In *Nebraska Symposium on Motivation* edited by C.B. Keasey 1977. Lincoln Nebr.: University of Nebraska Press.
Campbell, D.T. 1979 "Degrees of Freedom" and the case study. Pp. 49–67 in *Qualitative and Quantitative Methods in Evaluation Research* edited by T.D. Cook and C.S. Reichardt. Beverly Hills/London: Sage.
Cassirer, E. 1946. *Language and Myth.* New York: Dover.
Chiriboga, D.A. 1978. Evaluated Time: A Life Course Perspective. *J. Gerontol.,* 33:388–393.
Clayton, V.P. and J.E. Birren 1980. The Development of Wisdom across the Life-span: A Reexamination of an Ancient Topic. Pp. 103–35. in P.B. Baltes and O.G. Brim, Jr. (eds), *Life-span Development and Behavior,* 3. New York: Academic Press.
Cooper, P.A., L.E. Thomas, S.J. Stevens and D. Suscovich 1981. Subjective Time Experience in an Intergenerational Sample. *Int. J. Aging hum. Dev.,* 13:183–193.
Craik, F. June 1982. *Aging and Competency: Changes in Cognitive Processes.* Paper presented at the Invitational Research Symposium "Metaphors in the Study of Aging", Vancouver, University of British Columbia.
Cuilenburg, J.J. Van, J. Kleinnijenhuis and J.A. De Ridder 1985. Een Theorie over Evaluatieve Betogen. *Acta Politica,* 3, 291–330.
Cuilenburg, J.J. Van, J. Kleinnijenhuis and J.A. De Ridder 1986. *Artificial Intelligence and Content Analysis.* Paper presented at the XV Conference of the International Association for Mass Communiction Research IAMCR, New Delhi.
Erickson, E.H. 1963. *Childhood and Society* (2nd edition). New York: Norton.
Erickson, E.H. 1982. *The Life Cycle Completed.* New York: Norton.
Ericsson, K.A. and H.A. Simon 1980. Verbal reports as data. *Psychological Review,* 3:215–51.
Fischer, W. 1978. Struktur und Funktion erzählter Lebensgeschichten. Pp. 311–36 in: *Soziologie des Lebenslaufs* edited by M. Kohli Darmstadt: Luchterhand.
Gatz, M., C. Pearson and W. Weicker (in press). Older Persons and Health Psychology. In *Health Psychology: A Discipline and a Profession,* edited by G.C. Stone, S.M. Weiss, J.D. Matarazzo, N.E. Miller, J. Rodin, G.E. Schwartz, C.D. Belar, M.J. Follick and J.E. Singer.
Gentner, D. and J. Grudin 1985. The Evolution of Mental Metaphors in Psychology: A 90-year Retrospective. *American Psychologist,* 40:181–92.
Gesell, A. 1928. *Infancy and Human Growth.* New York: Macmillan.
Gould, R.L. 1978. *Transformations: Growth and Change in Adult Life.* New York: Simon and Schuster.
Hall, E.T. 1983. *The Dance of Life: The other dimension of time.* New York: Anchor Press/ Doubleday.
Holsti, O.R. 1968. Content Analysis. In *The Handbook of Social Psychology* edited by G. Lindsey and Aronson. London: Addison-Wesley.
Johnson, M.L. 1976. That was your Life: A Biographical Approach to Later Life. In *Dependency or independency in old age* edited by J.M.A. Munnichs & W. van den Heuvel, Den Haag: Nijhoff.
Johnson, M.L. 1982. *Personal Biography and Group Experience: A Methodological Innovation.* Mexico City: World Congress of Sociology.
Kohli, M. 1981. Biography: Account, Text, Method. Pp. 61–75 in *Biography and Society,* edited by D. Bertaux. London: Sage.
Krippendorff, K. 1980. *Content Analysis.* London: Sage.
Lakoff, G. and Johnson, M. 1980. *Metaphors We Live by.* Chicago: The University of Chicago Press.
Lasch, C. 1979. *The Culture of Narcissism.* New York: Norton.
Leatherdale, W.H. 1974. *The Role of Analogy, Model and Metaphor in Science.* Amsterdam: North-Holland Publishing Co.
Lecompte, M.D. and J.P. Goetz 1982. Problems of Reliability and Validity in Ethnographic Research. *Review of Educational Research,* 52, 31–60.

298 JOHANNES J.F. SCHROOTS and CORINE A. TEN KATE

Levinson, D.J. 1978. *The Season's of a Man's Life.* New York: Knopf.
Lowenthal, M.F., Thurner, M. and Chiriboga, D. 1975. *Four Stages of Life: A Comparative Study of Women and Men Facing Transitions.* San Fransisco: Jossey Boss.
Maas, H.S. and J.A. Kuypers 1974. *From Thirty to Seventy.* San Fransisco: Jossey Bass.
Mehrabian, A. 1968. *An Analysis of Personality Theories.* Englewood Cliffs, N.J.: Prentice Hall.
Miles, M.B. and A.M. Huberman 1984. *Qualitative Data Analysis.* London: Sage.
Nicholis, G. 1984. Symmetriebreuken en waarneming der vormen. In *Tijd, de vierde dimensie in de kunst.* Brussel: Vereniging van tentoonstellingen van het paleis voor schone kunsten.
Ordy, J.M. 1975. Principles of Mammalian aging. Pp. 1–22 in *Neurobiology of aging* edited by J.M. Ordy and K.R. Brizzee. New York: Plenum Press.
Ortony, A. (ed.) 1979. *Metaphor and Thought.* Cambridge: Cambridge University Press.
Osgood, Ch.E., Saporta, S. and Nunnally, J.C. 1956. Evaluative Assertion Analysis. *Litera*, 3:47–102.
Prigogine, I. 1979. *From Being to Becoming.* San Fransisco: W.H. Freeman.
Rakowski, W. and T. Hickey, 1981. A Brief Life-graph Technique for Work with Geriatric Patients. *J. Am. Geriatrics Soc.*, 29:373–378.
Raynor, J.E. and E.E. Entin (eds.) 1982. *Motivation, Career Striving, and Aging.* New York: Hemisphere.
Richards, I.A. 1936. *The Philosophy of Rhetoric.* London: Oxford University Press.
Rosenmayr, L. 1981. Objective and Subjective Perspectives of Life span Research. *Aging and Soc.*, 1:29–49.
Runyan, W.M. 1982. *Life Histories and Psychobiography.* New York, Oxford: Oxford University Press.
Runyan, W.M. 1984. *Life Histories and Psychobiography.* New York/Oxford: Oxford University Press.
Rust, H. 1983. *Inhaltsanalyse.* München: Urban & Schwarzenberg.
Sapir, E. 1921. *Language: An Introduction to the Study of the Speech.* New York: Harcourt, Brace and World.
Schön, D.A. 1979. Generative Metaphor: A Perspective on Problemsetting in Social Policy. Pp. 254–83 in *Metaphor and thought* edited by A. Ortony. Cambridge: Cambridge University Press.
Schroots, J.J.F. June 1982. *Metaphors of Aging: An Overview of their Nature and Implications.* Paper presented at the Invitational Research Symposium "Metaphors in the Study of Aging", Vancouver, University of British Columbia.
Schroots, J.J.F. 1984. The affective Consequences of Technological Change for Older Persons. Pp. 237–47 in *Aging and Technological Advances*, edited by P.K. Robinson, J. Livingston and J.E. Birren. New York: Plenum Press.
Schroots, J.J.F. 1986. *Metaforen, gerontologie en onderzoek; kroniek van een leerervaring.* Leiden, NIPG/TNO.
Schroots, J.J.F. (1988). On growing, formative change and aging. In *Emergent Theories of Aging* edited by J.E. Birren and V. Bengtson. New York: Springer.
Sheehy, G. 1977. *Passages: Predictable Crises of Adult Life.* New York: Datton.
Stoddart, J.B. 1980. *Content Analysis: A Study of the Implication of Differences in Verbal Behavior among Groups of Women.* Los Angeles: University of Southern California.
Valle, R.S. and R. Von Eckartsberg 1981. *The Metaphors of Consciousness.* New York: Plenum Press.
Vischer, A.L. 1961. *Seelische Wandlungen beim alternden Menschen.* Bazel/Stuttgart: Benno Schwabe.
Vygotski, L.S. 1978. *Mind in Society.* Cambridge: Harvard University Press.
Waddington, C.H. 1957. *The Strategy of the Genes.* London: Allen and Unwin.
Whorf, B.L. 1956. *Language, Thought and Reality: Selected Writings,* edited by J.B. Carroll. Cambridge, MA: MIT Press.

INDEX

Current Perspectives on Aging and the Life Cycle

Edited by **David Unruh,** *Office of Instructional Development, University of California, Los Angeles*

This series is intended to provide a forum for the publication of a broad spectrum of theoretical and empirical work on macro and micro levels pertaining to aging processes and the life cycle. Issues will be addressed such as the sources and consequences of continuities and discontinuities in earlier and later life stages, variations in aging processes in different socio-historical and national contexts, links between demographic and structural changes and age relevant social policy issues and conditions under which age groups function as political interest groups.

REVIEW: "This book can be well recommended as a sourcebook of recent research findings, based on national surveys on older workers, and retirement."
— *Contemporary Sociology*

Atchley, Miami University, Ohio. **Chinese Retirement: Policy and Practice,** Deborah Davis-Friedmann, Yale University. **The Vanishing Babushka: A Roleless Role for Older Soviet Women?,** Stephen Sternheimer, Central Intelligence Agency. **Employment Policy and Older Americans: A Framework and Analysis,** Stephen H. Sandell, National Commission for Employment Policy.

Volume 2, Family Relations in Life Course Perspective
1986, 280 pp. $63.50
ISBN 0-89232-522-4

Edited by **David I. Kertzer,** Department of Sociology, Bowdoin College

CONTENTS: Introduction. **A Life Course Perspective on Coresidence,** David I. Kertzer, Bowdoin College. **The Changing Place of Remarriage in the Life Course,** Peter Uhlenberg and Kenneth S.Y. Chew, University of North Carolina, Chapel Hill. **An Event History Analysis of the Process of Entry into First Marriage,** Annemette Sorensen and Aage Sorensen, Harvard University. **Markov Analyses of Family Event Histories,** Thomas Espenshade, The Urban Institute. **A Comparison of Statistical Models for Life Course Analysis with an Application to First Marriage,** Nancy B. Tuma, Stanford University and Robert T. Michael, The University of Chicago. **Maternal Influences on Adolescent Family Formation,** Dennis P. Hogan, The University of Chicago. **Data Base Management for Life Course Family Research,** Nancy Karweit, Johns Hopkins University and David I. Kertzer, Bowdoin College. **Kin Relationships and the Life Cycle: A Breton Village at the Turn of the Twentieth Century,** Martine Segalen, Centre d'Ethnologie Francaise, Paris. **Life Course: A Balkan Perspective,** Joel M. Halpern, University of Massachusetts. **Phrasing and Planning: A Rhetorical Analysis of Women's Statements about Family Formation,** Judith Modell, Colby College.

Volumes I and 2 of Current Perspectives on Aging and the Life Cycle were edited by Zena Smith Blau, Department of Sociology, University of Houston

Volume 3, Personal History Through the Life Course
1989, 298 pp. $63.50
ISBN 0-89232-739-1

Edited by **David Unruh,** Office of Instructional Development, University of California, Los Angeles and **Gail S. Livings,** Department of Sociology, University of California, Los Angeles

CONTENTS: List of Contributors. Introduction, David Unruh and Gail S. Livings, University of California, Los Angeles. **Social**

Construction of the Past: Autobiography and the Theory of G.H. Mead, *Denise D. Bielby, University of California, Santa Barbara and Hannah S. Kully, University of California, Los Angeles.* **Toward a Social Psychology of Reminiscence,** *David Unruh, University of California, Los Angeles.* **Changing Concepts of Self and Self Development in American Autobiographies,** *Diane Bjorklund, University of California, Davis.* **Treasured Possessions in Adulthood and Old Age,** *N. Laura Kamptner, Jean R. Kayano, and Joan L. Peterson, California State University, San Bernardino.* **Beginning with Life Histories: Interviewing in the Families of Welsh Steelworkers,** *W.R. Bytheway, University College of Swansea.* **Life Narratives: A Structural Model for the Study of Black Women's Cultures,** *Beverly J. Robinson, University of California, Los Angeles.* **Discovering the World of Twentieth Century Trade Union Waitresses in the West: A Nascent Analysis of Working Class Women's Meanings of Self and Work,** *Gail S. Livings, University of California, Los Angeles.* **Postsecondary Organizations as Settings for Life History: A Rationale and Illustration of Research Methods,** *Patricia J. Gumport, University of California, Los Angeles.* **Telling Women's Lives: "Slant," "Straight," and "Messy",** *Judy Long, Syracuse University.* **Continuities and Discontinuities in Elderly Women's Lives: An Analysis of Four Family Careers,** *Katherine R. Allen, Texas Women's University.* **Emotion Work and Emotive Discourse in the Alzheimer's Disease Experience,** *Jaber F. Gubrium, University of Florida.* **The Bereaved Parent's Position: Aspects of Life Review and Self Fulfillment,** *Meira Weiss, Tel Aviv University.* **Metaphors, Aging and the Life-Line Interview Method,** *Johannes J.F. Shroots and Corine A. ten Kate, TNO Institute of Preventive Health Care, The Netherlands.*

Volume 4, In preparation, Summer l990
ISBN 1-55938-106-X   Approx.$63.50

JAI PRESS INC.
55 Old Post Road - No. 2
P.O. Box 1678
Greenwich, Connecticut 06836-1678
Tel: 203-661-7602